A Journey Through the Wonders of Plane Geometry

Problem Solving in Mathematics and Beyond

Print ISSN: 2591-7234
Online ISSN: 2591-7242

Series Editor: Dr. Alfred S. Posamentier
Distinguished Lecturer
New York City College of Technology - City University of New York

There are countless applications that would be considered problem solving in mathematics and beyond. One could even argue that most of mathematics in one way or another involves solving problems. However, this series is intended to be of interest to the general audience with the sole purpose of demonstrating the power and beauty of mathematics through clever problem-solving experiences.

Each of the books will be aimed at the general audience, which implies that the writing level will be such that it will not engulfed in technical language — rather the language will be simple everyday language so that the focus can remain on the content and not be distracted by unnecessarily sophiscated language. Again, the primary purpose of this series is to approach the topic of mathematics problem-solving in a most appealing and attractive way in order to win more of the general public to appreciate his most important subject rather than to fear it. At the same time we expect that professionals in the scientific community will also find these books attractive, as they will provide many entertaining surprises for the unsuspecting reader.

Published

Vol. 34 *A Journey Through the Wonders of Plane Geometry*
by Alfred S Posamentier and Hans Humenberger

Vol. 33 *Geometrical Kaleidoscope*
by Boris Pritsker

Vol. 32 *Geometric Gems: An Appreciation for Geometric Curiosities*
Volume I: The Wonders of Triangles
by Alfred S Posamentier and Robert Geretschläger

Vol. 31 *Engaging Young Students in Mathematics through Competitions —*
World Perspectives and Practices
Volume III — Keeping Competition Mathematics Engaging in Pandemic Times
edited by Robert Geretschläger

For the complete list of volumes in this series, please visit www.worldscientific.com/series/psmb

Problem Solving in
Mathematics and Beyond — Volume **34**

A Journey Through the Wonders of Plane Geometry

Alfred S. Posamentier
City University of New York, USA

Hans Humenberger
University of Vienna, Austria

NEW JERSEY · LONDON · SINGAPORE · GENEVA · BEIJING · SHANGHAI · TAIPEI · CHENNAI

Published by

World Scientific Publishing Co. Pte. Ltd.
5 Toh Tuck Link, Singapore 596224
USA office: 27 Warren Street, Suite 401-402, Hackensack, NJ 07601
UK office: 57 Shelton Street, Covent Garden, London WC2H 9HE

Library of Congress Cataloging-in-Publication Data
Names: Posamentier, Alfred S., author. | Humenberger, Hans, author.
Title: A journey through the wonders of plane geometry / Alfred S. Posamentier,
　　City University of New York, USA, Hans Humenberger, University of Vienna, Austria.
Description: New Jersey : World Scientific, 2025. | Series: Problem solving in mathematics
　　and beyond, 2591-7234 ; vol. 34 | Includes index.
Identifiers: LCCN 2024022142 | ISBN 9789811292842 (hardcover) |
　　ISBN 9789811292859 (ebook for institutions) | ISBN 9789811292866 (ebook for individuals)
Subjects: LCSH: Geometry, Plane.
Classification: LCC QA474 .P67 2025 | DDC 516.22--dc23/eng20240805
LC record available at https://lccn.loc.gov/2024022142

British Library Cataloguing-in-Publication Data
A catalogue record for this book is available from the British Library.

Copyright © 2025 by World Scientific Publishing Co. Pte. Ltd.

All rights reserved. This book, or parts thereof, may not be reproduced in any form or by any means, electronic or mechanical, including photocopying, recording or any information storage and retrieval system now known or to be invented, without written permission from the publisher.

For photocopying of material in this volume, please pay a copying fee through the Copyright Clearance Center, Inc., 222 Rosewood Drive, Danvers, MA 01923, USA. In this case permission to photocopy is not required from the publisher.

For any available supplementary material, please visit
https://www.worldscientific.com/worldscibooks/10.1142/13828#t=suppl

Desk Editors: Kannan Krishnan/Rosie Williamson

Typeset by Stallion Press
Email: enquiries@stallionpress.com

About the Authors

Alfred S. Posamentier is currently Distinguished Lecturer at the New York City College of Technology of the City University of New York. Prior to that, he was Executive Director for Internationalization and Funded Programs at Long Island University, New York. This was preceded by five years as Dean of the School of Education and Professor of Mathematics Education at Mercy University, New York. For the prior 40 years, he was at The City College of the City University of New York, where he is now Professor Emeritus of Mathematics Education and Dean Emeritus of the School of Education. He is the author and co-author of more than 80 mathematics books for teachers, secondary and elementary school students, as well as the general readership. Dr. Posamentier is also a frequent commentator in newspapers and journals on topics related to education.

After completing his BA degree in mathematics at Hunter College of the City University of New York, he took a position as a teacher of mathematics at Theodore Roosevelt High School (Bronx, New York), where he focused his attention on improving the students' problem-solving skills and, at the same time, enriching their instruction far beyond what the traditional textbooks offered. During his six-year tenure there, he also developed the school's first mathematics teams (both at the junior and senior levels). He is still involved in working with mathematics teachers and supervisors, nationally and internationally, to help them maximize their effectiveness.

Immediately upon joining the faculty of the City College of New York in 1970 (after having received his master's degree there in 1966), he began to develop in-service courses for secondary school mathematics teachers, including such special areas as recreational mathematics and problem-solving in mathematics. As dean of the City College School of Education for 10 years, his scope of interest in educational issues covered the full gamut of educational issues. During his tenure as dean, he took the school from the bottom of the New York State rankings to the top with a perfect NCATE accreditation assessment in 2009. He also raised more than US$12 million from the private sector for educational innovative programs. Posamentier repeated this successful transition at Mercy College, where he enabled it to become the only college to have received both NCATE and TEAC accreditation simultaneously.

In 1973, Dr. Posamentier received his PhD from Fordham University (New York) in mathematics education and has since extended his reputation in mathematics education to Europe. He has been a visiting professor at several European universities in Austria, England, Germany, the Czech Republic, Turkey and Poland. In 1990, he served as Fulbright Professor at the University of Vienna.

In 1989, he was awarded an Honorary Fellow position at the South Bank University (London, England). In recognition of his outstanding teaching, the City College Alumni Association named him Educator of the Year in 1994 and in 2009. New York City had the day, May 1, 1994, named in his honor by the President of the New York City Council. In 1994, he was also awarded the *Das Grosse Ehrenzeichen für Verdienste um die Republik Österreich* (Grand Medal of Honor from the Republic of Austria), and in 1999, upon approval of Parliament, the President of the Republic of Austria awarded him the title of University Professor of Austria. In 2003, he was awarded the title of *Ehrenbürgerschaft* (Honorary Fellow) of the Vienna University of Technology, and in 2004, he was awarded the *Österreichisches Ehrenkreuz für Wissenschaft & Kunst 1. Klasse* (Austrian Cross of

About the Authors **vii**

Honor for Arts and Science, First Class) from the President of the Republic of Austria. In 2005, he was inducted into the Hunter College Alumni Hall of Fame, and in 2006, he was awarded the prestigious Townsend Harris Medal by the City College Alumni Association. He was inducted into the New York State Mathematics Educator's Hall of Fame in 2009, and in 2010, he was awarded the coveted Christian-Peter-Beuth Prize from the Technische Fachhochschule – Berlin. In 2017, Posamentier was awarded *Summa Cum Laude nemmine discrepante* by the Fundacion Sebastian, A.C., Mexico City, Mexico.

He has taken on numerous important leadership positions in mathematics education locally. He was a member of the New York State Education Commissioner's Blue Ribbon Panel on the Math-A Regents Exams and the Commissioner's Mathematics Standards Committee, which redefined the Mathematics Standards for New York State, and he also served on the New York City schools' Chancellor's Math Advisory Panel.

Dr. Posamentier is still a leading commentator on educational issues and continues his long-time passion of seeking ways to make mathematics interesting to teachers, students and the general public, as can be seen from some of his more recent books.

For more information and a list of his publications, see: https://en.wikipedia.org/wiki/Alfred_S._Posamentier.

Hans Humenberger studied mathematics and sports at the University of Vienna (Austria). In the 1990s, he was partly a high school teacher at several Viennese high schools and partly a graduate assistant at the University of Natural Resources and Life Sciences, Institute for Mathematics, Vienna. In 1993, he earned his PhD at the University of Vienna, and in 1998, he received his habilitation in the field of mathematics education. Between 2000 and 2005, he worked as an assistant professor of mathematics at the University of Dortmund (Germany). During this time, he spent a half year as a substitute professor at the University of Duisburg-Essen (Germany). He returned to Vienna in

2005 to accept the position as a full professor for mathematics, with special emphasis on mathematics education at the University of Vienna. Since then, he has served as the head of a working group, "Didactics of Mathematics and School Mathematics," and he was also responsible for the educational training of mathematics teachers for secondary schools.

He has written many papers in German and English on mathematics education and mathematics, as well as several books (mostly in German). One of his recent books, *Aschaulich Elementaregeometrie*, published in German, was coauthored with his former colleague in Dortmund, Berthold Schuppar. This book, which focuses on geometry, was written especially for student teachers in mathematics.

He has been interested in problem-solving for many years, as reflected in many of his papers and a seminar on the topic that he has conducted at several universities for preservice mathematics teachers.

In 2007, he established a program for secondary school students (between grades 5 and 8) to regularly attend seminars and workshops at the University of Vienna, providing them the opportunity to work on interesting and challenging problems. This successful program is funded by the Viennese Department of Education and continues to this day. Furthermore, he is particularly interested in making mathematics interesting for a general audience since this is important for the general perception of mathematics in our society.

Since 2007, he has been an editor of an Austrian school textbook series in mathematics for grades 5–8, and in 2022, he was appointed as a member of a committee that established a new syllabus for mathematics at primary and secondary school levels in Austria.

His main fields of interest include mathematics as a process, applications of mathematics, problem-solving, geometry and stochastics.

More details about his work and a complete list of his publications can be found at the following website: https://homepage.univie.ac.at/hans.humenberger/.

Contents

About the Authors		v
Introduction		xi
Chapter 1	The Golden Ratio in Geometry	1
Chapter 2	Unexpected Concurrencies	53
Chapter 3	Unexpected Collinearities	91
Chapter 4	Squares on Triangle Sides	127
Chapter 5	Similar Triangles on Triangle Sides	165
Chapter 6	Discovering Concyclic Points	189
Chapter 7	Circle Wonders	245
Chapter 8	Polygons and Polygrams	291
Chapter 9	Geometric Surprises	321
Chapter 10	Geometric Fallacies	361
Chapter 11	Homothety, Similarity, and Applications	401
Index		437

Introduction

Geometry is one of the most beautiful aspects of mathematics. This beauty is because you can "see" geometry at work. Most people are exposed to the very basic elements of geometry throughout their schooling with the most concentrated study in the secondary school curriculum. High schools in the United States offer one year of concentrated study of geometry that shows students how a mathematician functions, since everything that is accepted beyond the basic axioms must be proved. Unfortunately, as the course is only one year long, there is still very much in geometry left unexplored for the general audience. That is the challenge of this book, in which we will present a plethora of amazing geometric relationships. Perhaps the most glorious of these is a very special relationship appropriately named the Golden Ratio, which is best seen in the relationship between the length and width of a special rectangle, which is appropriately named the Golden Rectangle. This amazing relationship is seen in architecture, art, nature, and many other unexpected appearances.

We know that any two nonparallel lines intersect in a common point; however, when more than two lines intersect in a common point, it often amazes us. The second chapter of this book will present some highly unusual relationships of concurrency of three or more relatively unrelated lines. Analogously, it is accepted that any two points determine a unique line. However, when we find that more than two points lie on the same line, we have an opportunity to appreciate this phenomenon. Thus, we will exhibit unusual situations that result in three or more collinear points that arise when least expected.

Although the triangle could be considered the basic linear construct in geometry, when considering quadrilaterals, the square provides some highly unexpected beauty. In the fourth chapter, we can see some surprising results when squares are placed on the sides of a triangle. There, we experience amazing collinearity and concurrency. An analogous theme takes us to an extension of placing similar figures on the sides of the triangle to further explore unexpected relationships.

Our journey beyond linear figures leads us to the appreciation of circles. As we admired when three or more points are collinear, we know that three noncollinear points determine a unique circle, so we find a great deal of beauty in discovering when more than three noncollinear points lie on the same circle, which are referred to as concyclic points. Such relationships are truly amazing and support another reason to love geometry. This leads to an often-hidden aspect of geometry, which is the relationship between circles and various linear figures.

More advanced aspects of linear geometry take us to the consideration of multisided geometric figures commonly referred to as polygons. They, too, provide an aspect of geometry that is not too well known to the general audience, yet enhances our appreciation of many geometric wonders. When all these aspects of geometry are brought together, we marvel at their many geometric surprises. This chapter should leave most readers with an amazing appreciation for geometry.

As with most aspects of mathematics, not following all the "rules" leads to incorrect results. Oftentimes, these rule violations are so well concealed that they lead to geometric fallacies, which can certainly provide a level of entertainment and wonder as to where the error might be hidden. For example, one can prove, using simple secondary school geometry techniques, that all triangles are isosceles. Of course, this is a ridiculous conclusion. But to find the fault in the reasoning adds further to the beauty and appreciation for geometry. To explore some of these peculiarities, we devote a chapter to entertain the readership with geometric fallacies.

In our final chapter, we explore the concept of homothety, which is not taught in the American year-long course in geometry; however, it does provide a useful method for establishing geometric relationships. We now invite you to join us in our journey through geometry, a journey that will clearly exhibit the wonders of plane geometry. Enjoy!

Chapter 1
The Golden Ratio in Geometry

For many centuries, architects, artists, mathematicians, scientists, and everyday citizens have been fascinated by a ratio that is ubiquitous in nature and commonly found across many cultures. It was named *Golden Ratio* because of its prevalence as a design element and its seemingly universal aesthetic appeal. The famous German astronomer and mathematician Johannes Kepler (1571–1630) summarized this ratio's importance by saying that "geometry harbors two great treasures: one is the Pythagorean theorem, and the other is the Golden Ratio. The first can compare with a heap of gold and the second we simply call a priceless jewel." The human body possesses the Golden Ratio, as do the helix structure of DNA and the principles of Greek architecture, as well as many modern masterpieces. The Golden Ratio is a unique pattern that not only has wide-ranging and endless applications but also manifests itself in the most unexpected places. Perhaps the most popular appearance of the Golden Ratio is in the rectangle whose sides form the Golden Ratio. This rectangle is referred to as the *Golden Rectangle*.

As we begin our journey investigating this amazing Golden Ratio, manifested in the Golden Rectangle and beyond, we will first define it and then marvel over its various construction possibilities.

Defining the Golden Ratio

The Golden Ratio is defined as the ratio of the lengths of the two parts of a line segment that allows us to make the following equality of two ratios, namely, a proportion that the longer segment (L) is to the shorter segment (S) as the entire original segment ($L+S$) is to the longer segment (L). Symbolically, this is written as $\frac{L}{S} = \frac{L+S}{L}$, as shown in Figure 1-1:

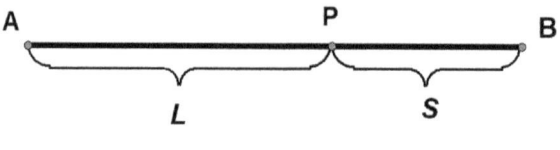

Figure 1-1

This is called the *Golden Ratio*, or the *Golden Section* (the latter term refers to the "sectioning" or partitioning of a line segment). The terms Golden Ratio and Golden Section were first introduced during the nineteenth century. We believe that the Franciscan friar and mathematician Fra Luca Pacioli (ca. 1445–1517) was the first to use the term *De divina proportione* (Divine Proportion) as the title of a book in 1509, while the German mathematician and astronomer Johannes Kepler (1571–1630) was the first to use the term *sectio divina* (Divine Section). Moreover, the German mathematician Martin Ohm (1792–1872) is credited for having first used the term *Goldener Schnitt* (Golden Section). In English, the term *Golden Section* was first used by James Sully in 1875 in an article on aesthetics in the 9th edition of the Encyclopedia Britannica.

You may be wondering what makes this ratio so outstanding that it merits the title "Golden." This designation, which it richly deserves, will be made clear throughout this chapter. We begin by considering its numerical value, which will bring us to its first unique characteristic.

To determine the numerical value of the Golden Ratio $\frac{L}{S}$, we will change the equation $\frac{L}{S} = \frac{L+S}{L}$, or $\frac{L}{S} = \frac{L}{L} + \frac{S}{L}$, to an equivalent, where $x = \frac{L}{S}$ to get $x = 1 + \frac{1}{x}$. This can be rewritten as $x^2 - x - 1 = 0$. We can now solve this equation for *x* by using the quadratic formula[1] to obtain the numerical value of the Golden Ratio (the negative solution needs not to be considered): $\frac{L}{S} = x = \frac{1+\sqrt{5}}{2}$. This value is commonly denoted by the Greek letter phi: ϕ.[2] Thus, $\phi = \frac{L}{S} = \frac{1+\sqrt{5}}{2} \approx 1.61803398874989485$. So, every time one ends up with the equation $x^2 - x - 1 = 0$ one can be sure, without the quadratic formula, that the only (positive) solution is ϕ. What makes this ratio $\frac{L}{S}$ mathematically unique is its reciprocal, namely, $\frac{S}{L} = \frac{1}{\phi} = \frac{2}{1+\sqrt{5}} \approx 0.61803398874989485$. Now, you should notice a very unusual relationship. The value of ϕ and $\frac{1}{\phi}$ differ by exactly 1. That is, $\phi - \frac{1}{\phi} = 1$. From the normal relationship of reciprocals, the product of ϕ and $\frac{1}{\phi}$ is also equal to 1, that is, $\phi \cdot \frac{1}{\phi} = 1$. Therefore, we have two numbers, ϕ and $\frac{1}{\phi}$, whose difference *and* product are equal to 1. It should be noted that these are the only two numbers for which this is true! By the way, you might have noticed that $\phi + \frac{1}{\phi} = \sqrt{5}$, since $\frac{\sqrt{5}+1}{2} + \frac{\sqrt{5}-1}{2} = \sqrt{5}$.

Constructing the Golden Section

Having now defined the Golden Ratio numerically, we shall now *construct* it geometrically. There are several ways to construct the Golden Section of a line segment. You may notice that we appear to be using the terms Golden Ratio and Golden Section interchangeably. To avoid confusion, we will use the term Golden Ratio to refer to the numerical

---

[1] The quadratic formula for the general quadratic equation $ax^2 + bx + c = 0$ yields $x = \frac{-b \pm \sqrt{b^2 - 4ac}}{2a}$.

[2] There is reason to believe that the letter ϕ was used because it is the first letter of the name of the celebrated Greek sculptor Phidias (ca. 490–430 BCE) (in Greek: (Pheidias) ΦΕΙΔΙΑΣ or Φειδίας), who produced the famous statue of Zeus in the Temple of Olympia and supervised the construction of the Parthenon in Athens, Greece. His frequent use of the Golden Ratio in this glorious building is likely the reason for this attribution. It must be said that there is no direct evidence that Phidias consciously used this ratio.
The American mathematician Mark Barr was the first using the letter ϕ in about 1909 – see Theodore Andrea Cook: *The Curves of Life*. Courier Dover Publications, 1914, p. 420 [New York, Dover Publications, 1979].

4 A Journey Through the Wonders of Plane Geometry

value of ϕ, and the term Golden Section to refer to the geometric division of a segment into the ratio ϕ.

Golden Section Construction 1

Our first method, which is perhaps the most popular, is to begin with a unit square[3] *ABCD*, with midpoint *M* of side *AB*, as shown in Figure 1-2. We draw a circular arc with radius *MC*, intersecting the extension of side *AB* at point *E*. We now can claim that line segment *AE* is partitioned into the Golden Section at point *B*, namely, $\frac{AB}{BE} = \frac{AE}{AB}$.

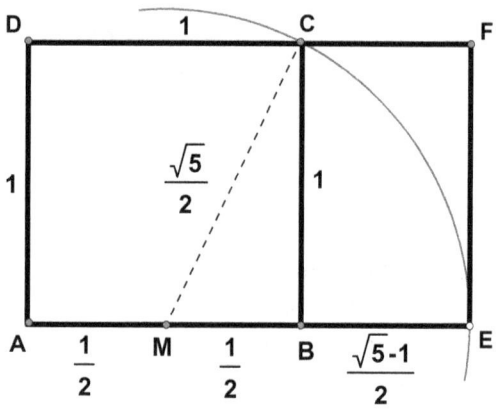

Figure 1-2

To verify that this is the Golden Ratio $\frac{AB}{BE} = \frac{AE}{AB}$, we begin by substituting the values obtained by applying the Pythagorean theorem to $\triangle MBC$ as shown in Figure 1-2, so that we get the following: $MC^2 = MB^2 + BC^2 = \left(\frac{1}{2}\right)^2 + 1^2 = \frac{1}{4} + 1 = \frac{5}{4}$; therefore $MC = \frac{\sqrt{5}}{2}$.

It thus follows that $BE = ME - MB = MC - MB = \frac{\sqrt{5}}{2} - \frac{1}{2} = \frac{\sqrt{5}-1}{2}$ and $AE = AB + BE = 1 + \frac{\sqrt{5}-1}{2} = \frac{2}{2} + \frac{\sqrt{5}-1}{2} = \frac{\sqrt{5}+1}{2}$. We then can find the value of $\frac{AB}{BE} = \frac{AE}{AB}$, that is, $\frac{1}{\frac{\sqrt{5}-1}{2}} = \frac{\frac{\sqrt{5}+1}{2}}{1}$, which turns out to be a true proportion, since the cross products are equal. That is, $\left(\frac{\sqrt{5}-1}{2}\right) \cdot \left(\frac{\sqrt{5}+1}{2}\right) = 1 \cdot 1 = 1$.

---

[3] A *unit square* is a square with side length of 1 unit.

We can also see from Figure 1-2 that point B can be said to divide the line segment AE into an *inner* Golden Section since $\frac{AB}{AE} = \frac{1}{\frac{\sqrt{5}-1}{2}} = \frac{\sqrt{5}+1}{2} = \phi$.

Meanwhile, point E can be said to divide the line segment AB into an *outer* Golden Section since $\frac{AE}{AB} = \frac{1+\frac{\sqrt{5}-1}{2}}{1} = \frac{\sqrt{5}+1}{2} = \phi$.

Notice the shape of the rectangle $AEFD$ in Figure 1-2. The ratio of the length to the width is the Golden Ratio: $\frac{AE}{EF} = \frac{\frac{\sqrt{5}+1}{2}}{1} = \frac{\sqrt{5}+1}{2} = \phi$. This appealing shape is called the *Golden Rectangle*, and plays a dominant role in architecture, art, and beyond.

Golden Ratio Construction 2

Another method for constructing the Golden Section is credited to Heron of Alexandria (fl. c. 62 CE) and begins with a right triangle with one leg BC of unit length and the other leg AB with length 2, as shown in Figure 1-3. Here, we will partition the line segment AB into the Golden Ratio.

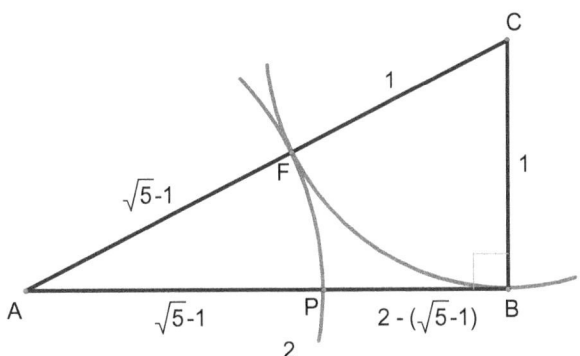

Figure 1-3

With $AB = 2$ and $BC = 1$, we apply the Pythagorean theorem to $\triangle ABC$. We find that $AC = \sqrt{2^2 + 1^2} = \sqrt{5}$. With the center at point C, we draw a circular arc with radius 1, intersecting line segment AC at point F. We then draw a circular arc with the center at point A and the radius AF, intersecting AB at point P. Because $AF = \sqrt{5} - 1$, we get $AP = \sqrt{5} - 1$. Therefore, $BP = 2 - (\sqrt{5} - 1) = 3 - \sqrt{5}$. To determine the

6   A Journey Through the Wonders of Plane Geometry

ratio $\frac{AP}{BP}$, we set up the ratio $\frac{\sqrt{5}-1}{3-\sqrt{5}}$, and to make this more manageable, we rationalize the denominator by multiplying the ratio by 1 in the form of $\frac{3+\sqrt{5}}{3+\sqrt{5}}$. We then find that $\frac{\sqrt{5}-1}{3-\sqrt{5}} \cdot \frac{3+\sqrt{5}}{3+\sqrt{5}} = \frac{3\sqrt{5}+5-3-\sqrt{5}}{3^2-(\sqrt{5})^2} = \frac{2\sqrt{5}+2}{9-5} = \frac{2(\sqrt{5}+1)}{4} = \frac{\sqrt{5}+1}{2} = \phi \approx 1.618033988$, which is the Golden Ratio! Therefore, point P partitions the line segment AB in the Golden Ratio.

Golden Ratio Construction 3

We have yet another way of constructing the Golden Section. Consider the two adjacent unit squares shown in Figure 1-4. We construct the angle bisector of ∠BHE. There is a convenient geometric relationship that will be very helpful to us here: the angle bisector in a triangle divides the side to which it is drawn proportionally to the two sides of the angle being bisected.[4] From Figure 1-4, we derive the relationship $\frac{BH}{EH} = \frac{BC}{CE}$. We apply the Pythagorean theorem to $\triangle HFE$ to get $HE = \sqrt{5}$. We can now evaluate the earlier proportion by substituting the values shown in Figure 1-4: $\frac{1}{\sqrt{5}} = \frac{x}{2-x}$. From this, we get $x = \frac{2}{\sqrt{5}+1}$, which is the reciprocal of $\frac{\sqrt{5}+1}{2} = \phi$. Therefore, $x = \frac{1}{\phi} \approx 0.61803398874989485$, which is the reciprocal of the Golden Ratio.

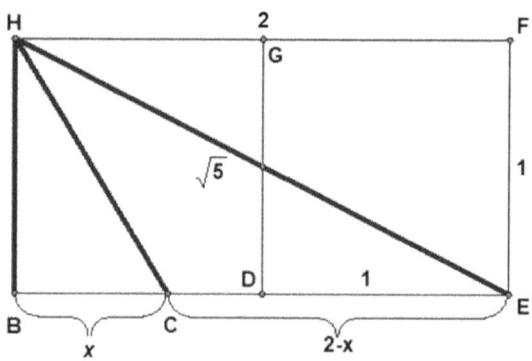

Figure 1-4

---

[4] This theorem was originally demonstrated by Euclid in § 3, Book VI of his *Elements*. The proof can also be found in A. S. Posamentier and R. Bannister: *Geometry: Its Elements and Structure*, 2nd Edition, New York: Dover, 2014.

Thus, we can conclude that point C divides the line segment BD into the Golden Section, since $\frac{BD}{BC} = \frac{1}{x} = \frac{\sqrt{5}+1}{2} = \phi \approx 1.61803398874989485$, the "approximate" value of the Golden Ratio.

Golden Ratio Construction 4

This construction is analogous to the previous one and begins with two congruent squares, as shown in Figure 1-5. A circle is drawn with its center at the midpoint M of the common side of the squares, and a radius half the length of the side of the square. The point of intersection C of the circle and the diagonal of the rectangle determines the Golden Section AC with respect to a side of the square AD.

With $AD = 1$ and $DM = \frac{1}{2}$, we get $AM = \frac{\sqrt{5}}{2}$ by applying the Pythagorean theorem to triangle AMD, shown in Figure 1-6. Since CM

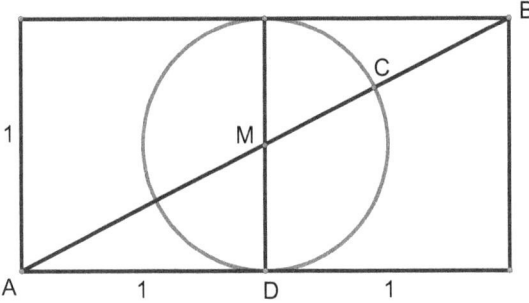

Figure 1-5

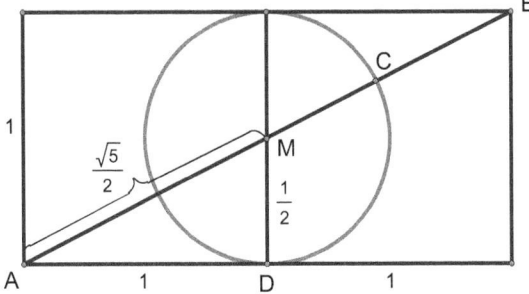

Figure 1-6

is also a radius of the circle, $CM = DM = \frac{1}{2}$. We can then conclude that $AC = AM + CM = \frac{\sqrt{5}}{2} + \frac{1}{2} = \frac{\sqrt{5}+1}{2} = \phi$. Furthermore, $BC = AB - AC = \sqrt{5} - \frac{\sqrt{5}+1}{2} = \frac{\sqrt{5}-1}{2} = \frac{1}{\phi}$. We have thus constructed the Golden Section and its reciprocal.

Golden Ratio Construction 5

In this rather simple construction, we will show that the semicircle on the side (extended) of a square whose radius is the length of the segment from the midpoint of the side of the square to an opposite vertex, creates a line segment where the vertex of the square determines the Golden Ratio. In Figure 1-7, we have square $ABCD$ and a semicircle on line AB with center at the midpoint M of AB and radius MC. We encountered a similar situation in Construction 1, where we concluded that $\frac{AB}{BE} = \phi$, and $\frac{AE}{AB} = \phi$.

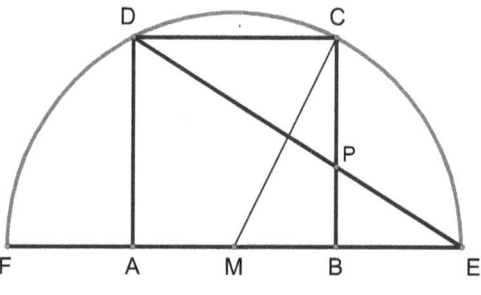

Figure 1-7

However, here we have an extra added attraction: DE and BC partition each other into the Golden Section at point P. This is easily justified by the triangles DPC and EBP being similar and their corresponding sides DC and BE being in the Golden Ratio. Hence, all the corresponding sides are in the Golden Ratio, which here is $\frac{CP}{PB} = \frac{DP}{PE} = \phi$.

Golden Ratio Construction 6

Some constructions of the Golden Section are rather creative. Consider the inscribed equilateral triangle ABC with line segment PT parallel to

BC and bisecting the two sides of the equilateral triangle at points Q and S, as shown in Figure 1-8.

We let the side length of the equilateral triangle equal 2, which provides us with the segment lengths as shown in Figure 1-8. The proportionality there gives us $\frac{RS}{CD} = \frac{AS}{AC}$, which, by substituting appropriate values, yields $\frac{RS}{1} = \frac{1}{2}$ and $RS = \frac{1}{2}$. A useful geometric theorem enables us to find the length of the segments $PQ = ST = x$, due to the symmetry of the figure. The theorem states that the products of the segments of two intersecting chords of a circle are equal. From that theorem, we find $PS \cdot ST = AS \cdot SC$, from which we get $(x+1) \cdot x = 1 \cdot 1$, which is $x^2 + x - 1 = 0$, leading to $x = \frac{\sqrt{5}-1}{2}$.

Therefore, the segment QT is partitioned into the Golden Section at point S, since $\frac{QS}{ST} = \frac{1}{x} = \frac{2}{\sqrt{5}-1} = \frac{\sqrt{5}+1}{2} \approx 1.61803398874989485$, which we recognize as the value of the Golden Ratio. We can generalize this construction by saying that the midline of an equilateral triangle extended to the circumcircle is partitioned into the Golden Section by the sides of the equilateral triangle.

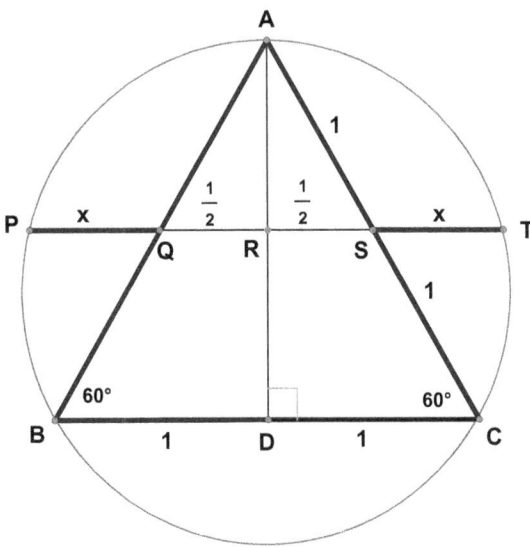

Figure 1-8

Golden Ratio Construction 7

This is a rather easy construction of the Golden Ratio because it simply requires constructing an isosceles triangle inside a square, as shown in Figure 1-9. The vertex E of $\triangle ABE$ lies on side DC of square $ABCD$, and altitude EM intersects the inscribed circle of $\triangle ABE$ at point H. The Golden Ratio appears in two ways here: When the side of the square is length 2, the radius of the inscribed circle $r = \frac{1}{\phi}$, and the point H partitions EM into the Golden Ratio as $\frac{EM}{HM} = \phi$.

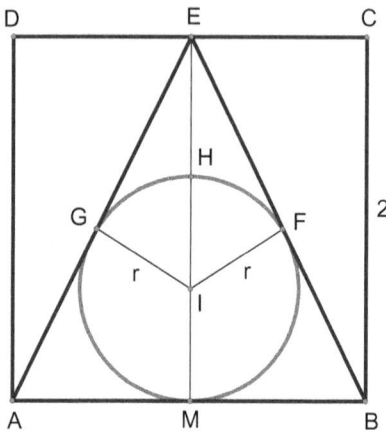

Figure 1-9

To justify this construction, we let the side of the square have length 2. This gives us $BM = 1$ and $EM = 2$. We apply the Pythagorean theorem to triangle MEB and derive $AE = BE = \sqrt{5}$, whereupon we recognize that $GE = \sqrt{5} - 1$, which is shown in Figure 1-10.

Once again, we apply the Pythagorean theorem, this time to $\triangle EGI$, giving us $EI^2 = GI^2 + GE^2$. Put another way: $(2-r)^2 = r^2 (\sqrt{5} - 1)^2$; therefore, $4 - 4r + r^2 = r^2 + 5 - 2\sqrt{5} + 1$. This determines the length of the radius of the inscribed circle $r = \frac{\sqrt{5}-1}{2} = \frac{1}{\phi}$. Now, with some simple substitution, we have $EM = 2$ and $HM = 2r$, yielding the ratio $\frac{EM}{HM} = \frac{2}{2r} = \frac{1}{r} = \phi$.

Golden Ratio Construction 8

A more contrived construction also yields the Golden Section of a line segment. To do this, we will construct a unit square with one vertex

The Golden Ratio in Geometry   11

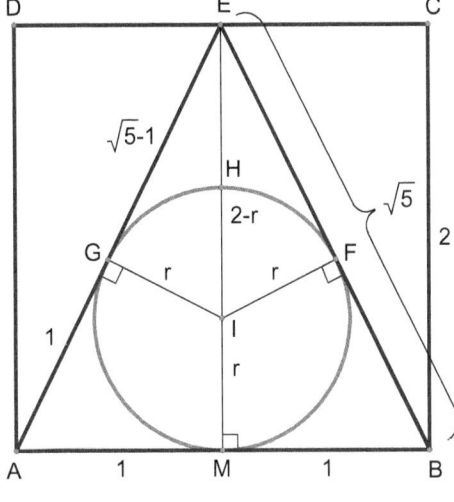

Figure 1-10

placed at the center of a circle whose radius is the length of the diagonal of the square. On one side of the square, we will construct an equilateral triangle, as we show in Figure 1-11.

Applying the Pythagorean theorem to triangle *ACD*, we get the radius of the circle as $\sqrt{2}$, which gives us the lengths of *AD*, *AG*, and *AJ*.

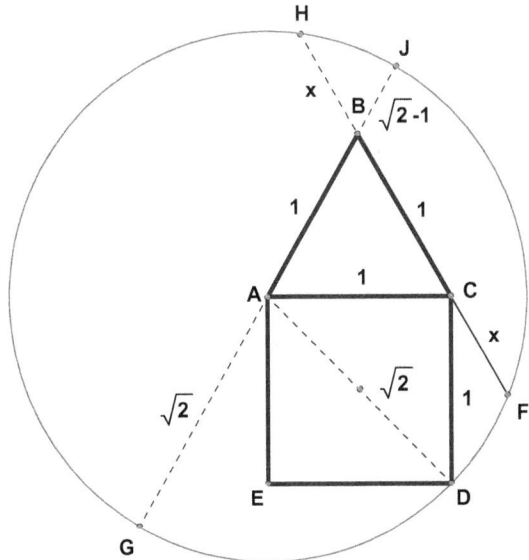

Figure 1-11

12 A Journey Through the Wonders of Plane Geometry

Because of symmetry, we have $BH = CF = x$. Again, applying the theorem involving intersecting chords of a circle (as in Construction 6), we get $GB \cdot BJ = HB \cdot BF$, which gives us $(\sqrt{2}+1)(\sqrt{2}-1) = x(x+1)$, and therefore, $x = \frac{\sqrt{5}-1}{2}$. Once again, we find the segment BF is partitioned into the Golden Section at point C, since $\frac{BC}{CF} = \frac{1}{x} = \frac{2}{\sqrt{5}-1} = \frac{\sqrt{5}+1}{2} = \phi \approx$ 1.61803398874989485, which we recognize as the "approximate" value of the Golden Ratio.

Golden Ratio Construction 9

We can derive the equation $x^2 + x - 1 = 0$, the so-called *Golden Equation*, in several other ways, one of which involves a circle with a chord AB. Suppose we can extend AB beyond A to a point P so that when a tangent from P is drawn to the circle, its length equals that of AB.[5] We can see this in Figure 1-12, where $PT = AB = 1$.

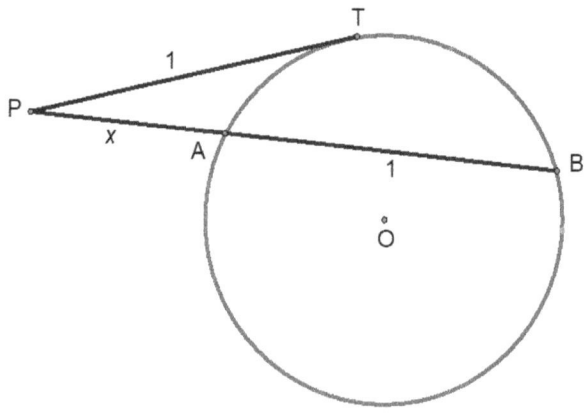

Figure 1-12

Here, we apply a geometric theorem that states that when, from an external point P, a tangent (PT) and a secant (PB) are drawn to a circle, the tangent segment is the mean proportional between the

---

[5] Since we do not say how to construct this constellation, this "construction" is rather a visualization.

entire secant and the external segment, $\frac{PB}{PT} = \frac{PT}{PA}$. This yields $PT^2 = PB \cdot PA$, or $PT^2 = (PA + AB) \cdot PA$. If we let $PA = x$, then $1^2 = (x + 1)x$, or $x^2 + x - 1 = 0$. As before, we can conclude that point A determines the Golden Section of line segment PB, since the solution to this equation is the Golden Ratio.

Golden Ratio Construction 10

This time we will construct the Golden Section with three equal circles. Consider the three tangent congruent circles with radius $r = 1$ shown in Figure 1-13.

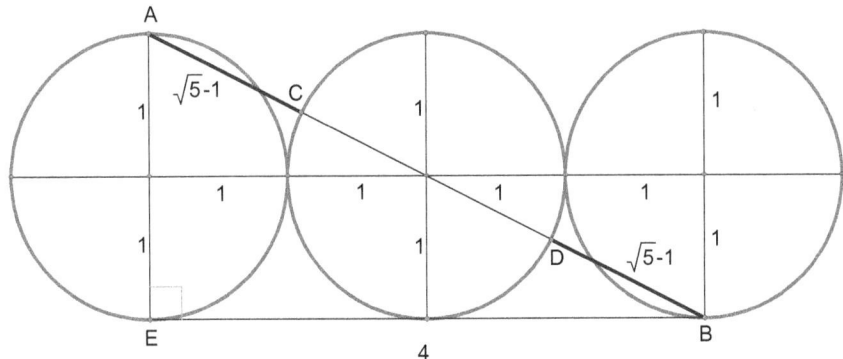

Figure 1-13

In Figure 1-13, we have $AE = 2$ and $BE = 4$. We apply the Pythagorean theorem to $\triangle ABE$ to get $AB = \sqrt{2^2 + 4^2} = \sqrt{20} = 2\sqrt{5}$. Because of the symmetry $AC = BD$, and $CD = 2$, we then have $AB = AC + CD + BD = 2AC + BD = 2AC + 2$. Therefore, $2AC + 2 = 2\sqrt{5}$. It follows that $AC = \sqrt{5} - 1$ and $AD = AB - BD = AB - AC = 2\sqrt{5} - (\sqrt{5} - 1) = \sqrt{5} + 1$.

The ratio $\frac{AD}{CD} = \frac{\sqrt{5}+1}{2} \approx 1.61803398874989485$ again denotes the Golden Ratio. You may notice that each time we used a unit measure as our basis, we could have used a variable, such as x, and we would

14 A Journey Through the Wonders of Plane Geometry

have gotten the same result; however, using 1 rather than x is just a bit simpler.

Golden Ratio Construction 11

When we place the three equal unit circles tangent to each other and tangent to the semicircle, as shown in Figure 1-14 (note that the points K and L can easily be found by extending ME and MF), we have the makings for another construction of the Golden Section.

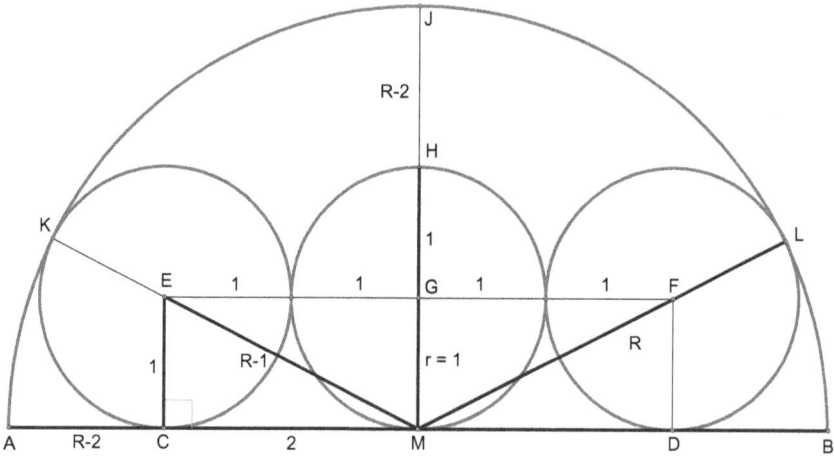

Figure 1-14

First, we note that $AM = BM = JM = KM = LM = R$, and $GH = GM = CE = DF\ (= r) = 1$ (and also $CM = DM = EG = FG = 2$), and $EM = R - r = R - 1$. When we apply the Pythagorean theorem to $\triangle CEM$ in Figure 1-14, we get $(EM)^2 = (CM)^2 + (CE)^2$, or $(R - 1)^2 = 2^2 + 1^2$.

When we solve this equation for R, we get $R^2 - 2R + 1 = 5$, and $R = 1 \pm \sqrt{5}$.

Since a radius cannot be negative, we only use the positive value of R; therefore, $R = 1 + \sqrt{5}$.

We take the ratio $\frac{R}{r} = \sqrt{5} + 1$. Yet, half of this ratio will give us the Golden Ratio: $\frac{1}{2}\left(\frac{R}{r}\right) = \frac{\sqrt{5}+1}{2}$; and therefore, $\frac{LM}{HM} = \frac{R}{2r} = \frac{R}{2} = \frac{\sqrt{5}+1}{2} \approx$

1.61803398874989485. Additionally, the ratios $\frac{HM}{HJ}$ and $\frac{CM}{AC}$ also produce the Golden Ratio, since with $R - 2r = R - 2 = 1 + \sqrt{5} - 2 = \sqrt{5} - 1$ we have $\frac{HM}{HJ} = \frac{CM}{AC} = \frac{2r}{R-2r} = \frac{2}{\sqrt{5}-1} = \frac{\sqrt{5}+1}{2}$.

Golden Ratio Construction 12

The last in our collection of constructions of the Golden Section is one that may look a bit overwhelming but is actually very simple, as it uses only a compass! All we need is to draw five circles.[6]

In Figure 1-15, we begin by constructing circle c_1 with center M_1 and radius $r_1 = r$. Then, with a randomly selected point M_2 on circle c_1,

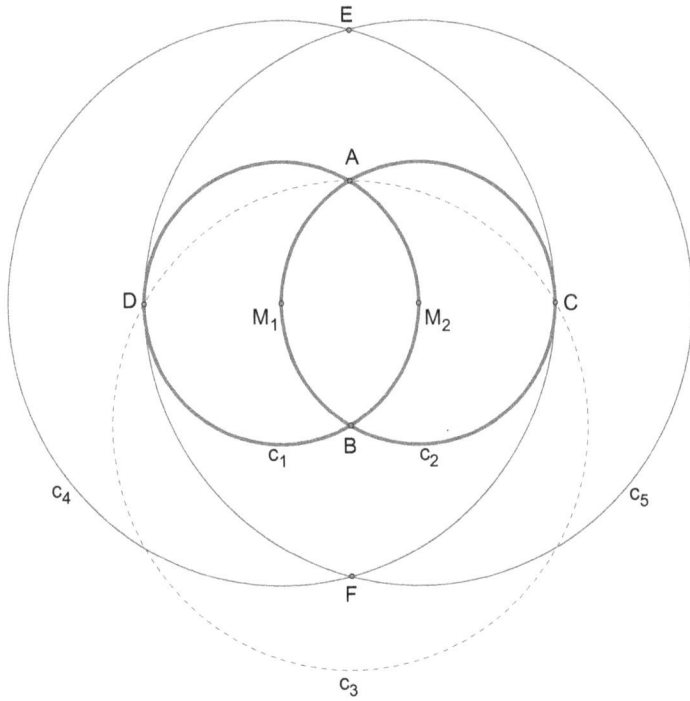

Figure 1-15

---

[6] Kurt Hofstetter: A Simple Construction of the Golden Section. *Forum Geometricorum* 2 (2002), 65–66.

16 *A Journey Through the Wonders of Plane Geometry*

we construct circle c_2 with center M_2 and radius $r_2 = r$; naturally, $M_1M_2 = r$. We denote the points of intersection of the two circles c_1 and c_2 as A and B. Circle c_3 with center B and radius $AB = r_3$ will intersect circles c_1 and c_2 at points C and D. We now construct circle c_4 with center at M_1 and radius $M_1C = r_4 = 2r$. Lastly, circle c_5 with center M_2 and radius $M_2D = r_5 = r_4 = 2r$ is constructed so that it intersects circle c_4 at points E and F.

From Figure 1-16, as a result of obvious symmetry, we have $AE = BF$, $AF = BE$, $AM = BM$, $EM = FM$, $CM = DM$, and $MM_1 = MM_2$. We can then get $\frac{AB}{AE} = \frac{BE}{AB} = \phi$ (or analogously, $\frac{AB}{BF} = \frac{AF}{AB} = \phi$).

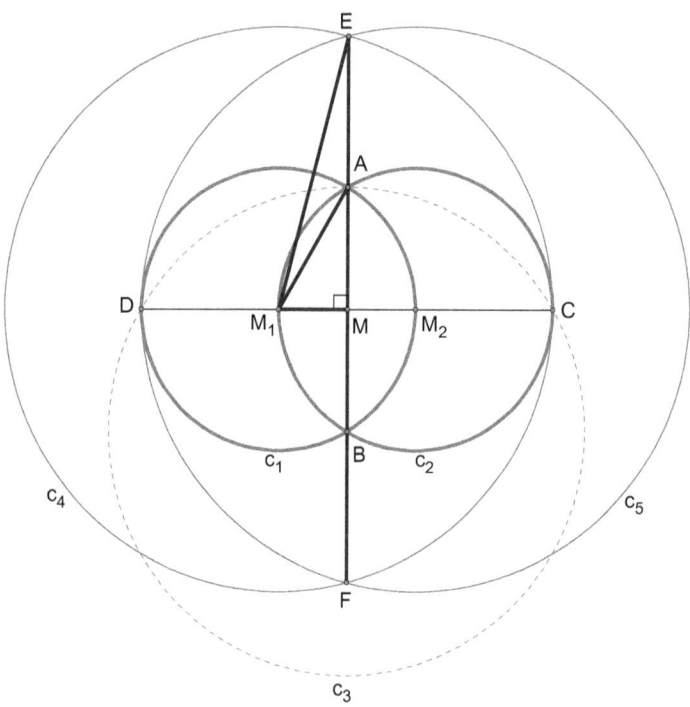

Figure 1-16

This can be justified simply by inserting a few line segments. The radius of the first circle is $r_1 = r = AM_1$, and the radius of the fourth circle is $r_4 = 2r = CM_1 = EM_1$. We can apply the Pythagorean theorem to $\triangle AMM_1$ to get $AM_1^2 = AM^2 + MM_1^2$, or $r^2 = AM^2 + (\frac{r}{2})^2$, which determines $AM = \frac{r}{2}\sqrt{3}$. By applying the Pythagorean theorem to $\triangle EMM_1$ we get $EM^2 + MM_1^2 = EM_1^2 = CM_1^2$, or $(2r)^2 = EM^2 + (\frac{r}{2})^2$, whereupon $EM = \frac{r}{2}\sqrt{15}$.

We now need to show that the ratio we asserted above is in fact the Golden Ratio.

$$\frac{AB}{AE} = \frac{AM+BM}{EM-AM} = \frac{2AM}{EM-AM} = \frac{2 \cdot \frac{r}{2}\sqrt{3}}{\frac{r}{2}\sqrt{15} - \frac{r}{2}\sqrt{3}} = \frac{2\sqrt{3}}{\sqrt{3}(\sqrt{5}-1)}$$

$$= \frac{2}{\sqrt{5}-1} \cdot \frac{\sqrt{5}+1}{\sqrt{5}+1} = \frac{\sqrt{5}+1}{2} = \phi.$$

Now, the second ratio that we must check is

$$\frac{BE}{AB} = \frac{EM+BM}{AM-BM} = \frac{EM+BM}{2AM} = \frac{\frac{r}{2}\sqrt{15} + \frac{r}{2}\sqrt{3}}{2 \cdot \frac{r}{2}\sqrt{3}} = \frac{\sqrt{3}(\sqrt{5}+1)}{2\sqrt{3}} = \frac{\sqrt{5}+1}{2} = \phi.$$

In both cases we have shown that the Golden Ratio is, in fact, determined by the five circles we constructed.

The Golden Rectangle

For centuries, artists and architects have identified what they believed to be the most perfectly shaped rectangle. This ideal rectangle, often referred to as the "Golden Rectangle", has also proved to be the most pleasing to the eye. As mentioned earlier, the Golden Rectangle is one that has the following ratio of its length and width: $\frac{w}{l} = \frac{l}{w+l}$.

Numerous psychological experiments have borne out the desirability of this rectangle. For example, German experimental psychologist Gustav Fechner (1801–1887), inspired by Adolf Zeising's book *Der goldene Schnitt*,[7] began a serious inquiry to see if the Golden

---

[7] Adolf Zeising (1810–1876), a German philosopher, *Neue Lehre von den Proportionen des menschlichen Körpers* (New theories about the proportions of the human body), Leipzig, Germany: R. Weigel, 1854. The book *Der goldene Schnitt* (The Golden Section) was published posthumously by the Leopoldinisch-Carolinische Akademie, Halle, Germany, 1884.

18 *A Journey Through the Wonders of Plane Geometry*

Rectangle had a special psychological aesthetic appeal. His findings were published in 1876.[8] Fechner made thousands of measurements of commonly seen rectangles, such as playing cards, writing pads, books, windows, etc. He found that most had a ratio of length to width that was close to ϕ. He also tested people's preferences and found most people preferred the shape of the Golden Rectangle.

Gustav Fechner asked 228 men and 119 women which of the following rectangles, shown in Figure 1-17, is aesthetically the most pleasing. Which rectangle would you choose as the most pleasing to

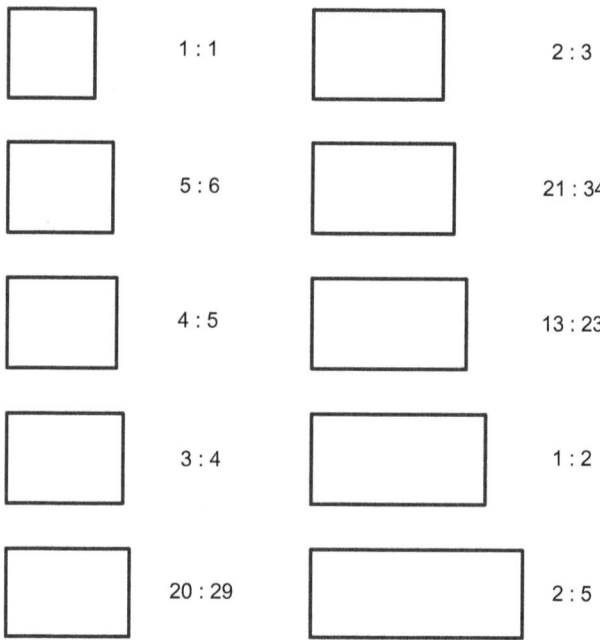

Figure 1-17   Fechner's rectangles

---

[8] Gustav Theodor Fechner: *Zur experimentalen Ästhetik* (On Experimental Aesthetics), Leipzig, Germany: Breitkopf & Haertel, 1876.

look at? Rectangle 1:1 is actually a square, considered by the general public as not representative of a "general rectangle." On the other hand, rectangle 2:5 (the other extreme) is uncomfortable to look at since it requires the eye to scan it horizontally. The rectangle 21:34 seemed to have been most appreciated at a single glance. Fechner's findings bear this out. Here are the results that Fechner reported:

Ratio of sides of rectangle	Percent response for best rectangle	Percent response for worst rectangle
1:1 = 1.00000	3.0	27.8
5:6 = .83333	.02	19.7
4:5 = .80000	2.0	9.4
3:4 = .75000	2.5	2.5
20:29 = .68966	7.7	1.2
2:3 = .66667	20.6	0.4
21:34 = .61765	**35.0**	**0.0**
13:23 = .56522	20.0	0.8
1:2 = .50000	7.5	2.5
2:5 = .40000	1.5	35.7
	100.00	100.00

Figure 1-18

Fechner's experiment has been repeated many times with variations in methodology, and his results have been further supported. For example, in 1917, American psychologist and educator Edward Lee Thorndike (1874–1949) carried out similar experiments with analogous results.

As we can see, the rectangle with the ratio of 21:34 was most preferred. They are the Fibonacci numbers, which we will discuss a bit later in this chapter. The ratio $\frac{21}{34}$ = 0.61764705882352941 approaches the value of ϕ, and gives us a good approximation of the so-called "Golden Rectangle."

Consider the rectangle shown in Figure 1-19, where the length l and the width w are in the following proportion: $\frac{w}{l} = \frac{l}{w+l}$.

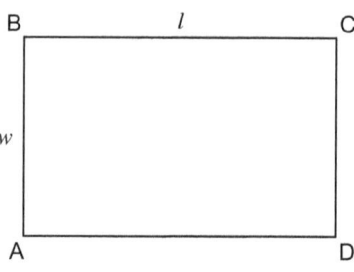

Figure 1-19

By multiplying the means and extremes of this proportion, we get $w^2 + wl = l^2$, or $w^2 + wl - l^2 = 0$. If we let $l = 1$, then $w^2 + w - 1 = 0$. Using the quadratic formula,[9] we get $w = \frac{-1 \pm \sqrt{5}}{2}$. Because we have lengths, the negative value is irrelevant. Therefore, $w = \frac{-1+\sqrt{5}}{2} = \frac{\sqrt{5}-1}{2} = \frac{1}{\phi}$, and the Golden Ratio again emerges.

This is the same equation that gave us $\frac{1}{\phi}$ above. So we now know that the ratio of the rectangle's dimensions is $\frac{w}{l} = \frac{l}{w+l} = \frac{1}{\phi} = \frac{\sqrt{5}-1}{2}$, or $\frac{l}{w} = \frac{w+l}{l} = \phi = \frac{\sqrt{5}+1}{2}$, giving us a Golden Rectangle.

Let's see how this rectangle may be constructed using the traditional Euclidean tools: an unmarked straightedge and compasses. Another way would be to use a geometric construction program such as Geometer's Sketchpad or GeoGebra. With a width of 1 unit, our objective is to get the length to be $\frac{1+\sqrt{5}}{2}$, so that the ratio of the length to the width will be ϕ, which equals $\frac{1+\sqrt{5}}{2}$. Perhaps one of the simpler ways to construct this Golden Rectangle is to begin with a square, as we had in Construction 1 (see Figure 1-2).

As we like to relate mathematical phenomena to each other, we shall mention that Underwood Dudley[10] drew a "cute" relationship

---

[9] The quadratic formula presented in the secondary school algebra course says that the solution of the general quadratic equation $ax^2 + bx + c = 0$ is $x = \frac{-b \pm \sqrt{b^2 - 4ac}}{2a}$.

[10] *Mathematical Cranks*, Washington, D.C.: Mathematical Association of America, 1992.

between the Golden Ratio and the value of π. He showed that the following is just a good approximation, but nothing more: $3.1415926535897932384 \approx \pi \approx \frac{6}{5}\phi^2 \approx 3.1416407864998738178$.

Golden/Fibonacci Spirals

As we return to the Golden Rectangle *ABCD* shown in Figure 1-20, we should recall the relationship of the sides, where $AF = 1$, $AD = \phi$, and $FD = \phi - 1 = \frac{1}{\phi}$. In rectangle *CDFE*, $FD = \frac{1}{\phi}$ and $CD = 1$. When we take the ratio of the length to the width of rectangle *CDFE*, we get $\frac{EF}{FD} = \frac{1}{\frac{1}{\phi}} = \phi$, and it is thus also a Golden Rectangle.

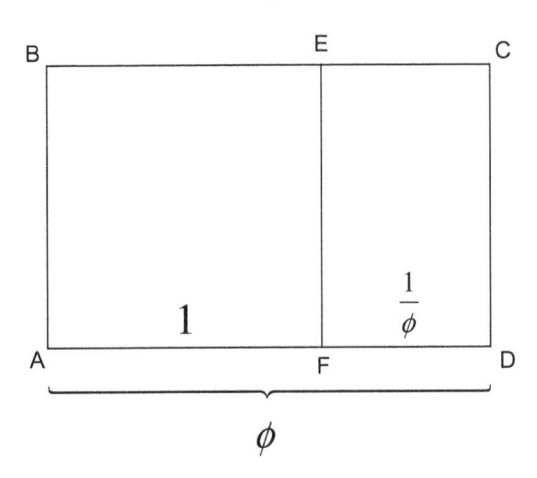

Figure 1-20

Now that we have established that rectangle *CDFE* is a Golden Rectangle, when we construct square *DFGH*, as we show in Figure 1-21, we find that $CH = 1 - \frac{1}{\phi} = \frac{1}{\phi^2}$, so the ratio of the length to width of rectangle *CHGE* is $\frac{\frac{1}{\phi}}{\frac{1}{\phi^2}} = \phi$. Thus, rectangle *CHGE* is also a Golden Rectangle.

We can now continue this process by constructing square *CHKJ* in Golden Rectangle *CHGE*. We find that $EJ = \frac{1}{\phi} - \frac{1}{\phi^2} = \frac{\phi - 1}{\phi^2} = \frac{\frac{1}{\phi}}{\phi^2} = \frac{1}{\phi^3}$.[11] We

---
[11] We showed earlier that $\phi - \frac{1}{\phi} = 1$, therefore, $\phi - 1 = \frac{1}{\phi}$.

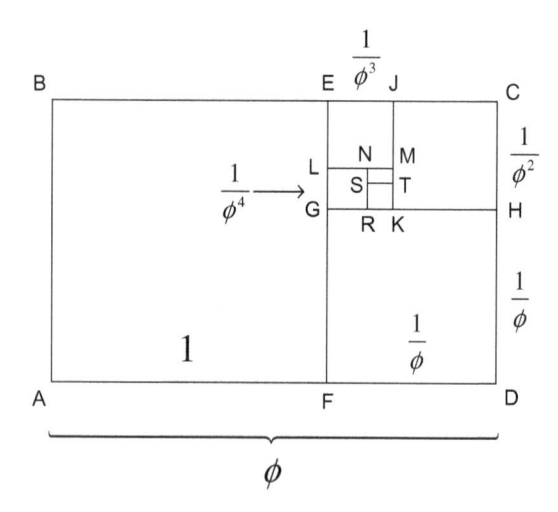

22 A Journey Through the Wonders of Plane Geometry

Figure 1-21

now inspect the ratio of the dimensions of rectangle *EJKG*. This time, the ratio of length to width is $\frac{\frac{1}{\phi^2}}{\frac{1}{\phi^3}} = \phi$. This produces another Golden Rectangle *EJKG*. By continuing this process, we get Golden Rectangles *GKML, NMKR, MNST,* and so on.

Suppose we now draw the following quarter circles with the respective centers and radii as indicated in the following:

- center *E*, radius *EB*
- center *G*, radius *GF*
- center *K*, radius *KH*
- center *M*, radius *MJ*
- center *N*, radius *NL*
- center *S*, radius *SR* etc. (see Figure 1-22).

By connecting these quarter circles, as shown in Figure 1-22, we have essentially created the approximation of a logarithmic spiral, sometimes referred to as a Golden Ratio spiral.

Since all Golden Rectangles have the same shape, rectangle *ABCD* in Figure 1-23 is similar to rectangle *CEFD*. This implies that $\triangle ECD \sim \triangle CDA$. Therefore, $\angle CED = \angle DCA$, and $\angle DCA$ is complementary

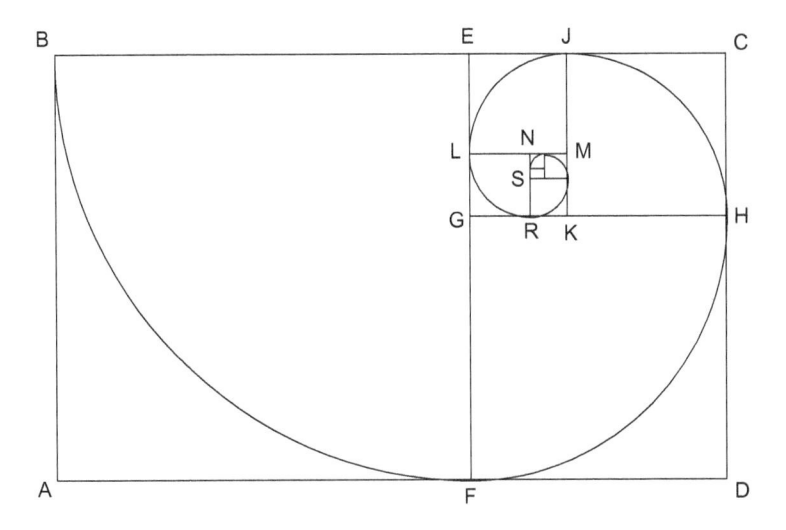

Figure 1-22

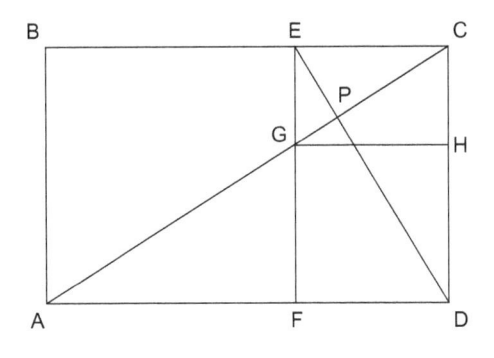

Figure 1-23

to ∠ECA. Therefore, ∠CED is complementary to ∠ECA. Thus, ∠EPC must be a right angle, equivalent to AC⊥ED.

If the width of one rectangle is the length of the other and the rectangles are similar, then the rectangles are said to be *reciprocal rectangles* with a ratio of similitude[12] of ϕ. Therefore, in Figure 1-24,

---

[12] The ratio of similitude is the ratio of the corresponding sides of the two similar figures, in this case, rectangles.

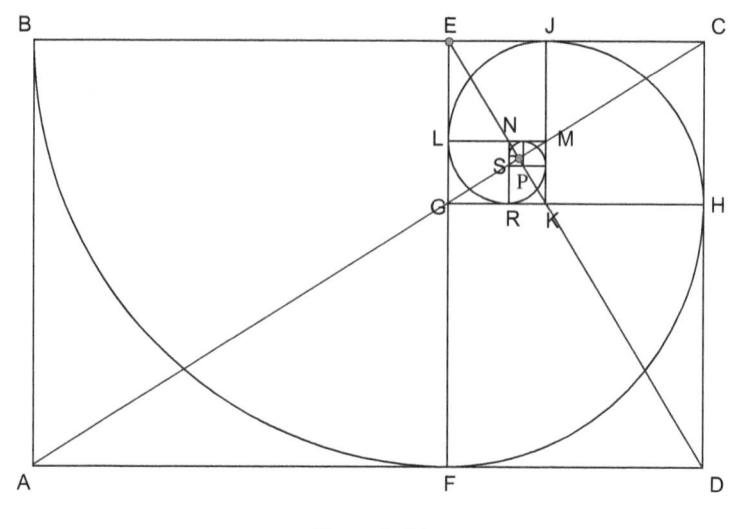

24 A Journey Through the Wonders of Plane Geometry

Figure 1-24

the two Golden Rectangles *ABCD* and *CEFD* are reciprocal and have perpendicular corresponding diagonals. Similarly, rectangles *CEFD* and *CEGH* are also reciprocal, with diagonals $ED \perp CG$.

The Golden Ratio in Architectural History

The more we analyze structures and artwork in early parts of civilization the more we find the ubiquity of the Golden Ratio or Golden Section. From the Western world's point of view, the earliest use of the Golden Ratio occurred in the ancient Egyptians' construction of the Great Pyramid at Giza, the only one of the seven wonders of the ancient world that still exists today. This by no means implies that earlier sightings of the Golden Ratio cannot be found. What is still unknown to this day is whether the architect of this structural wonder, Hemiunu (ca. 2570 BCE), consciously chose the dimensions that yield the Golden Ratio as he strove to achieve beauty in this structure, or if it arose simply by chance. This and other questions about the structure of the pyramid have prompted numerous books, and yet the issue still lacks a definitive conclusion.

The Golden Ratio in Geometry 25

This colossal structure, built about 2560 BCE, is the oldest and largest of three pyramids in the Giza Necropolis near modern-day Cairo, Egypt, which is shown in Figure 1-25.

Figure 1-25   Giza Necropolis, Egypt
Photo courtesy of Free Range Stock, by David McEachan

Although the structure may be analyzed from many perspectives, our interest is largely its exterior dimensions. We will use the cubit[13] as the unit measure, since that was what was used at the time of construction. The diagram of the pyramid (Figure 1-26) shows its height to be 280 cubits, its base length as 440 cubits, and its slant height as 356 cubits.[14]

The ratio of the slant height to half the base length is $\frac{h_{\Delta}}{\frac{a}{2}} = \frac{AB}{BC} = \frac{356}{220} = \frac{89}{55} \approx 1.61818$, which is very close to the Golden Ratio of 1.61803398874989485.... We can take this search for a relationship

---

[13] A *cubit* is the first recorded unit of length used in ancient times, and is the measure from the elbow to the tip of the middle finger, and was assumed to be a length of 52.25 cm.

[14] The British Egyptologist W.M.F. Petrie (1853–1942) established these measurements.

26 A Journey Through the Wonders of Plane Geometry

A

280 cubits

356 cubits

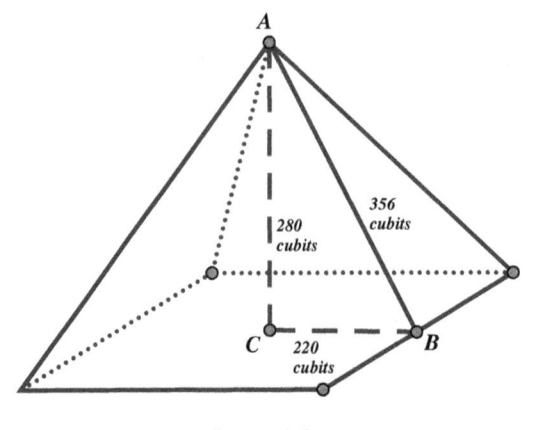

C 220 cubits *B*

Figure 1-26

to the Golden Ratio (ϕ) a step further by noting that the ratio of the height of the pyramid to half the base length, namely, $\frac{280}{220} = \frac{14}{11} = 1.27272\ldots$, is extremely close to $\sqrt{\phi} \approx 1.2720196$. Were we to divide each of the dimensions of triangle *ABC*, shown in Figure 1-26, by 220, we would get a triangle with the dimensions shown in Figure 1-27.

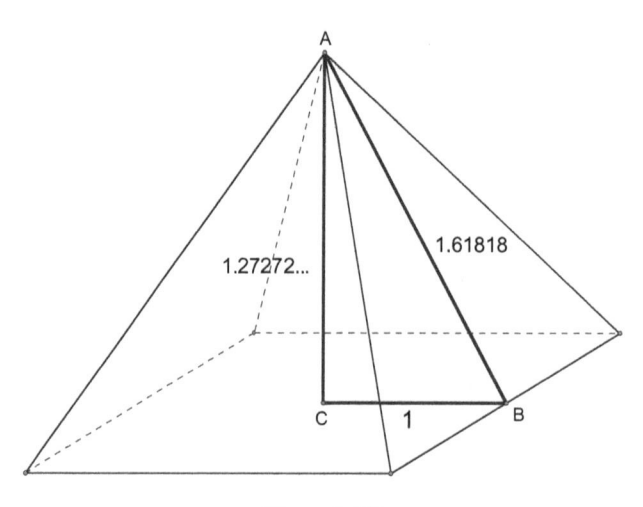

A

1.27272... 1.61818

C 1 B

Figure 1-27

The Golden Ratio in Geometry 27

These values approximate the Golden Ratio in various forms, as we can see in Figure 1-28, where we have, in terms of ϕ, the dimensions of a similar right triangle.

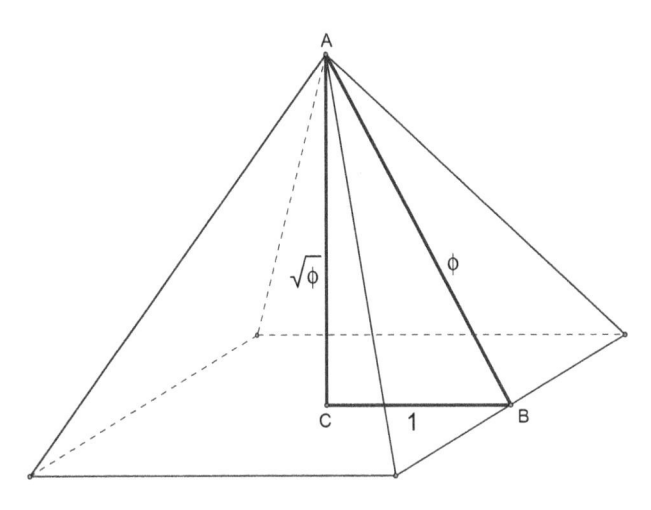

Figure 1-28

According to the Greek historian Herodotus (ca. 484–424 BCE), the Pyramid of Khufu (Cheops) at Giza was constructed in such a way that the square of the height of the pyramid is equal to the area of one of the lateral sides – a relationship that yields some curious results.

When we apply the Pythagorean theorem to triangle ABC in Figure 1-29, we get $h_\Delta^2 = \frac{a^2}{4} + h_P^2$. The area (A) of one of the lateral triangles is $A = \frac{a}{2} h_\Delta$. Using the curious relationship mentioned above that Herodotus ascribed to the Giza Pyramid, we get $h_P^2 = h_\Delta^2 - \frac{a^2}{4} = A = \frac{a}{2} h_\Delta$. If we divide both sides of the equation $\frac{a}{2} h_\Delta = h_\Delta^2 - \frac{a^2}{4}$ by $\frac{a}{2} h_\Delta$, we get $1 = \frac{h_\Delta}{\frac{a}{2}} - \frac{\frac{a}{2}}{h_\Delta}$. By letting $x = \frac{h_\Delta}{\frac{a}{2}}$, then taking the reciprocal $\frac{\frac{a}{2}}{h_\Delta} = \frac{1}{x}$, and substituting in the equation above, we get a simplified equation $1 = x - \frac{1}{x}$. This generates the Golden Equation $x^2 - x - 1 = 0$, whose solution is the Golden Ratio $x_1 = \phi$ and $x_2 = -\frac{1}{\phi}$. We will not consider x_2, since it is negative and holds no real meaning for us geometrically. Using today's measurement capabilities, this great pyramid has the dimensions shown in Figure 1-30.

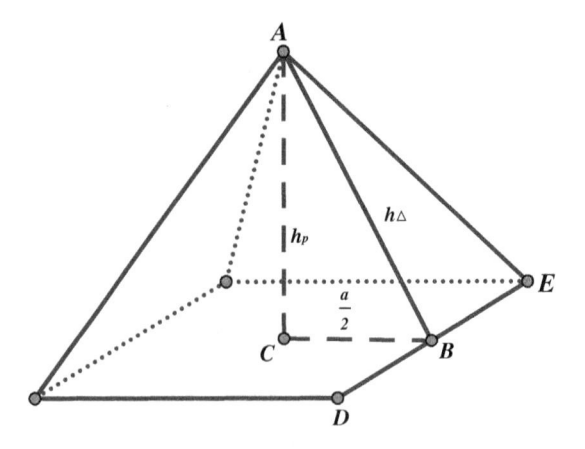

28 A Journey Through the Wonders of Plane Geometry

Figure 1-29

Cheops pyramid	Length of the side of the base: a	Height of lateral triangle: $h_\triangle$	Pyramid height h_P	$\dfrac{h_\triangle}{\frac{a}{2}}$
Measurements	230.56 m	186.54 m	146.65 m	1.61813471 ($\approx \phi$)

Figure 1-30

Surprisingly, the ratio of the height of the lateral triangle to half its base is $\dfrac{h_\triangle}{\frac{a}{2}} = 1.61813471$. This is amazingly close to the Golden Ratio and leads us to wonder if this was done by design or by chance.

Further investigations of the history of the Golden Ratio take us to Euclid's *Elements*, which was a compilation of everything the author knew about mathematics at the time of its writing around 300 BCE. This monumental work consisted of thirteen books, in which there are two references made to the Golden Ratio. In Book II, Proposition 11, Euclid constructs a straight line (segment) that is cut so that the whole segment forms a rectangle and so one of its parts (segments) is equal to the square of the remaining segment. This can be demonstrated pictorially, as in Figure 1-31. Consider the line segment ACB, where point C cuts the segment so that CB is used to form rectangle $ABHF$, where $CB = HB$. The square $ACGD$ would then have the same area as the rectangle *ABHF*. This equality of areas can be expressed as $AC^2 = AB \times CB$, which then can be converted to $\frac{AB}{AC} = \frac{AC}{CB}$, as cited in the *Elements*.[15]

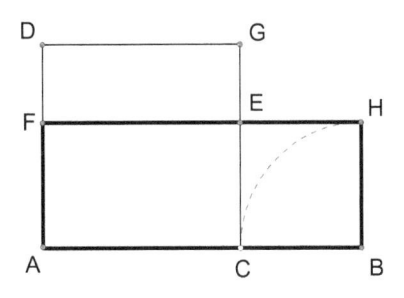

Figure 1-31

There, Euclid refers to a given straight line (segment) that is cut – or sectioned – into a mean and extreme ratio.[16] That is, for the line segment *AB* containing point *C*, we get $\frac{AB}{AC} = \frac{AC}{CB}$, which is the definition of the Golden Section, a name that seems to have been used since the early part of the nineteenth century.[17]

We find the Golden Ratio's next prominent display in the works of the great Greek sculptor Phidias (ca. 490–430 BCE). His design for the construction of the Parthenon in Athens, Greece, (Figure 1-32) as well as the famous statue of Zeus at Olympia, are said to be reflective of this beautiful ratio. As a matter of fact, the Greek letter ϕ is used by many mathematicians today to represent the Golden Section because it is the first letter of Phidias' name when written in the Greek alphabet (Φειδίας). As we see in Figure 1-32, the Parthenon fits nicely into a Golden Rectangle. Furthermore, Figure 1-33 demonstrates several additional Golden Ratios. Yet even today, no one can say with

---

[15] Euclid's *Elements*, Book VI, Definition 3.
[16] This "mean and extreme ratio" Euclid defines (Book VI, Definition 3) as follows: "A straight line [segment] is said to have been cut in *extreme and mean ratio* when, as the whole line [segment] is to the greater, so is the greater to the lesser."
[17] The first use of the term "Golden Section" is attributed to Martin Ohm (1792–1872), who referred to this partitioning of a line segment as the "Goldener Schnitt" in his book *Pure Elementary Mathematics* in 1835.

30 *A Journey Through the Wonders of Plane Geometry*

Figure 1-32   The Parthenon in Athens, Greece

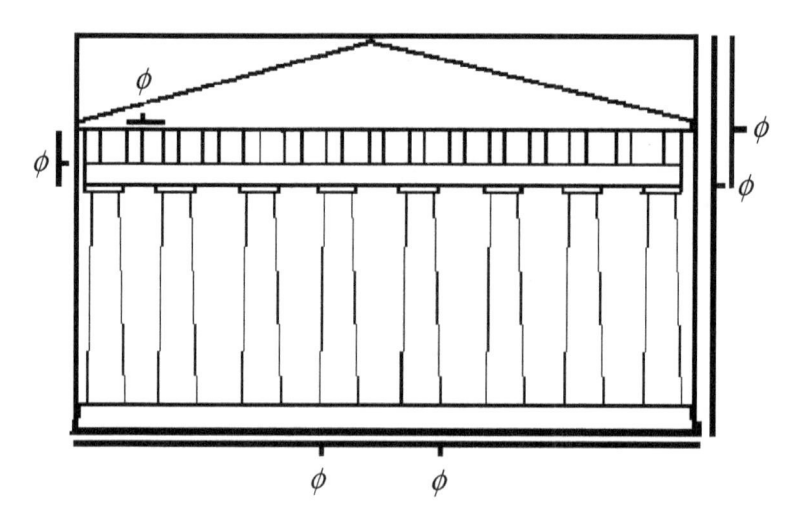

Figure 1-33

certainty that Phidias had the Golden Ratio in mind when he designed the structure.

The oldest stone structure in Germany, dating to the post-Roman period (764 CE), is the "Königshalle" in the town of Lorsch (Figure 1-34). It was erected by the Frankish Count Candor and his mother Williswinda as a proprietary church and monastery. This magnificent example of architecture from the early Middle Ages has an impressive open-air ground floor. The interior space of this building exhibits a practically perfect Golden Rectangle.

Figure 1-34 Lorsch Abbey, Germany
CC-BY SA 3.0 (https://creativecommons.org/licenses/by-sa/3.0/)

The Cathedral in Chartres, France, (built 1194–1260) clearly exhibits the Golden Rectangle. In Figure 1-35, you can see the front portal exhibiting this famous rectangle.

The famous Santa Maria del Fiore Cathedral in the city of Florence, Italy, exhibits the Golden Rectangle throughout its construction. In Figure 1-36, we show two places where the Golden Rectangle can be seen.

32 A Journey Through the Wonders of Plane Geometry

Figure 1-35 Chartres Cathedral, France
CC BY-SA 2.0 (https://creativecommons.org/licenses/by-sa/2.0/)

When speaking of the Golden Rectangle and the Golden Ratio, we are led to the natural connection to the Fibonacci numbers. Leonardo of Pisa (today known as Fibonacci)[18] first presented Fibonacci numbers in chapter 12 of his 1202 book *Liber abaci*, in which they aided his solution to a problem about the regeneration of rabbits. The *Fibonacci numbers* begin with 1 and 1, and each succeeding number is the sum of the two preceding numbers. Thus, the Fibonacci sequence looks like this: 1, 1, 2, 3, 5, 8, 13, 21, 34, 55, 89, 144, ...

The relationship between the Golden Ratio and the Fibonacci numbers is quite astounding. As the Fibonacci numbers increase, the

---

[18] *Fibonacci* is a short version of *Filius Bonacci*, which means "son of Bonacci" in Latin, telling us that his father was named Bonacci.

The Golden Ratio in Geometry **33**

Figure 1-36   Santa Maria del Fiore Cathedral in Florence, Italy
CC BY 3.0 (https://creativecommons.org/licenses/by/3.0/)

ratio of two consecutive Fibonacci numbers approaches the Golden Ratio. The larger the selected consecutive Fibonacci numbers are, the closer the ratio is to the Golden Ratio. In other words, although $\frac{89}{55} = 1.618181818181\overline{818}$ is very close to the Golden Ratio, when we choose a pair of larger consecutive Fibonacci numbers, such as $\frac{6765}{4181} \approx 1.6180339631667065\ldots$, the ratio is even closer to the Golden Ratio, $\phi \approx 1.618033988749894848\ldots$ (Incidentally, if we take the

34 *A Journey Through the Wonders of Plane Geometry*

reciprocal $\frac{4181}{6765}$, we get 0.618033998 . . . , which essentially differs from its reciprocal by 1.) Therefore, as we search for mathematical appearances in architecture and art, we can seek Golden Rectangles, Golden Triangles, or Fibonacci numbers.

Throughout the Renaissance, the designs of many construction projects used the Fibonacci numbers or the Golden Ratio. The Golden Rectangle can be found on the front of the Santa Maria del Fiore Cathedral in Florence, Italy, shown in Figure 1-36. The rough sketch (Figure 1-37) by Giovanni di Gherardo da Prato (1426) exhibits the Fibonacci numbers 55, 89, and 144, as well as 17 (which is half the Fibonacci number 34) and 72 (which is half the Fibonacci number 144).

The remainder of the Florence Cathedral further demonstrates the Fibonacci numbers. But the most dramatic proportions are those shown in Figure 1-37, where the ratio of 89:55 (= 1.6181818 . . .) has an extremely close approximation to the Golden Ratio 1.618033988 . . . .

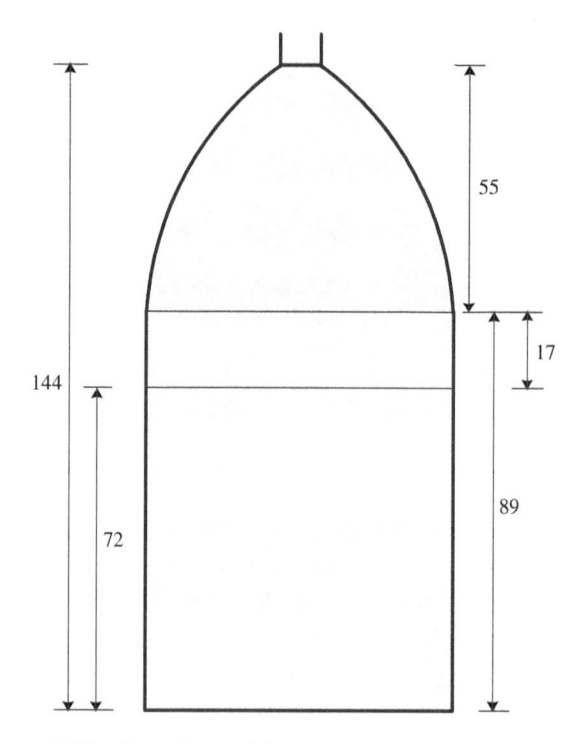

Figure 1-37 Santa Maria del Fiore Cathedral in Florence, Italy

The Golden Ratio in Geometry   35

More recently, the Swiss-French architect Le Corbusier (1887–1965)[19] designed the *unités d'habitation* in Marseille, France, (Figure 1-38) between 1946 and 1952, which again exhibits the Golden Ratio. Here, it is exhibited in the tower that divides the remainder of the building into the Golden Ratio, as indicated by the bold horizontal lines at the top of the building. This was not done by chance, but rather by specific design! Le Corbusier displays the Golden Section throughout his designs.

Figure 1-38   *Unités d'habitation* in Marseille, France

In 1947, Le Corbusier was a member of the architectural commission that planned the building of the U.N. headquarters in New York, shown in Figure 1-39. He was ultimately responsible for the basic concept of the plans for the 39-story Secretariat building. As you might expect, the height and the width of the building are close to the ratio of 1.618 : 1. Once again, we see the Golden Ratio.

Last, but not least, an obvious exhibit of the Golden Ratio can be found in an aerial view of the Pentagon in Arlington, Virginia

---

[19] His actual name was Charles Édouard Jeanneret-Gris.

36 A Journey Through the Wonders of Plane Geometry

Figure 1-39   United Nations Headquarters in New York City, USA

(see Figure 1-40). By now you might (rightly) expect that the intersection point of two diagonals divides each of the diagonals in the Golden Ratio, as is the case with any regular pentagon.

Figure 1-40   Pentagon building in Arlington, Virginia, USA
Combined Military Service Digital Photographic Files released to the public

In Figure 1-41, we see the various places where the diagonals of the pentagon intersect to form the Golden Ratio in a number of ways: $d : a = \phi$, $a : e = \phi$, and $e : f = \phi$.

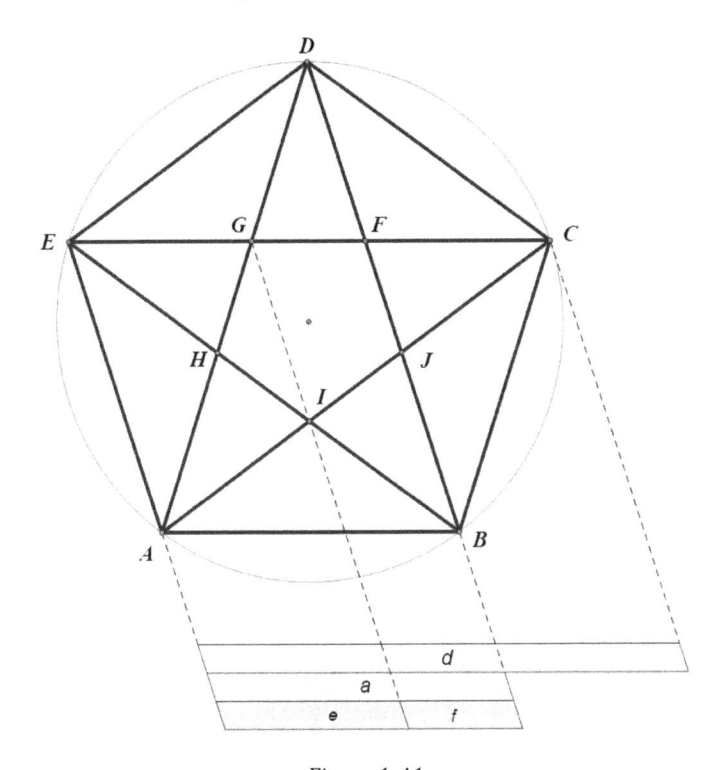

Figure 1-41

In the field of architecture there are countless illustrations of the use of the Golden Ratio, which, of course, employ the Fibonacci numbers. The lingering question persists: Are these coincidences, or are these Golden Sections deliberate? In the case of some buildings, such as those designed by Le Corbusier, we have enough evidence to know that the use of the Golden Ratio was intentional. As for ancient architecture, we can sometimes speculate or assume that if an architect had a connection with a mathematician, then he might have used it by design. Still, there is always the notion that this ratio delivers the most beautifying partition in geometry, and therefore the "trained eye" might have simply landed on the Golden Ratio through artistic prowess.

Golden Triangles

The isosceles triangles that can be seen in the diagonals of the regular pentagon are called *Golden Triangles*. There are two types of such isosceles triangles: The *acute* ones have the ratio $\frac{\text{leg}}{\text{base}} = \phi$, and the *obtuse* triangles have the ratio $\frac{\text{base}}{\text{leg}} = \phi$. Both types can also be determined by their vertex angle, shown in Figure 1-42.

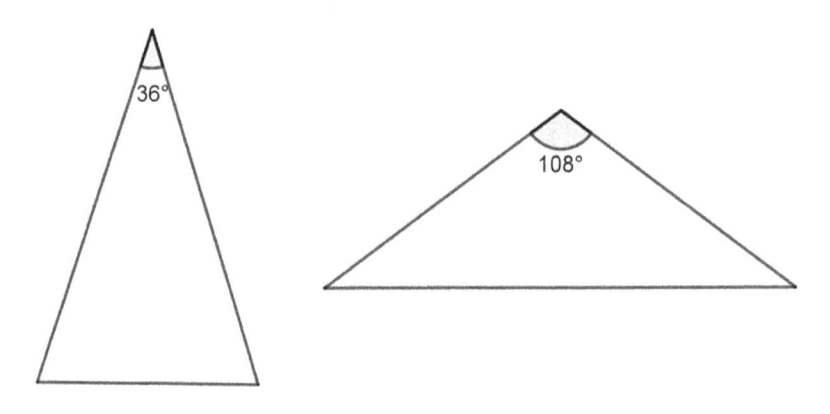

Figure 1-42   Golden triangles

The vertex angle of 108° emanates from the 108° interior angle of a regular pentagon. The vertex angle in the other case is 36° since, as we will see, the diagonals trisect the angles of a regular pentagon (see Figure 1-41). Due to symmetry, the isosceles triangles $\triangle AED \cong \triangle BCD$, we have $\angle ADE = \angle BDC = \frac{180°-108°}{2} = 36°$ and thus also $\angle ADB = 108 - (2 \times 36°) = 36°$. Consequently, we have the trisection $\angle EDA = \angle CDB = \angle ADB = 36°$. Since AED is isosceles, we have $\angle EDA = \angle DAE$ and $\angle ADB = \angle DAE$, so that $AE \| BD$, and we can show that $AEDJ$ is a parallelogram, even a rhombus (note that, analogously, we have $AC \| ED$).

Golden Parallelograms

A *Golden Parallelogram* has sides in the ratio $\phi : 1$ and an acute angle of 60°. It has some striking properties. Consider cutting off two equilateral triangles, as shown in Figure 1-43, where the remaining

The Golden Ratio in Geometry 39

parallelogram is still golden. This is justified because the interior angles of the reduced parallelogram are 60° and 120°, and its side lengths are $\phi - 1$ and 1, and for the corresponding ratio we have $(\phi-1):1=\frac{1}{\phi}$. (Recall the crucial property of ϕ that $\phi-1=\frac{1}{\phi}$, or equivalently $\phi-\frac{1}{\phi}=1$ determines ϕ.)

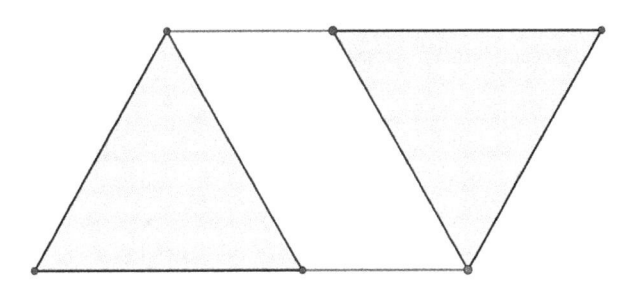

Figure 1-43 Equilateral triangles cut off in a Golden Parallelogram

Furthermore, the diagonals of the parallelogram have the ratio $\phi : 1$ and determine an angle of 60°. This can be justified by using the *law of cosines* to calculate the lengths of the diagonals of a parallelogram with side lengths 1 and $\phi = \frac{1+\sqrt{5}}{2}$, and angles of 60° and 120°. For the shorter diagonal f, seen in Figure 1-44, we apply the law of cosines in the triangle ABD to get $f^2 = 1^2 + \left(\frac{1+\sqrt{5}}{2}\right)^2 - \left(1+\sqrt{5}\right)\cdot\underbrace{\cos 60°}_{\frac{1}{2}} =$ $1 + \frac{\sqrt{5}+1}{2}\cdot\frac{\sqrt{5}-1}{2} = 2$, where then $f = \sqrt{2}$. Analogously, the longer diagonal is $e = \phi\cdot\sqrt{2}$. Using the law of cosines in triangle AMD, we get

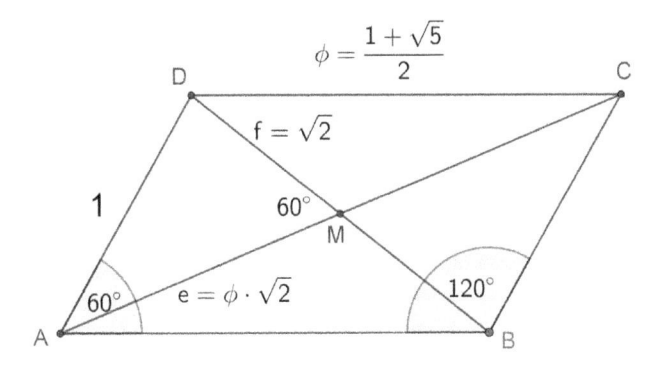

Figure 1-44 Diagonals in the Golden Ratio make an angle of 60°

40 A Journey Through the Wonders of Plane Geometry

$1^2 = \left(\frac{f}{2}\right)^2 + \left(\frac{e}{2}\right)^2 - \frac{ef}{2}\cos\angle AMD = \frac{1}{2} + \frac{\phi^2}{2} - \phi\cos\angle AMD$, which is equivalent to $\cos\angle AMD = \frac{\phi^2 - 1}{2\phi} = \frac{1}{2}$, and, thus, $\angle AMD = 60°$.

The medial quadrilateral, shown in Figure 1-45, is again a Golden Parallelogram and is thus similar to the initial parallelogram. This follows immediately since the sides of the medial quadrilateral are parallel to, and half the length of, the diagonals.

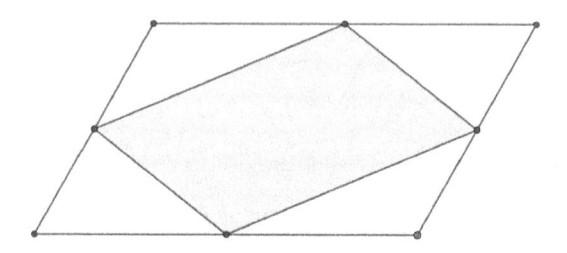

Figure 1-45 The medial quadrilateral is also a Golden Parallelogram

The Golden Ratio in Platonic Solids

There are only five *regular* polyhedra – that is, those three-dimensional shapes that have congruent polygon faces, vertices, and edges, shown in Figure 1-46. These *Platonic solids* are named after the Greek philosopher Plato (ca. 424–348 BCE), who described them as representing the elements of the universe in his work *Timaeus*. Plane geometry also plays an important role in appreciating three-dimensional geometry as we see in this section.

Tetrahedron	Hexahedron (Cube)	Octahedron	Dodecahedron	Icosahedron
4 vertices 6 edges 4 (*tetra*) faces (equilateral triangles)	8 vertices 12 edges 6 (*hexa*) faces (squares)	6 vertices 12 edges 8 (*okta*) faces (equilateral triangles)	20 vertices 30 edges 12 (*dodeka*) faces (regular pentagons)	12 vertices 30 edges 20 (*eikosi*) faces (equilateral triangles)

Figure 1-46

The Greek mathematician Euclid (ca. 300 BCE) in his *Elements* (Book XIII, 13–17) describes the construction of these five polyhedra using only a straightedge and compasses, and then proves that these are the only such regular polyhedra.

Among their many properties, each of the five Platonic solids can be inscribed in a sphere and can have a sphere inscribed in them. There is a famous formula discovered by the Swiss mathematician Leonhard Euler (1707–1783) that connects the number of vertices (v), edges (e), and faces (f) of any convex polyhedron: $v + f = e + 2$.[20] Curiously, the octahedron (Figure 1-47) is the only Platonic solid that can be colored with four colors so that no same colors share a common edge.

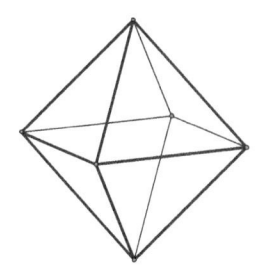

Figure 1-47

Of the five Platonic solids, only three of these, namely, the octahedron, the dodecahedron and the icosahedron, demonstrate the Golden Ratio. These are shown from various aspects in Figure 1-48.

Further investigation of these Platonic solids leads us to some interesting aspects hiding the Golden Ratio. The icosahedron can produce Golden Rectangles by connecting the 12 vertices to form three congruent Golden Rectangles that, in pairs, also happen to be perpendicular to one another, as we can see in Figures 1-49 and 1-50. This was first discovered by Luca Pacioli (ca. 1446–1517) in his book *De divina proportione*.

---

[20] A proof of this formula can be found in H.S.M. Coxeter: *Introduction to Geometry*. 2nd ed., New York: Wiley, 1989.

42 *A Journey Through the Wonders of Plane Geometry*

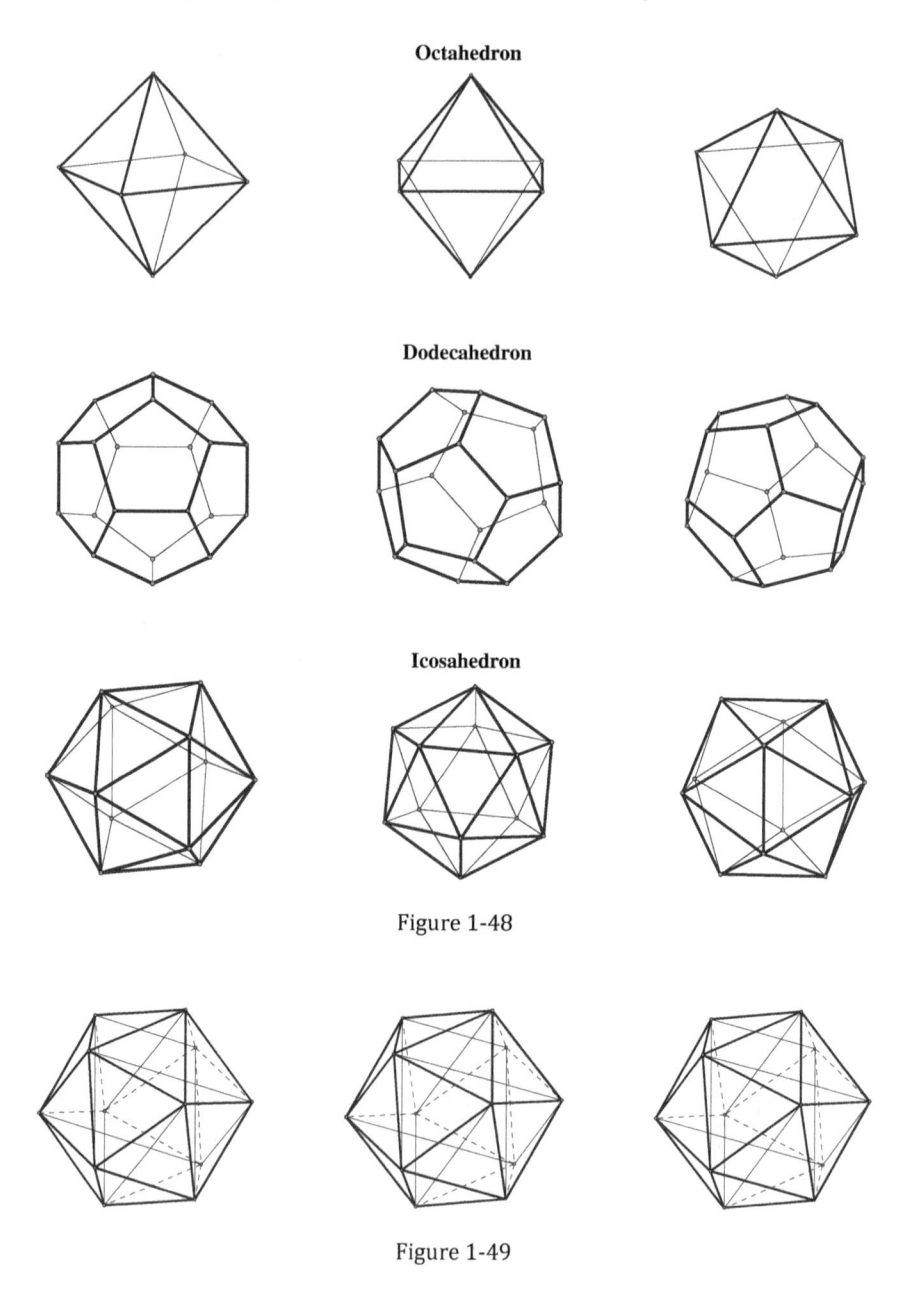

Octahedron

Dodecahedron

Icosahedron

Figure 1-48

Figure 1-49

The Golden Ratio in Geometry **43**

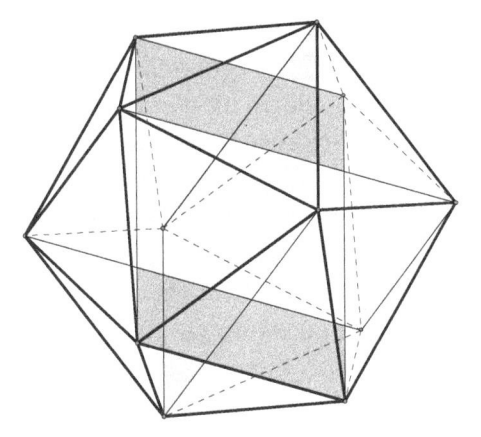

Figure 1-50

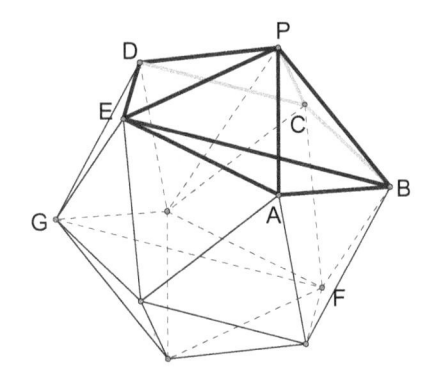

Figure 1-51

In Figure 1-49, we show the three congruent and perpendicular Golden Rectangles separately, and in Figure 1-50, we show them together in one icosahedron. To better understand this, consider the five equilateral triangles that share vertex *P*, as shown in Figure 1-51. They form a pyramid with the regular pentagonal base *ABCDE*.[21]

It can be shown that *BEGF* is a rectangle with opposite sides *BF* and *EG*, which are also edges of the icosahedron, as can be seen in Figure 1-51. The longer side *BE* of this rectangle is also a diagonal of

---

[21] Hans Walser: *The Golden Section*. Washington, DC: Mathematical Association of America, 2001.

44 A Journey Through the Wonders of Plane Geometry

the pentagon *ABCDE*. In a regular pentagon, the ratio of the diagonal to a side is the Golden Ratio, $\frac{BE}{AE} = \phi$. Therefore, $\frac{BE}{BF} = \frac{BE}{AE} = \phi$. Analogously, the other two rectangles shown in Figure 1-50 are also Golden Rectangles. This can be more clearly seen by constructing three congruent Golden Rectangles with a slit in each piece of cardboard down the middle and parallel to the longer side. The slit should be long enough to allow for the width of another of the pieces to fit. You may have to extend one slit to the side of the rectangle. Then place the rectangles so that each one goes through the slit of another, as shown in Figure 1-52. By carefully connecting the vertices of this structure with a string, we can construct the icosahedron, which is shown in Figure 1-53. In a regular icosahedron it is possible to erect five such individual structures.

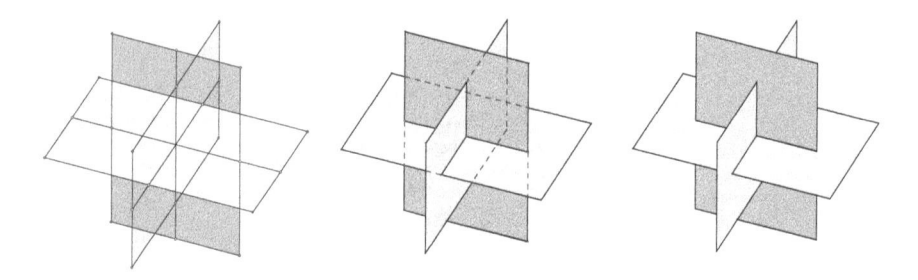

Figure 1-52

Since the pyramids formed at each vertex of the icosahedron have a base in the shape of a pentagon – for example, *ABCDE* in Figure 1-56 – we notice the close connection that the Golden Ratio has with the icosahedron.

In the interior of the icosahedron, there are numerous regular pentagons, and five equilateral triangles meet at each vertex. This should facilitate finding other parts of this figure that show the Golden Ratio.

As we move along to the octahedron, we recall that it has 12 edges, while the icosahedron has 12 vertices. This allows us to "encase" a regular icosahedron in a regular octahedron by having each edge of the octahedron contain one vertex of the icosahedron, as is shown in Figure 1-53. Moreover, the edges of the octahedron are partitioned by the vertices of the icosahedron in the Golden Ratio.

The Golden Ratio in Geometry **45**

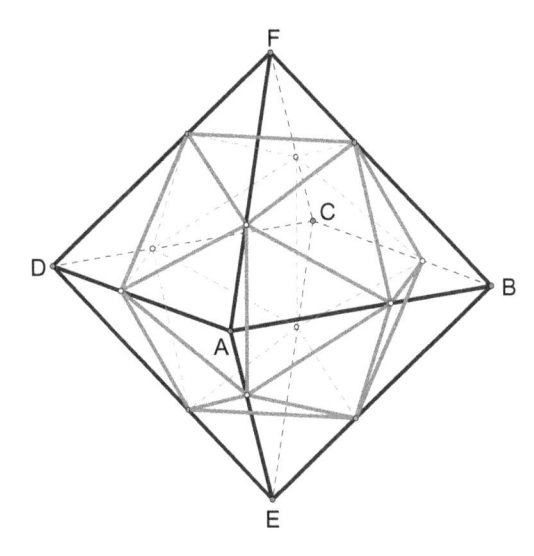

Figure 1-53

To justify this unexpected relationship, we must first notice that the six vertices of the octahedron, *A*, *B*, *C*, *D*, *E*, and *F*, are also the vertices of three mutually perpendicular squares *ABCD*, *AECF*, and *BEDF*. Furthermore, the sides of the squares are the edges of the octahedron. This can be seen in Figure 1-54.

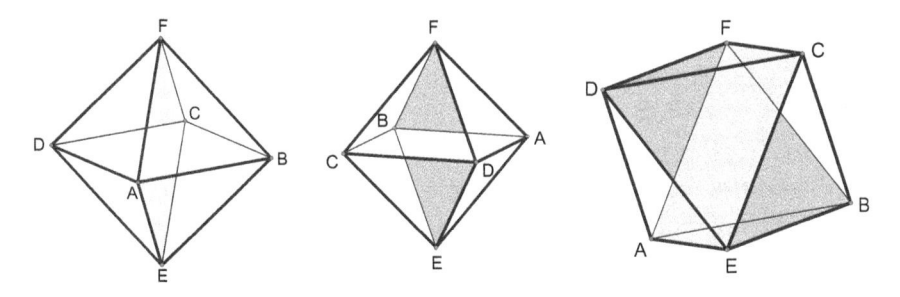

Figure 1-54

If we select the points on the 12 sides of the octahedron that cut each side into the Golden Ratio, as shown in Figures 1-55 and 1-56, we will determine the vertices of an icosahedron. This produces the perpendicular Golden Rectangles that we identified earlier in Figure 1-52, which then justifies their perpendicularity.

Figure 1-55

Figure 1-56

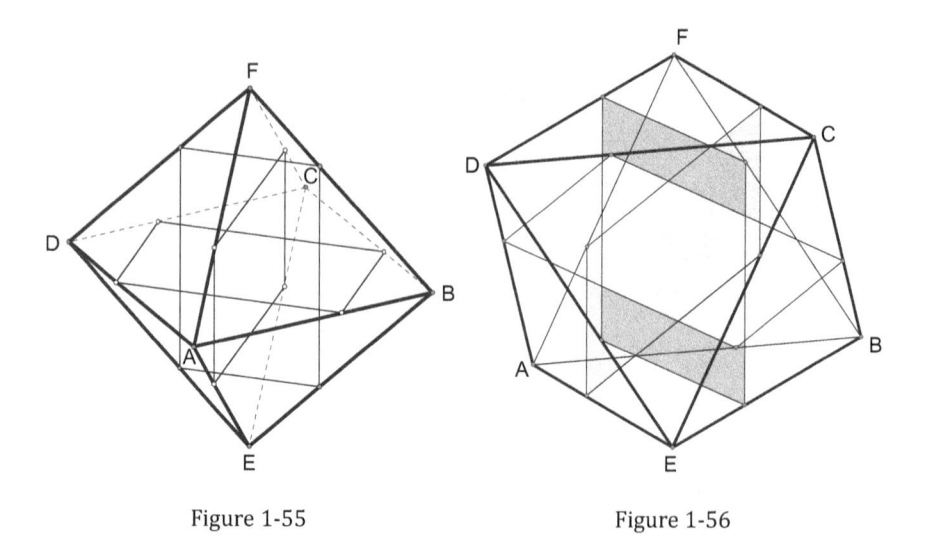

The Golden Ratio in Art

As we continue to trace the history of the Golden Ratio, we find a significant sighting in the aforementioned book *Divina Proportione*. The book contains drawings by the Italian painter, sculptor, architect, and mathematician Leonardo da Vinci (1452–1519) of the five Platonic solids and, as a frontispiece, the Vitruvian Man (Figure 1-57), drawn about 1487. This is a picture of a man's body – supposedly Marcus Vitruvius Pollio (ca. 75 BCE–15 BCE), a Roman writer, architect, and engineer – that clearly exhibits a very close approximation to the Golden Ratio. Vitruvius had a great influence on Roman architecture largely through his multivolume *De architectura, libri decem* (The Ten Books on Architecture), which influenced the building of structures for many centuries. It resurfaced in the 15th century and is still available in modified form today. It is believed that the Roman Pantheon's incredible preservation is due to Vitruvius' influence. This might be the reason that da Vinci named this picture for this great architect.

Da Vinci provided notes based on the work of Vitruvius. The drawing, which is in the possession of the Gallerie dell'Accademia in

The Golden Ratio in Geometry 47

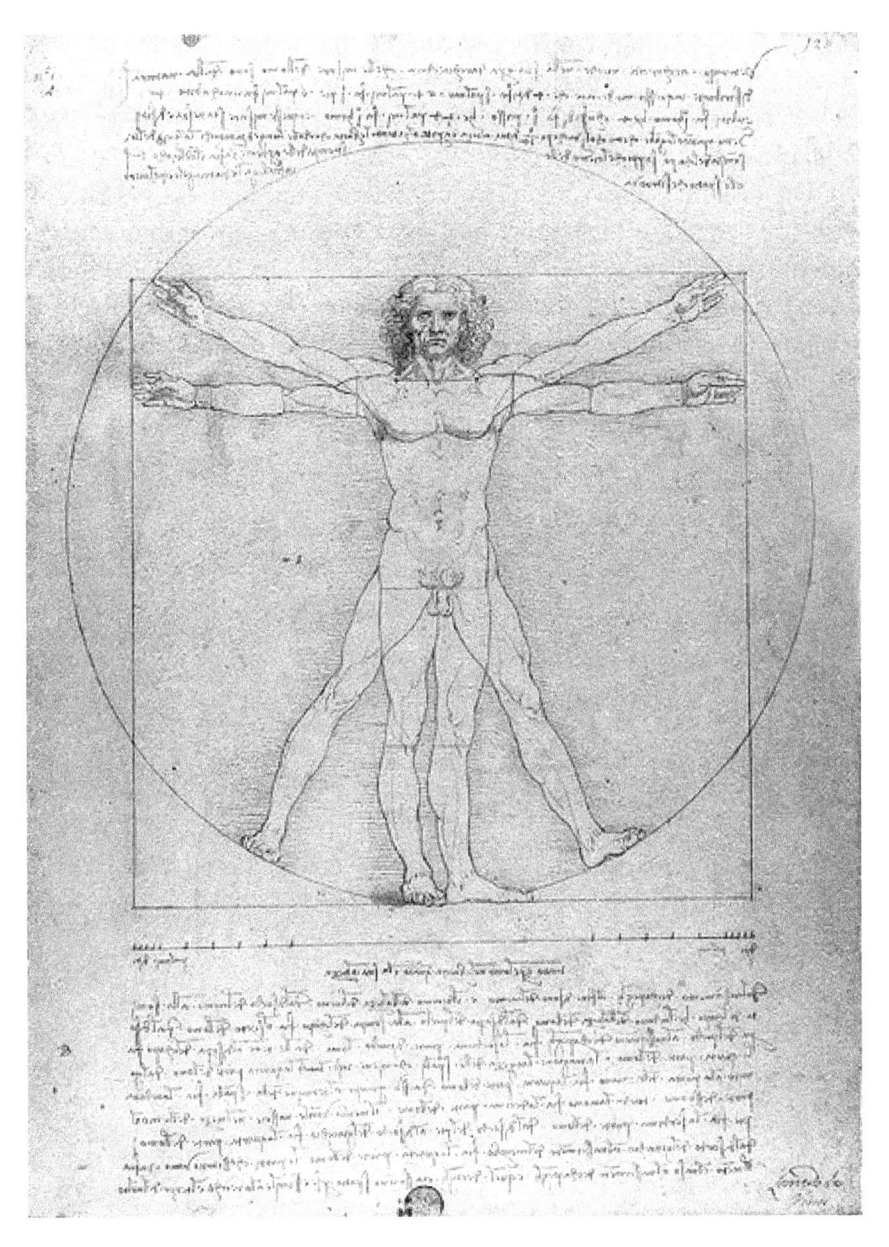

Figure 1-57   The Vitruvian Man

Venice, Italy, is often considered one of the early breakthroughs of pictorially depicting a perfectly proportioned human body. Apparently, da Vinci derived these geometric proportions from Vitruvius' treatise *De Architectura*, Book III.

The drawing shows a male figure in two superimposed positions with his arms and legs apart and inscribed in a circle and square, which are tangent at only one point. The Golden Ratio is exhibited when the distance from the soles of the man's feet to his navel (which appears to be the center of the circle, as shown in Figure 1-58) is

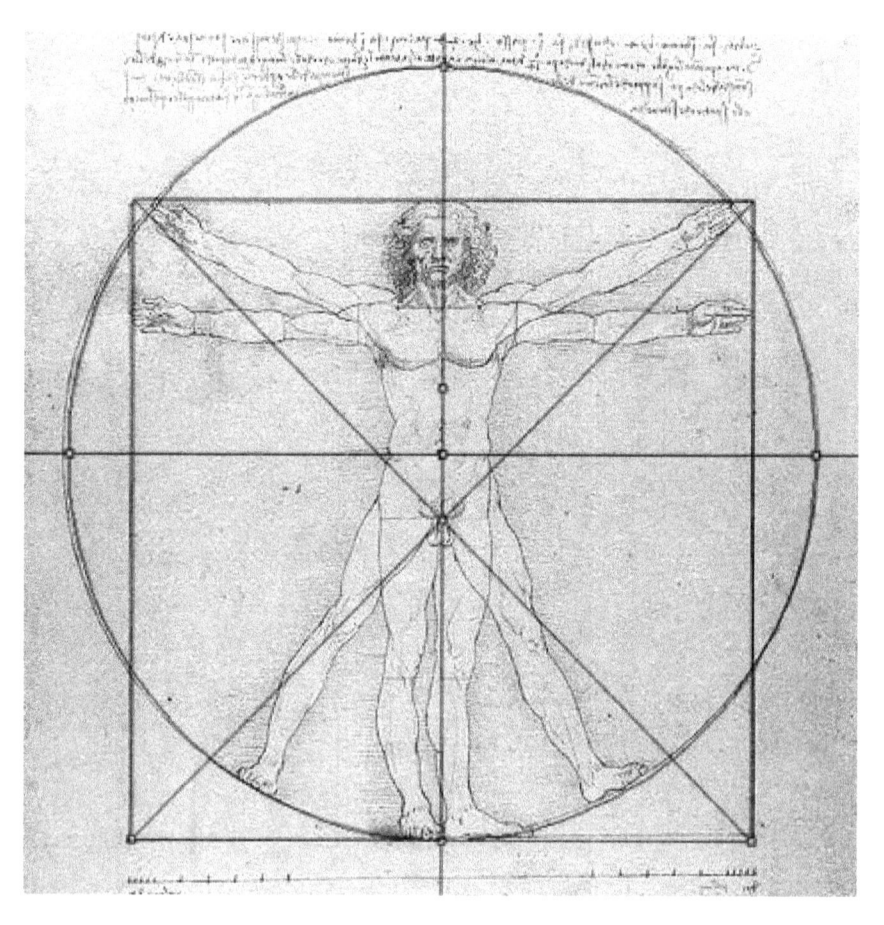

Figure 1-58

divided by the distance from the navel to the top of his head. The quotient is about 0.656, and approximates the Golden Ratio (which we know is 0.618...).

Had the square's upper vertices been somewhat closer to the circle, then the Golden Ratio would have been attained. This can be seen in Figure 1-59, where the radius of the circle is selected to be 1, and the side of the square is 1.618, approximately equal to ϕ, the Golden Ratio.

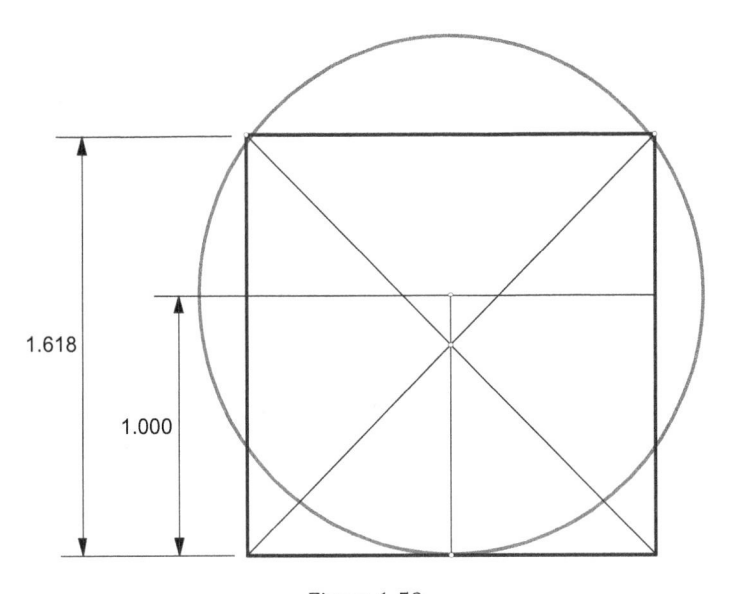

Figure 1-59

Since Leonardo da Vinci was aware of the Golden Ratio, we might wonder where it might appear in his works. The most famous work by Leonardo da Vinci is *Mona Lisa*, (Figure 1-60) which was painted between 1503 and 1506 and is on exhibit in the Louvre Museum in Paris. King François I of France paid 15.3 kilograms of gold for this painting. Let's look at this masterpiece from the point of view of the Golden Ratio. When a rectangle is drawn around Mona Lisa's face, it seems to measure exactly as a Golden Rectangle.

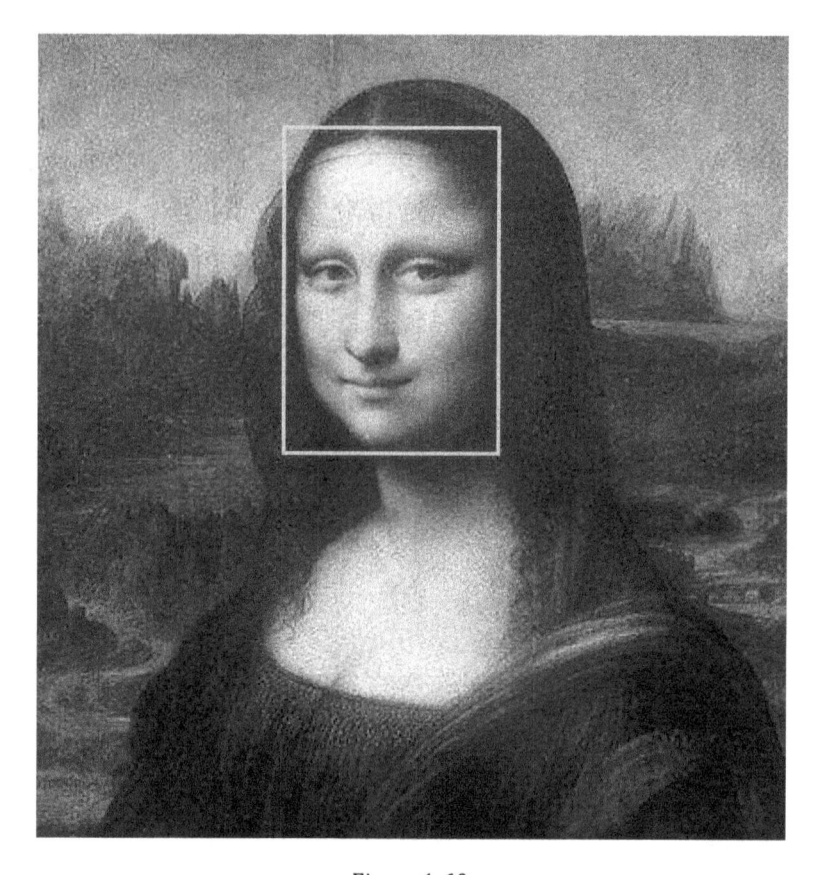

Figure 1-60

Since da Vinci illustrated Pacioli's book *Da Divina Proportione*, which thoroughly discussed the Golden Ratio, it can be assumed that he was guided by this magnificent ratio.

In 1500, one of Germany's most famous artists, Albrecht Dürer (1471–1528), produced a self-portrait. He drew himself with a head of wavy hair, the outlines of which create an equilateral triangle, as can be seen in Figure 1-61, where Dürer superimposes the triangle and several other guidelines over the self-portrait. The base of the equilateral triangle divides the height of the entire picture into the Golden Ratio. The chin also divides the height of the entire picture into the Golden Ratio – in the other direction.

The Golden Ratio in Geometry **51**

Figure 1-61

Many artists have consciously used the Golden Ratio in their artwork and said so in their descriptions. The Golden Ratio appears throughout the artwork discussed here, though we are not always certain whether it was done intentionally or subconsciously. In any case, the Golden Ratio seems to manifest itself geometrically in so many different areas.

Chapter 2

Unexpected Concurrencies

We know that two nonparallel lines create a unique point of intersection. However, three nonparallel lines need not necessarily intersect at a unique point. In fact, when three or more lines share a common point, this is a special situation, and we have a concurrency that is worth noting. We will start with the four main centers of a triangle, all of them points of concurrency of three particular lines. One can prove three of these concurrencies by a famous theorem called *Ceva's theorem*. However, we will introduce this very useful theorem a bit later in this chapter. Rather, we will discuss the concurrencies as presented in most high school textbooks, since Ceva's theorem is, unfortunately, not included in the typical high school curriculum.

Recall that all the points along the perpendicular bisector of a line segment AB are equidistant from points A and B, as we can see in Figure 2-1, where $CD \perp AB$ at point E, and $AC = BC$ and $AD = BD$. Also, all points along the bisector of the angle with vertex A are equidistant from the two sides of the angle. In Figure 2-2, where AG bisects angle BAC, any point on AG is equidistant from the two rays of the angle, so that $FE = FD$. In both cases, the converse is also true: that is, if a point has equal distances to two points A and B, then it must lie on the perpendicular bisector of AB; and if a point has equal distances from two rays of an angle, it must lie on the angle bisector.

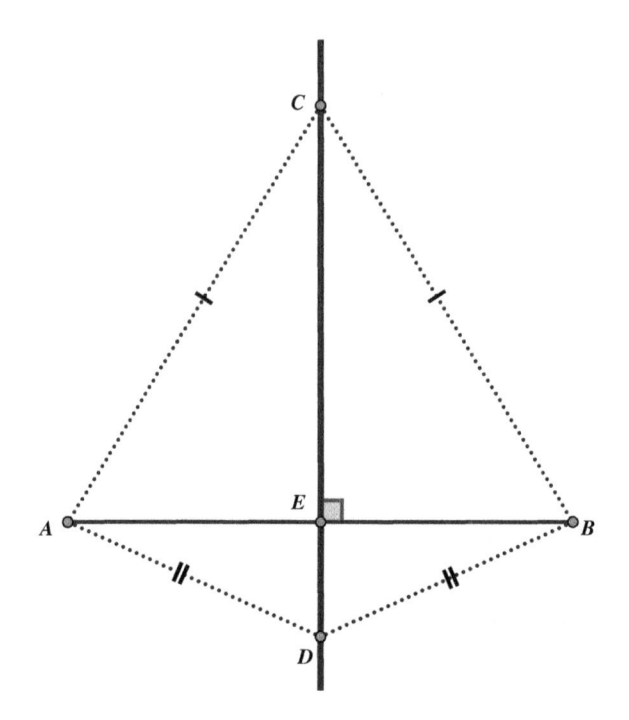

54 A Journey Through the Wonders of Plane Geometry

Figure 2-1

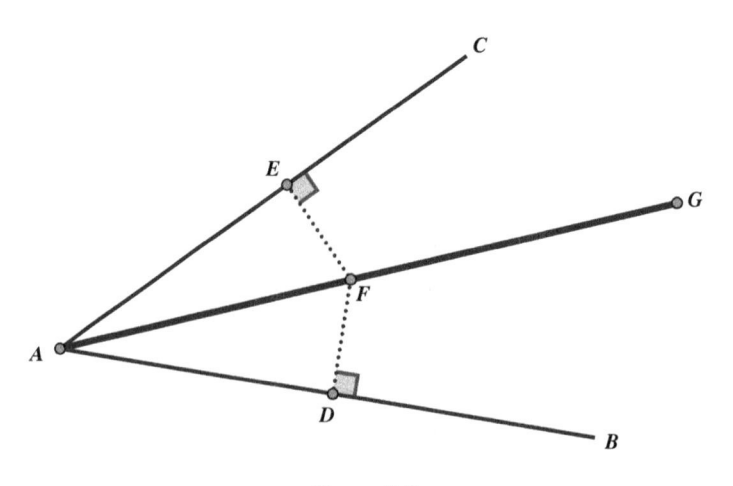

Figure 2-2

Concurrency of the Perpendicular Bisectors

The perpendicular bisectors of the triangle sides are concurrent at the center of the circumcircle of the triangle. In Figure 2-3, the point of intersection O of the perpendicular bisectors DE, KL, and FJ of the sides of triangle ABC is the center of the circle circumscribed around triangle ABC, known as the circumcircle. This point O is inside the triangle when the triangle is an acute triangle and outside the triangle when the triangle is an obtuse triangle. If, and only if, point O is on one of the sides of the triangle, the triangle is a right triangle, with the hypotenuse being the diameter of the circumcircle.

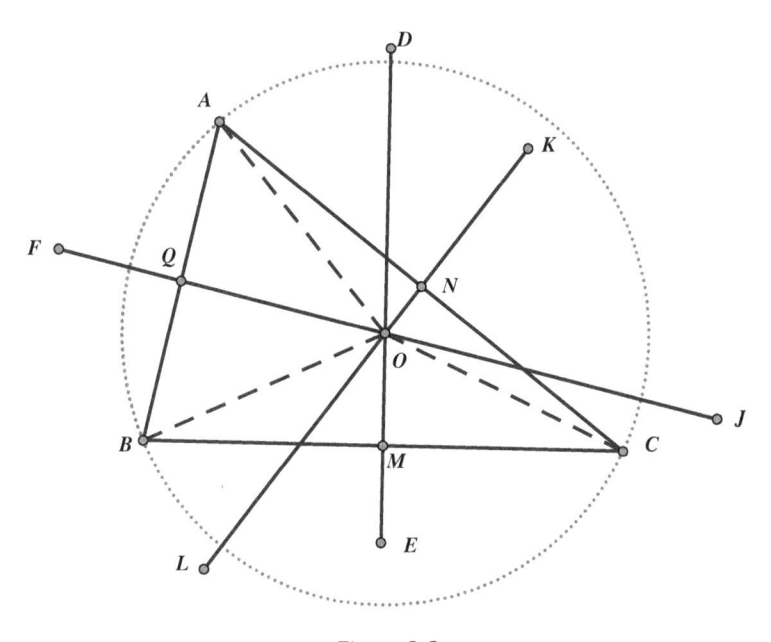

Figure 2-3

Proof

Let O be the intersection point of the perpendicular bisectors of BC and AC. Thus, we have $BO = CO$ and $AO = CO$. This gives us $BO = AO$, so that point O is equidistant from the vertices of triangle ABC and thus, determines the center of the triangle's circumcircle.

56 *A Journey Through the Wonders of Plane Geometry*

Furthermore, *O* also lies on the third perpendicular bisector, that is, on the perpendicular bisector of *AB*.

Concurrency of the Angle Bisectors

The angle bisectors of the triangle are concurrent at *I*, the center of the circle inscribed in triangle *ABC*, as can be seen in Figure 2-4. This circle is called *incircle*, and *I* is called *incenter*. Of course, the incenter *I* must lie in the interior of the triangle.

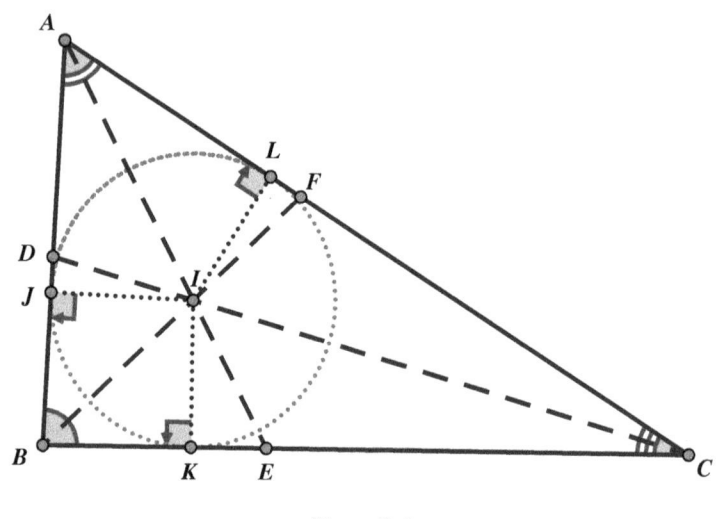

Figure 2-4

Proof

The proof for the incenter *I* works analogously to that of the circumcenter. The only difference is that in the case of the angle bisectors, we have equal distances to the sides of the triangle instead of equal distances to the vertices of the triangle, which we can see in Figure 2-4.

Excircles and Excenters

Every triangle also has three *excircles* that are tangent to one side of the triangle (externally) and to two of the triangle's side extensions, and whose centers *S*, *V*, and *M* are exterior to the triangle. The construction of these *excenters* is analogous to the construction of the incenter, the intersections of angle bisectors. The difference is that in the case of an excenter, such as point *S* in Figure 2-5, the intersection point is where one angle bisector of an interior angle (in this case angle *A*) intersects the two angle bisectors of exterior angles at the other two vertices *B* and *C*. The excircles are tangent to the triangle sides from the exterior of the triangle.

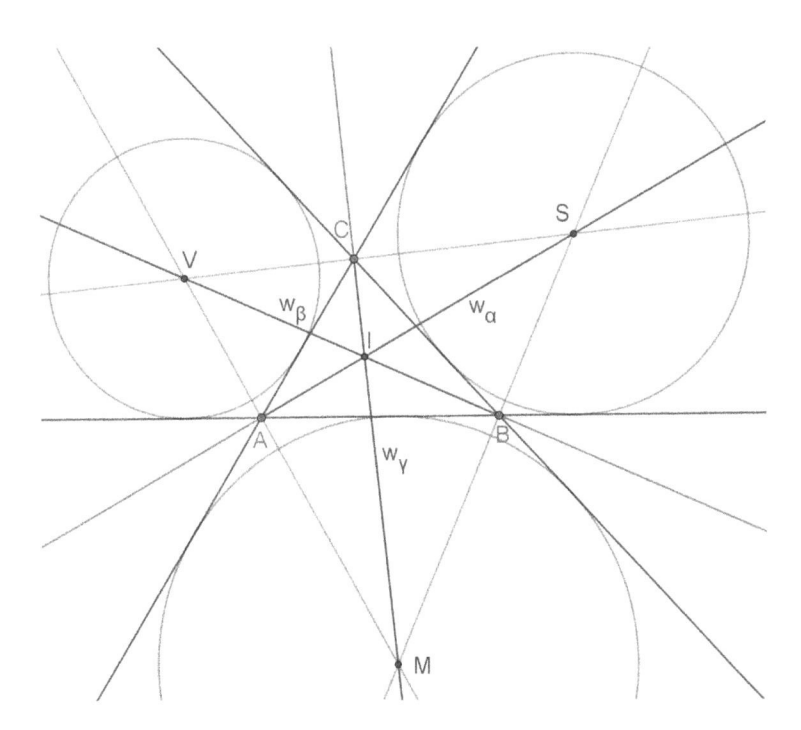

Figure 2-5 Excircles and excenters

Concurrency of the Altitudes

The altitudes of the triangle are also concurrent, namely at the orthocenter H of the triangle ABC, as shown in Figure 2-6.

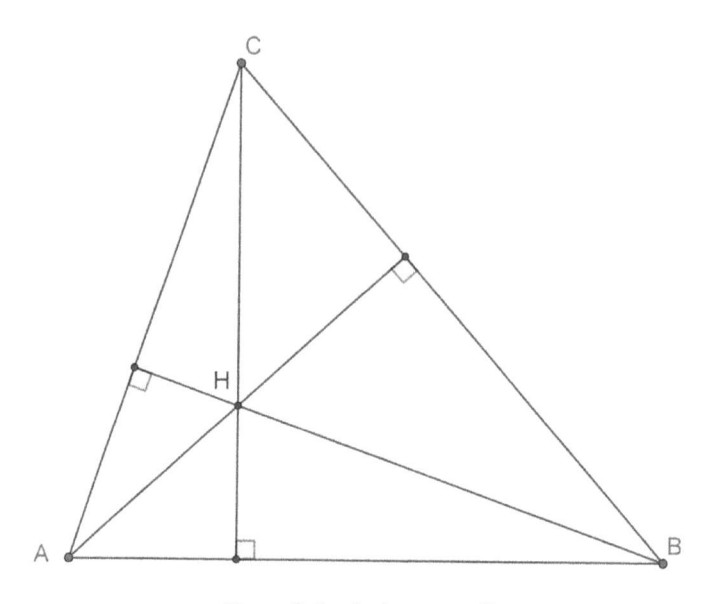

Figure 2-6   Orthocenter H

Proof

As the altitudes do not have a locus property, we require a different approach to prove their concurrency. We will consider the theorem on the concurrency of perpendicular bisectors at the circumcenter of the triangle.

Given the triangle $\triangle ABC$, we construct a second triangle $\triangle A'B'C'$, which is shown in Figure 2-7, where through point C, the line $A'B'$ is drawn parallel to AB so that $AB = A'C = B'C$. Similarly, a line is drawn through point A parallel to BC so that $CA' = AC' = AB'$. We then draw line $A'BC'$, which turns out to be parallel to AC and twice its length. We see that the altitudes of $\triangle ABC$ are the perpendicular bisectors of $\triangle A'B'C'$, which we have shown to be concurrent.

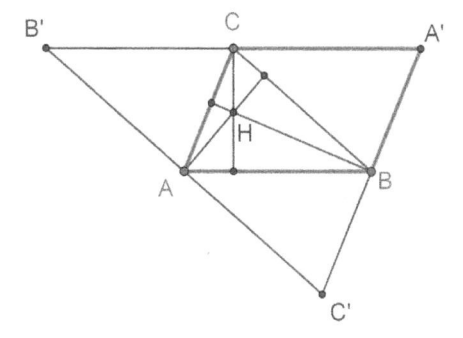

Figure 2-7   Orthocenter

Concurrency of the Medians

The three medians of triangle ABC are concurrent at G, the centroid, or balancing point, of the triangle. Furthermore, G divides the medians in the ratio 2:1, as shown in Figure 2-8.

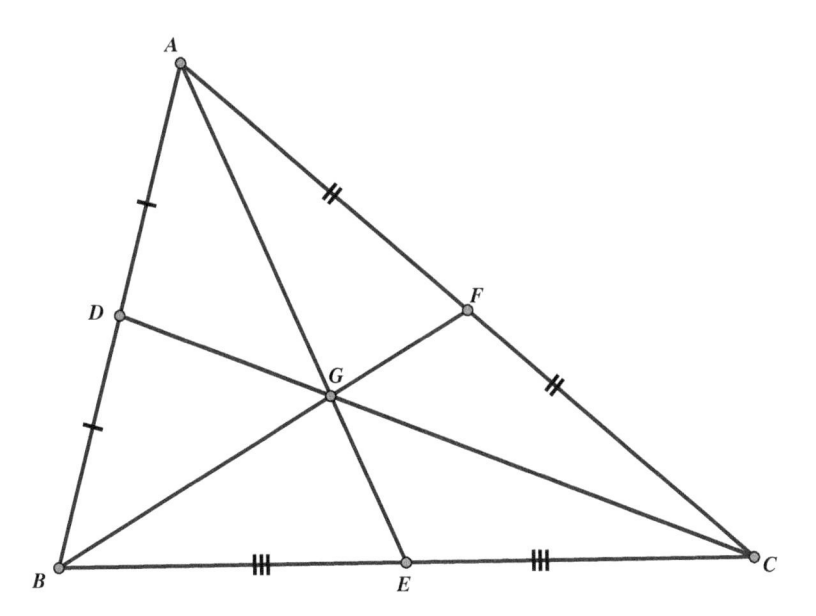

Figure 2-8

Proof

We begin the proof by drawing triangle *ABC* with medians *CD* and *BF* to meet at point *G*, as shown in Figure 2-9. We draw *AG* to intersect *BC* at point *E* and extend line *AGE* to point *J*, so that *AG* = *GJ*. When we consider triangle *JAC*, we find that *GF* connects the midpoints of its two sides so that it is parallel to the third side, namely, *GF* ∥ *JC*. Similarly, in triangle *BAJ* we find *DG* ∥ *BJ*. Therefore, quadrilateral *BGCJ* is a parallelogram where the diagonals bisect each other. That is, point *E* is the midpoint of *BC* and the line *AGE* is a median of triangle *ABC*, which contains the centroid point *G*. Of particular note is the fact that the medians intersect at a trisection point for each of the medians.

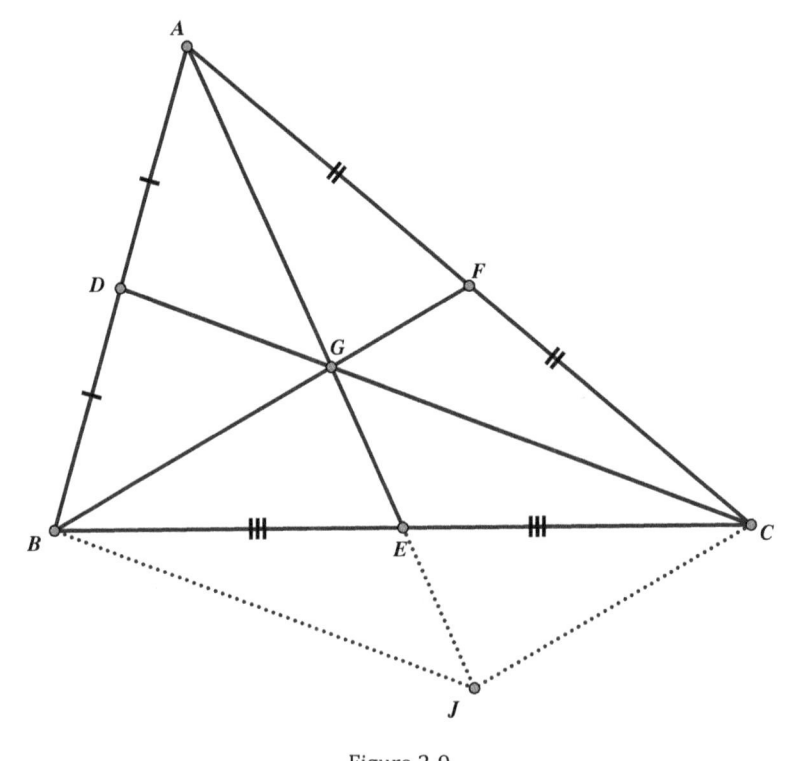

Figure 2-9

Introducing the Famous Ceva's Theorem

One of the common relationships that occurs in triangles involves lines joining a vertex to a point on the opposite side of the triangle. We often find ourselves pleasantly surprised when these three lines – one from each vertex – intersect at a common point. Perhaps the most powerful tool in determining concurrency of three lines is a theorem that we use extensively to explore the wonders of concurrency in geometry. We often refer to these lines as *Cevians*, which are named for their discoverer, the Italian mathematician Giovanni Ceva (1647–1734), who in 1678 developed the theorem about their concurrency in his work *De lineis invicem secantibus statica constructio*. This theorem is very powerful and helps us more easily establish many geometric wonders by facilitating proving the concurrency of these three lines. In particular, Ceva stated that when three Cevians are concurrent, the product of the alternate segments determined by the points of contact along the sides of the triangle are equal. This can be best demonstrated in Figure 2-10, where the concurrent Cevians *AL*, *BM*, and *CN* of triangle *ABC* intersect at point *P* and intersect the sides *BC*, *AC*, and *AB* in points *L*, *M*, and *N*, respectively. According to Ceva's theorem, the following relationship holds: $AN \cdot BL \cdot CM = NB \cdot LC \cdot MA$.

The converse of this relationship is also true. If the products of these alternate segments are equal, then the Cevians determined by these points are concurrent. We will prove both directions below.

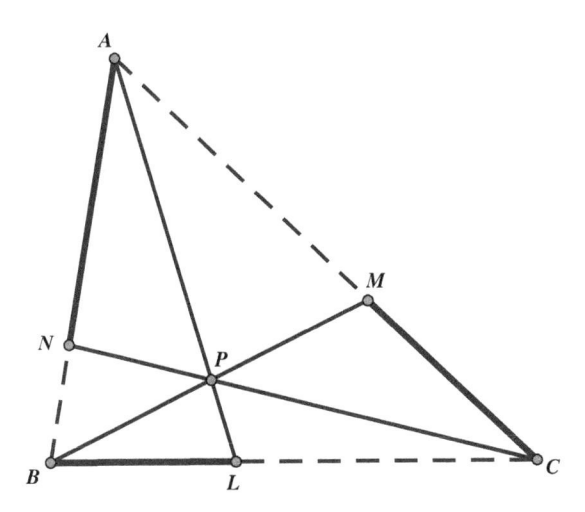

Figure 2-10

Ceva's Theorem

If three Cevians (lines containing the vertices of triangle *ABC*, shown in Figure 2-10) are concurrent (at a point *P*) and intersect the opposite sides in points *L*, *M*, and *N*, respectively, then $\frac{AN}{NB} \cdot \frac{BL}{LC} \cdot \frac{CM}{MA} = 1$.

Proof

It is perhaps easier to follow the proof by considering Figure 2-11 as well as verifying the validity of each of the statements in Figure 2-12, as the following statements in this proof hold for *both* figures.

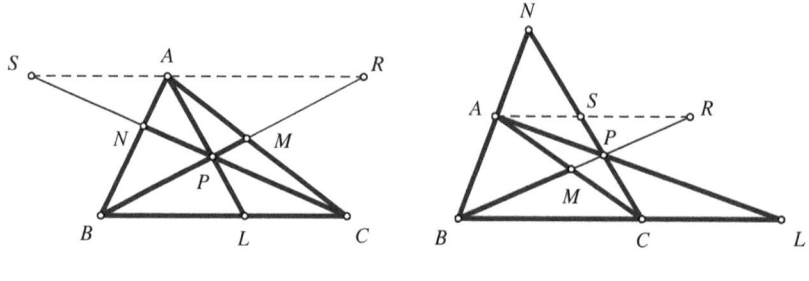

Figure 2-11

Figure 2-12

Consider Figure 2-11, where we have triangle *ABC* with a line (*SR*) containing *A* and parallel to *BC* intersecting *CP* extended at *S*, and *BP* extended at *R*. The parallel lines enable us to establish the following pairs of similar triangles:

$$\triangle ASN \sim \triangle BCN, \text{ therefore, } \frac{AN}{NB} = \frac{SA}{BC}, \tag{I}$$

$$\triangle LCP \sim \triangle ASP, \text{ therefore, } \frac{LC}{SA} = \frac{LP}{AP}, \tag{II}$$

$$\triangle BLP \sim \triangle RAP, \text{ therefore, } \frac{BL}{AR} = \frac{LP}{AP}. \tag{III}$$

From (II) and (III) we get $\frac{BL}{LC} = \frac{AR}{SA}$. \hfill (IV)

$$\triangle AMR \sim \triangle CMB, \text{ therefore, } \frac{CM}{MA} = \frac{BC}{AR}. \tag{V}$$

Now by multiplying (I), (IV), and (V) we obtain our desired result: $\frac{AN}{NB} \cdot \frac{BL}{LC} \cdot \frac{CM}{MA} = \frac{SA}{BC} \cdot \frac{AR}{SA} \cdot \frac{BC}{AR} = 1$. This can also be written as $AN \cdot BL \cdot CM = NB \cdot LC \cdot MA$.

A nice way to read this theorem is that the product of the alternate segments along the sides of the triangle, determined by the concurrent line segments (the *Cevians*) emanating from the triangle's vertices and ending at the opposite side, are equal.

There is also another, even shorter proof for Ceva's theorem, based on *triangle areas* instead of similar triangles. Consider the following.

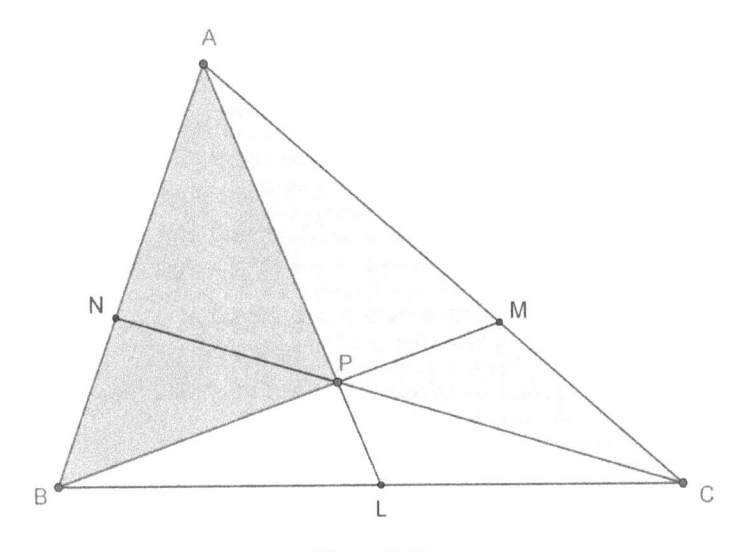

Figure 2-13

First, to simplify the notation, let us denote with $[\triangle ABC]$ the area of $\triangle ABC$. It is easy to see in Figure 2-13 that $\frac{BL}{LC} = \frac{[\triangle BPA]}{[\triangle CPA]}$: Note that $\frac{[\triangle BLA]}{[\triangle LCA]} = \frac{BL}{LC} = \frac{[\triangle BLP]}{[\triangle LCP]}$ because in each case the involved triangles have equal altitudes; furthermore, $[\triangle BPA] = [\triangle BLA] - [\triangle BLP]$ and $[\triangle CPA] = [\triangle LCA] - [\triangle LCP]$. Analogously, we have $\frac{AN}{NB} = \frac{[\triangle CPA]}{[\triangle BCP]}$ and $\frac{CM}{MA} = \frac{[\triangle BCP]}{[\triangle BPA]}$, and by multiplication of the three equalities, we get

$$\frac{AN}{NB} \cdot \frac{BL}{LC} \cdot \frac{CM}{MA} = \frac{[\triangle CPA]}{[\triangle BCP]} \cdot \frac{[\triangle BPA]}{[\triangle CPA]} \cdot \frac{[\triangle BCP]}{[\triangle BPA]} = 1.$$

Converse of Ceva's Theorem

We often find that the converse of this theorem is particularly valuable as it establishes a concurrency. That is, if the products of the alternate segments along the sides of the triangle are equal, then the Cevians determined by these points are concurrent.

Proof

We shall now prove that if the lines containing the vertices of $\triangle ABC$ intersect the opposite sides in points L, M, and N, respectively, so that $AM \cdot BN \cdot CL = MC \cdot NA \cdot BL$ (which is equivalent to $\frac{AM}{MC} \cdot \frac{BN}{NA} \cdot \frac{CL}{BL} = 1$), then these Cevian lines AL, BM, and CN are concurrent.

Suppose BM and AL intersect at P. Draw PC and call its intersection with AB point N', as shown in Figure 2-14.

Now because AL, BM, and CN' are concurrent, we can use Ceva's theorem (proved above) to state $\frac{AM}{MC} \cdot \frac{BN'}{N'A} \cdot \frac{CL}{BL} = 1$. But our hypothesis stated that $\frac{AM}{MC} \cdot \frac{BN}{NA} \cdot \frac{CL}{BL} = 1$. Therefore, $\frac{BN'}{N'A} = \frac{BN}{NA}$ so that N and N' must coincide, which, therefore, proves the concurrency.

For convenience, we can again restate this relationship as: If $AM \cdot BN \cdot CL = MC \cdot NA \cdot BL$, then the three lines AL, BM, and CN are concurrent.

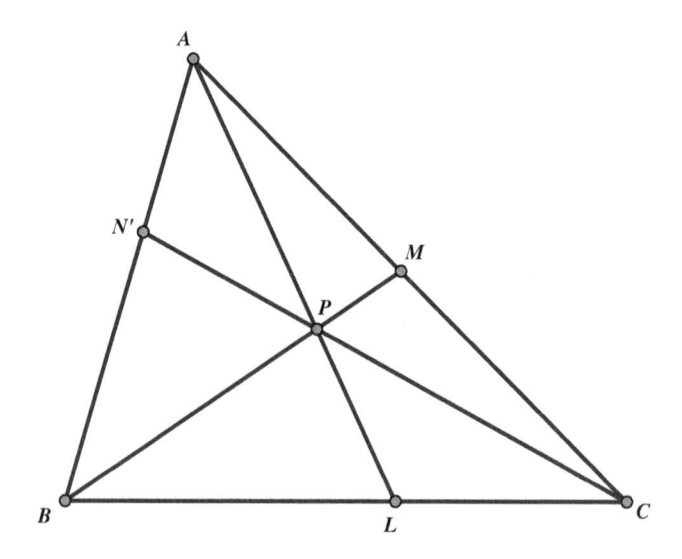

Figure 2-14

There is an interesting trigonometric version of Ceva's theorem and its converse, discovered by the French mathematician Lazare Carnot (1753–1823). Here, the concurrency of the three Cevians is specified by the partitioned angles at the triangle's vertices. In Figure 2-15, we have Cevians AL, BM, and CN that partition the angles: $\angle A$ into α_1 and α_2, $\angle B$ into β_1 and β_2, and $\angle C$ into γ_1 and γ_2. They will intersect at a common point P if, and only if, $\frac{\sin\alpha_1}{\sin\alpha_2} \cdot \frac{\sin\beta_1}{\sin\beta_2} \cdot \frac{\sin\gamma_1}{\sin\gamma_2} = 1$.

Proof

Let $BC = a$, $AC = b$, and $AB = c$, as shown in Figure 2-15. By the law of sines applied to triangles $\triangle BLA$ and $\triangle LCA$, we get $\frac{BL}{\sin\alpha_1} = \frac{c}{\sin\angle ALB}$ and $\frac{LC}{\sin\alpha_2} = \frac{b}{\sin\angle ALC}$, and since the supplementary angles on either side of point L have the same sine, we have $\sin\angle ALB = \sin\angle ALC$. This gives us $\frac{BL}{LC} = \frac{b \cdot \sin\alpha_1}{c \cdot \sin\alpha_2}$. Analogously, we get $\frac{CM}{MA} = \frac{c \cdot \sin\beta_1}{a \cdot \sin\beta_2}$, and $\frac{AN}{NB} = \frac{a \cdot \sin\gamma_1}{b \cdot \sin\gamma_2}$. Multiplying these three expressions together yields $\frac{\sin\alpha_1}{\sin\alpha_2} \cdot \frac{\sin\beta_1}{\sin\beta_2} \cdot \frac{\sin\gamma_1}{\sin\gamma_2} = \frac{BL}{LC} \cdot \frac{CM}{MA} \cdot \frac{AN}{NB} = 1$, as claimed.

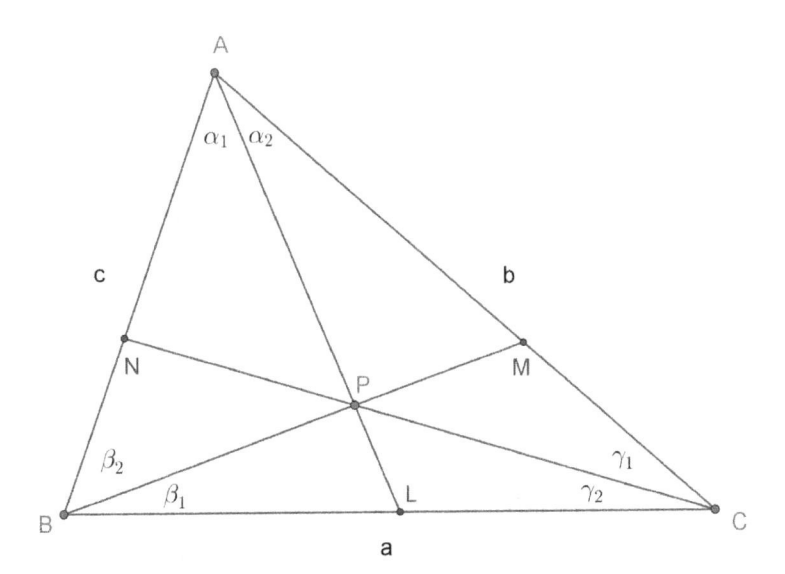

Figure 2-15

Applying the Converse of Ceva's Theorem to Prove Familiar Triangle Concurrencies

The converse of Ceva's theorem enables us to prove some familiar concurrencies, such as the concurrencies of the medians, the altitudes, and the angle bisectors of a triangle, in a simple way.

Concurrency of Medians of a Triangle

For example, it is almost trivial to use the converse of Ceva's theorem to prove that the medians are concurrent since by definition the Cevians in triangle *ABC*, shown in Figure 2-16, determine midpoints along the sides of the triangle so that $AD = DB$, $BE = EC$, and $CF = FA$. Therefore, according to the converse of Ceva's theorem, since $AD \cdot BE \cdot CF = DB \cdot EC \cdot FA$, the medians *AE*, *BF*, and *CD* are concurrent.

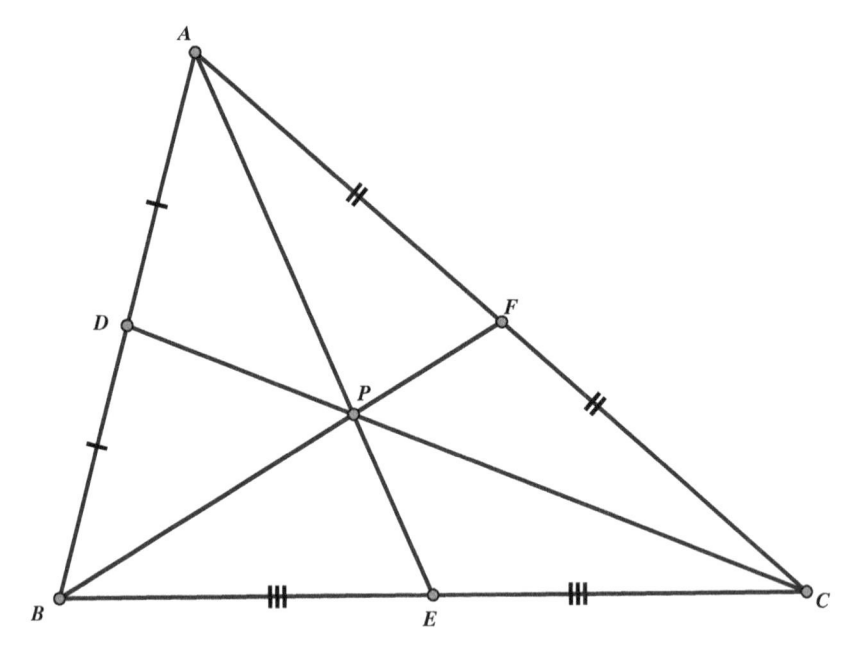

Figure 2-16

Concurrency of the Altitudes of a Triangle

We can use the converse of Ceva's theorem to show that the altitudes *AE*, *BF*, and *CD* are concurrent by establishing three pairs of similar triangles and their accompanying ratios as follows. In Figure 2-17, we have:

$$\triangle ABE \sim \triangle BDC, \text{ therefore, } \frac{BE}{DB} = \frac{AB}{BC},$$

$$\triangle BFC \sim \triangle AEC, \text{ therefore, } \frac{CF}{EC} = \frac{BC}{AC},$$

$$\triangle ADC \sim \triangle AFB, \text{ therefore, } \frac{AD}{FA} = \frac{AC}{AB}.$$

When we multiply these three equations, we get $\frac{BE}{DB} \cdot \frac{CF}{EC} \cdot \frac{AD}{FA} = \frac{AB}{BC} \cdot \frac{BC}{AC} \cdot \frac{AC}{AB} = 1.$

Therefore, $BE \cdot CF \cdot AD = DB \cdot EC \cdot FA$, which by the converse of Ceva's theorem tells us that the three altitudes are concurrent.

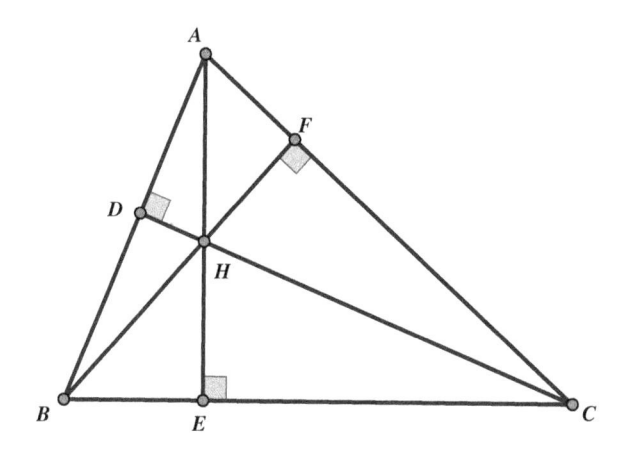

Figure 2-17

Concurrency of the Angle Bisectors of a Triangle

We can also use the converse of Ceva's theorem to prove that the angle bisectors of a triangle are concurrent by using a well-known relationship: the angle bisector of the triangle divides the side to which it is drawn proportional to the two adjacent sides. So that in triangle ABC, which is shown in Figure 2-18, we get the following proportions:

$$\text{for angle bisector } AE, \frac{BE}{BC} = \frac{AB}{AC},$$

for angle bisector BF, $\dfrac{CF}{FA} = \dfrac{BC}{AB}$,

for angle bisector CD, $\dfrac{AD}{DB} = \dfrac{AC}{BC}$.

Multiplying these three equations gives us $\dfrac{BE}{EC} \cdot \dfrac{CF}{FA} \cdot \dfrac{AD}{DB} = \dfrac{AB}{AC} \cdot \dfrac{BC}{AB} \cdot \dfrac{AC}{BC} = 1$. Therefore, $BE \cdot CF \cdot AD = EC \cdot FA \cdot DB$, which, according to the converse of Ceva's theorem, establishes that the three angle bisectors are concurrent.

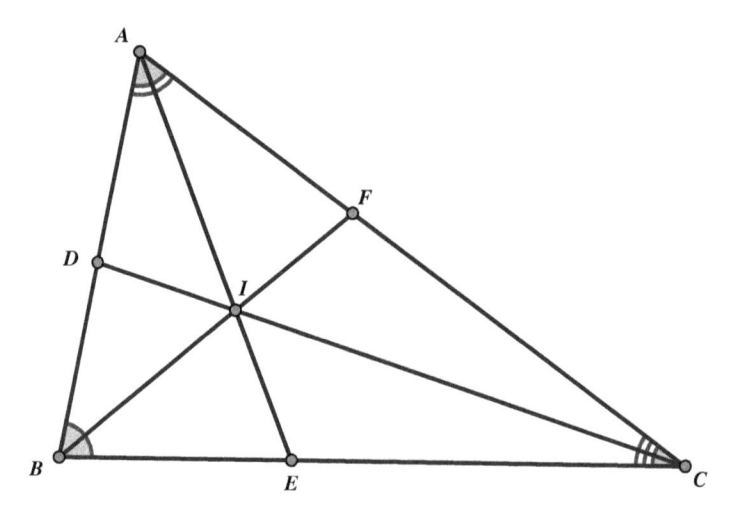

Figure 2-18

The Nagel Point of a Triangle

Escribed circles with centers M, V, and S are tangent externally to the sides of triangle ABC and their extensions at points R, Q, and S. The lines joining these points of tangency with the opposite vertices, AP, BQ, and CR, are concurrent at point N, as shown in Figure 2-19. This point is called the *Nagel point*, named after the German mathematician Christian Heinrich von Nagel (1803–1882).

Unexpected Concurrencies   69

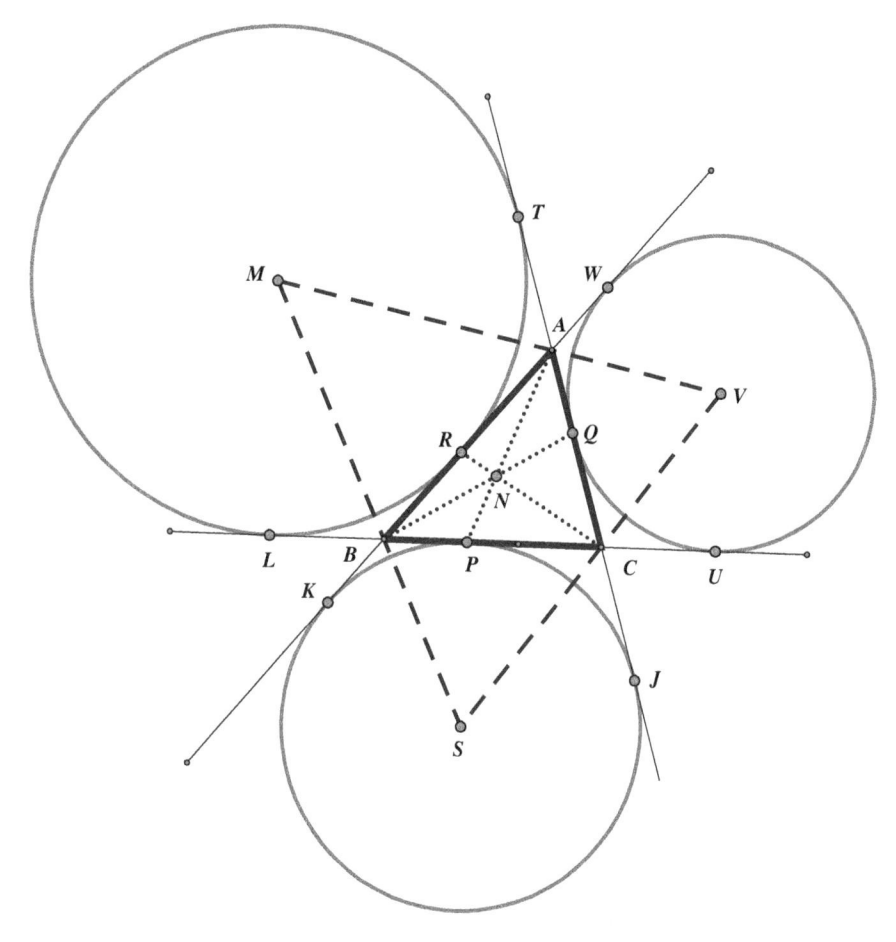

Figure 2-19

Proof

To prove this concurrency, we will use Ceva's theorem. However, to simplify matters, we first need to establish some equalities. Since tangents from an external point to the same circle are equal, we have $AK = AJ$, or we can write that as $AB + BK = AC + CJ$. We also know that $BK = BP$ and $CJ = CP$. Therefore, $AB + BP = AC + CP$. In a similar fashion, with tangent point Q on AC, we get $BC + CQ = AQ + AB$, and with tangent point R on AB, we get $AC + AR = BR + BC$. To simplify this proof, we will focus on triangle ABC, as shown in Figure 2-20.

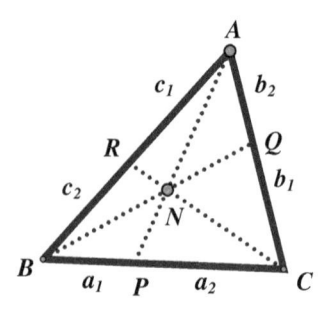

Figure 2-20

Let a_1, a_2, b_1, b_2, c_1, and c_2 denote BP, PC, CQ, QA, AR, and RB, respectively, and let p denote the perimeter of the triangle as shown in Figure 2-20. We then have $a_1 + a_2 + b_1 + b_2 + c_1 + c_2 = BC + AC + AB = p$. By making the proper substitutions, we get the following:

$$AB + BP = AC + CP \Rightarrow c_1 + c_2 + a_1 = a_2 + b_1 + b_2 = \frac{p}{2}. \qquad (I)$$

$$BC + CQ = AQ + AB \Rightarrow a_1 + a_2 + b_1 = b_2 + c_1 + c_2 = \frac{p}{2}. \qquad (II)$$

$$AC + AR = BR + BC \Rightarrow b_1 + b_2 + c_1 = c_2 + a_1 + a_2 = \frac{p}{2}. \qquad (III)$$

From equations (I) and (II), we get $a_2 + b_1 + b_2 = \frac{p}{2} = a_1 + a_2 + b_1$ so that we have $a_1 = b_2$. Similarly, from equations (II) and (III), we now obtain $b_1 = c_2$ and from equations (I) and (III), we can conclude that $c_1 = a_2$. Therefore, $\frac{AR}{RB} \cdot \frac{BP}{PC} \cdot \frac{CQ}{QA} = \frac{c_1}{c_2} \cdot \frac{a_1}{a_2} \cdot \frac{b_1}{b_2} = \frac{c_1}{a_2} \cdot \frac{a_1}{b_2} \cdot \frac{b_1}{c_2} = 1$, and by the converse of Ceva's theorem, AP, BQ, and CR are concurrent.

The Hidden Length of a Cevian

Finding line segments in a triangle, the sum of which is the same length as a given Cevian, provides a surprising wonder. Consider the Cevian AH in triangle ABC, which is shown in Figure 2-21. From point H, we construct a line parallel to side AB intersecting side AC at point E. Similarly, we construct a line through point H parallel to side AC intersecting side AB at point D. We then draw line DF parallel to AH intersecting BC at point F. In a similar manner, we draw the line EG parallel to AH, intersecting side BC at point G. Surprisingly, we find that $DF + EG = AH$.

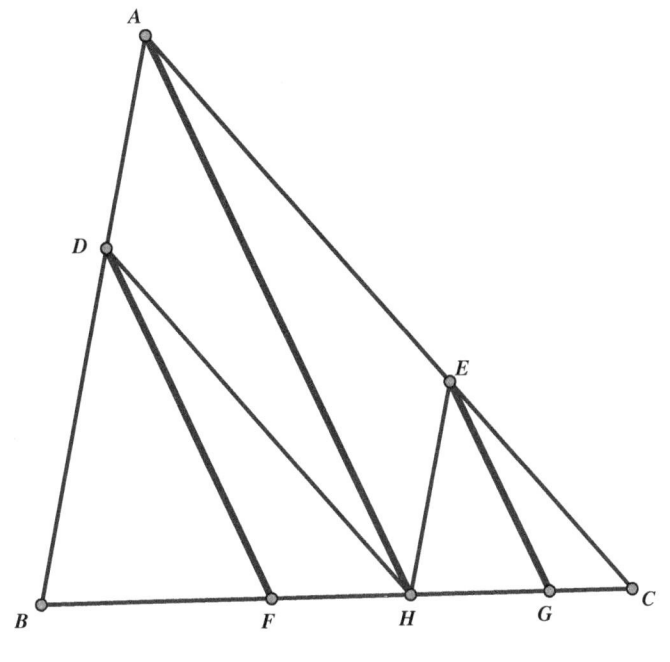

Figure 2-21

Proof

We notice in Figure 2-22 that since both pairs of opposite sides of quadrilateral $ADHE$ are parallel, quadrilateral $ADHE$ is a parallelogram. Next, we draw the parallel to AB through D and get the intersection points J and K. Also, $FHJD$ and $HCKD$ are parallelograms, and by the side-angle-side congruence theorem, we have $\triangle HGE \cong \triangle DJA$. This gives us $AJ = EG$ and $DF = JH$. Therefore, $DF + EG = AJ + HJ = AH$.

Another Surprising Aspects of Concurrent Cevians

Randomly drawn concurrent Cevians can also present a rather surprising result when, through the feet of each of these Cevians, we construct parallel lines to each of the other two Cevians. This can be seen in Figure 2-23, where we begin with the three Cevians AF, BE, and CD, which intersect at point U. We draw the following parallel lines: $AF \parallel DQ \parallel ER$, $BE \parallel DX \parallel FY$, and $CD \parallel EV \parallel FW$. The intersections

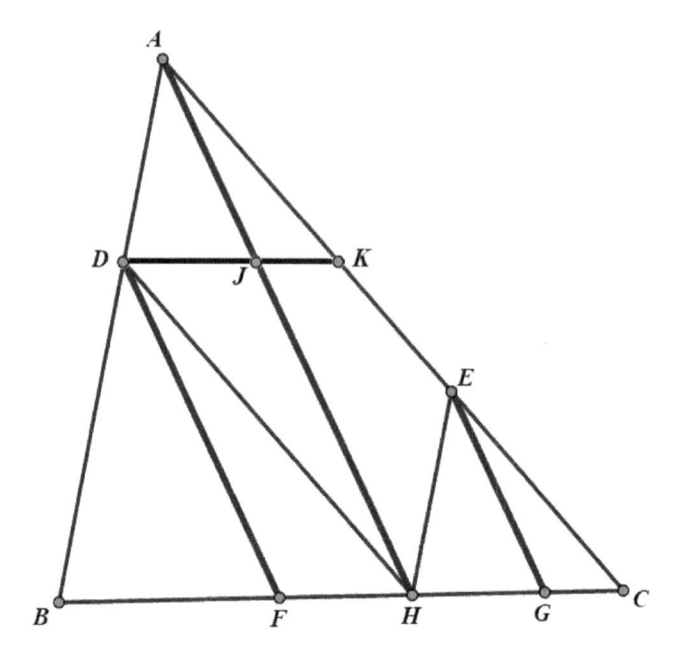

72 A Journey Through the Wonders of Plane Geometry

Figure 2-22

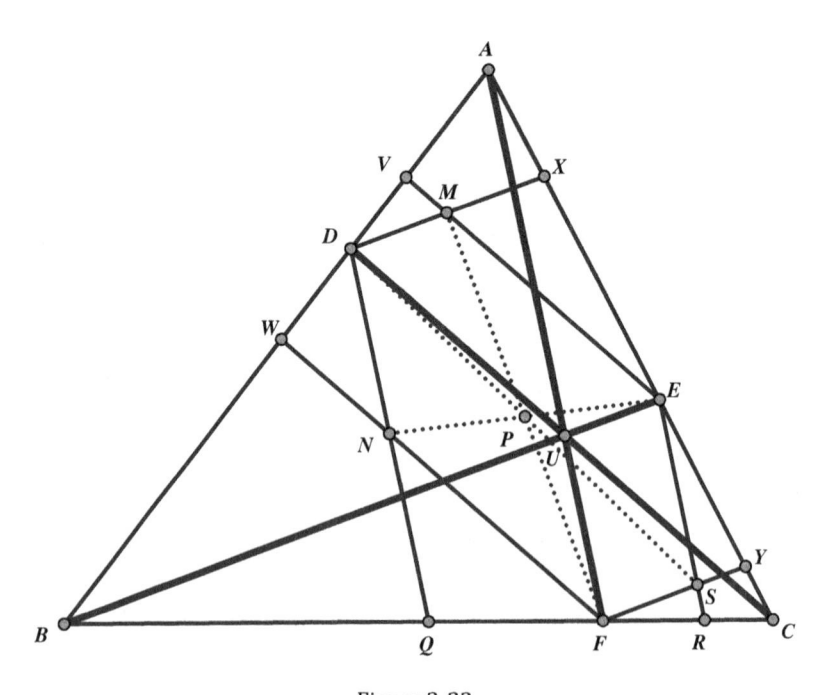

Figure 2-23

of these lines are as follows: *DX* intersects *VE* at point *M, DQ* intersects *FW* at point *N*, and *ER* intersects *FY* at point *S*. We find that the three lines joining these points of intersection, namely, *MF, NE,* and *SD,* are concurrent at point *P*, and they bisect each other as well! Although this may appear to be a bit complicated, it continues to show the many hidden relationships in the world of geometry.

Proof

This configuration has lots of parallelograms based on pairs of parallel opposite sides. We begin by noticing that quadrilaterals *DNFU* and *DMEU* are both parallelograms since both pairs of opposites sides are parallel. Consequently, we know that *NF = ME, NF* ∥ *ME,* and *UF = ND*. Therefore, we can show that quadrilateral *MEFN* is also a parallelogram, whose diagonals *NE* and *MF* bisect each other at point *P*, as shown in Figure 2-24. Furthermore, we know that

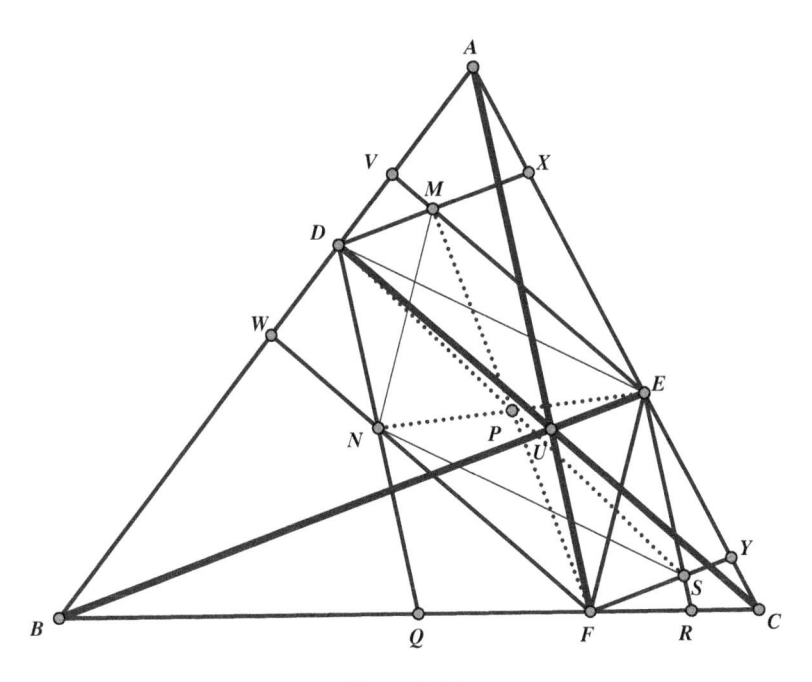

Figure 2-24

74 *A Journey Through the Wonders of Plane Geometry*

quadrilateral *FSEU* is also a parallelogram, and *ES* = *UF*; thus, *ND* = *ES*. Therefore, quadrilateral *DESN* is also a parallelogram, and its diagonals *NE* and *DS* bisect each other at the common point *P*. Thus, we have shown that the three lines *MF*, *NE*, and *SD* are concurrent at point *P* and bisect each other.

Gergonne's Discovery Involving the Inscribed Circle of a Triangle

We have established that the inscribed circle of the triangle has its center at the point of intersection of the angle bisectors of the triangle. However, this circle determines another point of concurrency in the triangle. In Figure 2-25, when we draw the lines from each vertex of triangle *ABC* to the points of tangency, *L*, *M*, and *N*, on the opposite sides, these three lines are concurrent at point *J*. This point is called the *Gergonne point*, named for the French mathematician Joseph-Diaz Gergonne (1771–1859), who first published this amazing concurrency. The triangle *LMN*, formed by joining these three points of tangency, is called the *Gergonne triangle*.

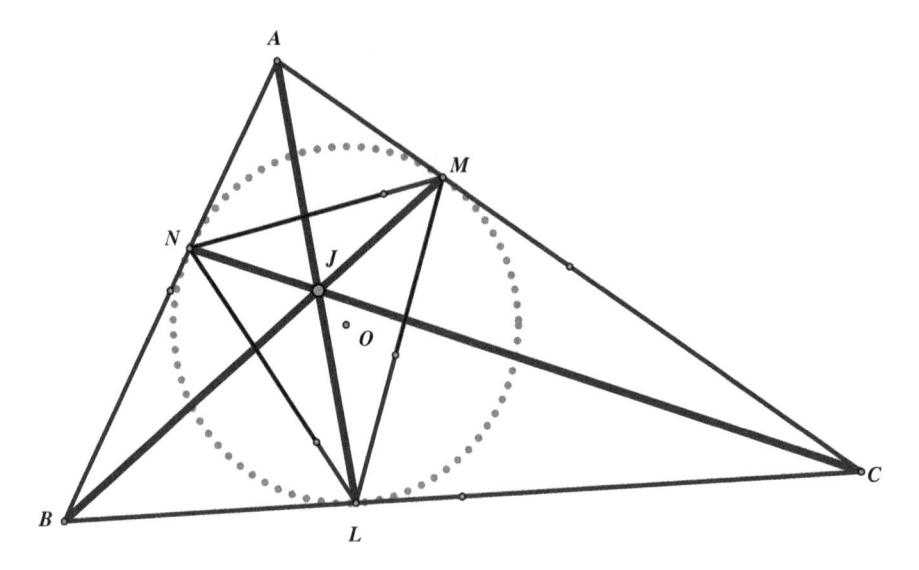

Figure 2-25

Proof

To prove that the three lines joining vertices to points of tangency of the inscribed circle are concurrent at the Gergonne point, we shall use the converse of Ceva's theorem, which states that if three lines are drawn from the vertices of a triangle to the opposite sides creating six segments along the sides, when the products of the alternate segments are equal, then the three lines are concurrent. In Figure 2-25, we have $MA = AN$, $BL = NB$, and $CM = LC$ since the tangent segments from external points to the same circle or equal. Therefore, since $AN \cdot BL \cdot CM = NB \cdot LC \cdot MA$, the lines AL, BM, and CN are concurrent.

Another Concurrency for the Gergonne Triangle

Fortunately, there is much more that can be found in this configuration. If we draw the perpendicular line segments from the midpoints R, S, and Q, of each of the sides of triangle ABC to each of the sides of the Gergonne triangle LMN, we find that these lines RE, SD, and QF are concurrent at point P, as shown in Figure 2-26.

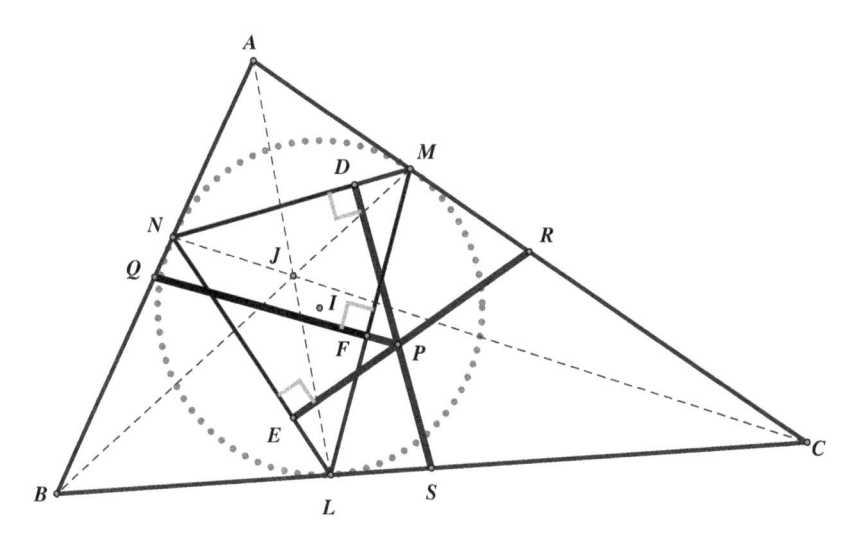

Figure 2-26

Proof

As we indicated earlier, triangle *NAM* is isosceles, so *AX*, bisector of ∠*NAM*, is perpendicular to *MN*, as we can see in Figure 2-27. Consider the medial triangle *RQS*, that is, the triangle formed by joining the midpoints of the sides of the original triangle *ABC*. We know that the sides of the medial triangle *RQS* are parallel to the sides of the original triangle *ABC*. This creates the parallelogram *AQSR*, where the bisectors of the opposite angles of the parallelogram are also parallel so that *AX* ∥ *DS*, and therefore, *DS* ⊥ *MN*. This tells us that the three lines we seek to show are concurrent are actually the bisectors of the three angles of the medial triangle *QRS*, as each of them is perpendicular to the Gergonne triangle. Since we know that the angle bisectors of a triangle (in this case, the medial triangle) are concurrent, we have shown that the perpendiculars to the sides of the Gergonne triangle are concurrent. This point *P*, the incenter of the medial triangle, is called the *Spieker center*[1] of the initial triangle *ABC* and is the center of gravity of a homogeneous wire frame in the shape of △*ABC*.

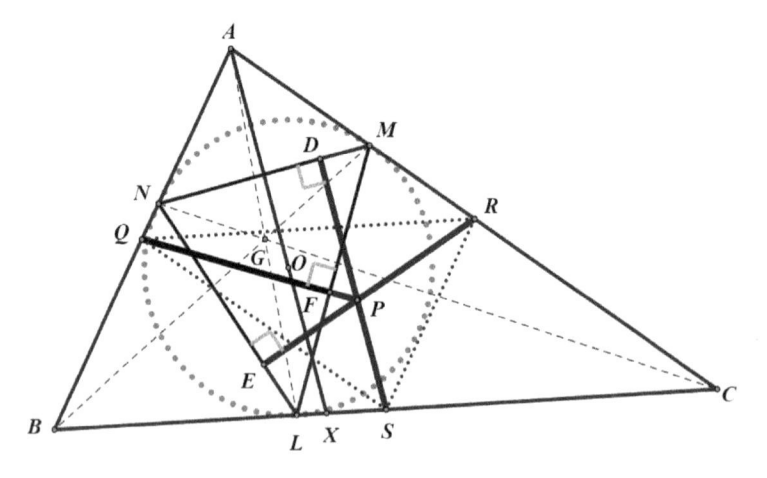

Figure 2-27

---

[1] Named after the German mathematician Theodor Spieker (1823–1913) who published it in 1888.

Circumscribed and Inscribed Circle Curiosities

We are now prepared for truly amazing counterintuitive relationships that involve surprising concurrencies and collinearities. For example, we draw the perpendicular bisectors *OS*, *OT*, and *OR* of each of the sides of triangle *ABC*, as shown in Figure 2-28. These perpendicular bisectors meet the circumscribed circle at points *D*, *F*, and *E*, respectively, which are the midpoints of the arcs *AC*, *AB*, and *BC*. We connect these latter arc midpoints to the points of tangency of the inscribed circle of triangle *ABC*, namely points *M*, *N*, and *L*. Surprisingly, we find that these segments *EN*, *DM*, and *FL* are concurrent at point *Q*.

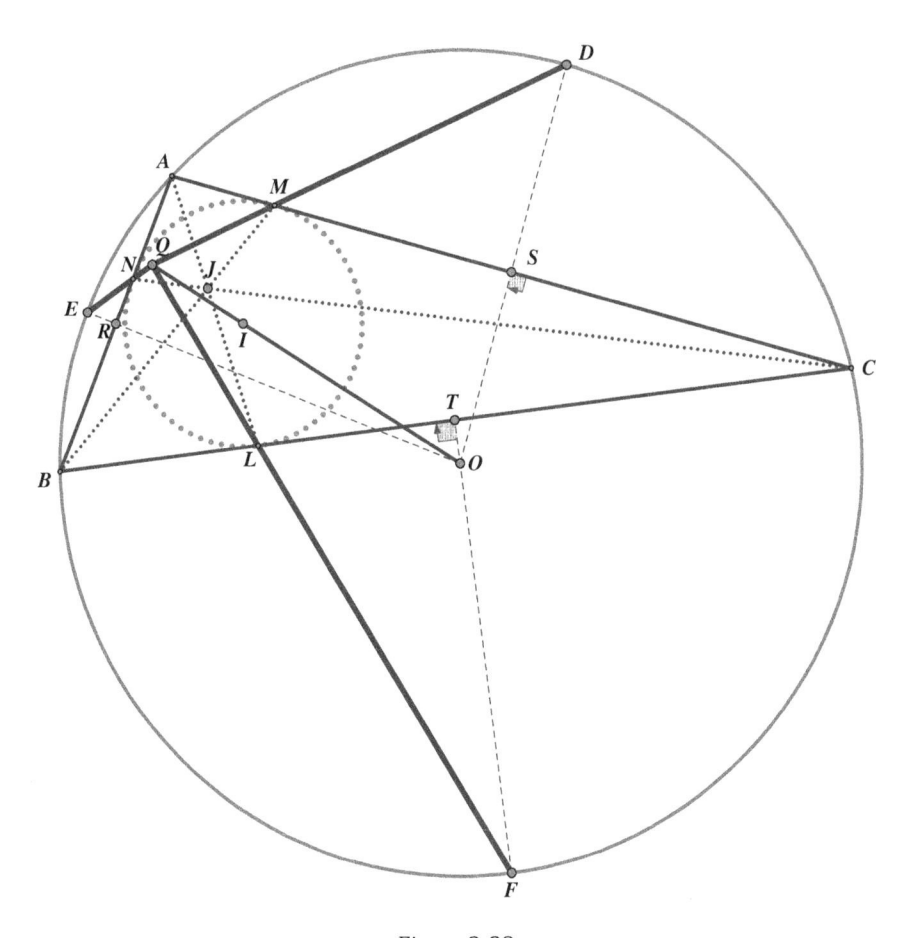

Figure 2-28

Proof

We begin in Figure 2-29 by noticing the radii $DO = EO = FO = x$ of the circle with center O. Similarly, we have $MI = NI = LI = y$ as the radii of the inscribed circle. We now define point Q as the intersection of the extension of OI and the extension of FL. Since IL and OF are each perpendicular to BC, we have $IL \parallel OF$ and $\triangle QIL \sim \triangle QOF$. We then have $\angle QLI = \angle QFO$. This allows us to indicate that QLF is a straight line. Similar arguments can be made that QMD and QNE are also straight lines. This proves the concurrency we set out to establish.

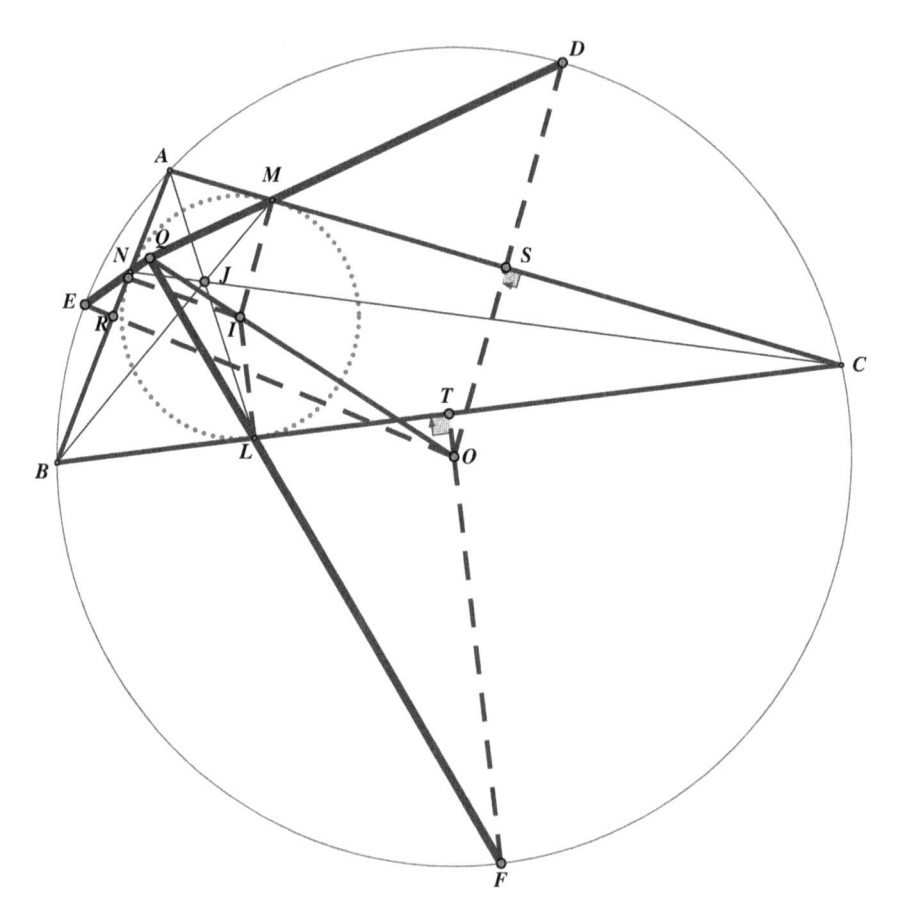

Figure 2-29

Two Triangles Related with a Common Point: Circumcenter – Centroid

Although a bit cumbersome, there are times when two triangles are related by the fact that they share a common point of concurrency; in this case, we consider that the center of the circumscribed circle of one triangle is the centroid of the other triangle. In Figure 2-30, the center of the circumscribed circle of triangle *ABC* is point *O*. We draw the medians of triangle *ABC*, which are *AE*, *BF*, and *CD* and meet at the centroid *G*. The perpendicular bisectors of *AG*, *BG*, and *CG* intersect their respective sides at points *Q*, *M*, and *N*, and we find that these perpendicular bisectors meet at points *X*, *Y*, and *Z*. The medians of triangle *XYZ*, namely, *XL*, *YJ*, and *ZK*, have their point of intersection (centroid) at point *O*. Therefore, point *O* is the centroid of triangle *XYZ* and the center of the circumscribed circle of triangle *ABC*.

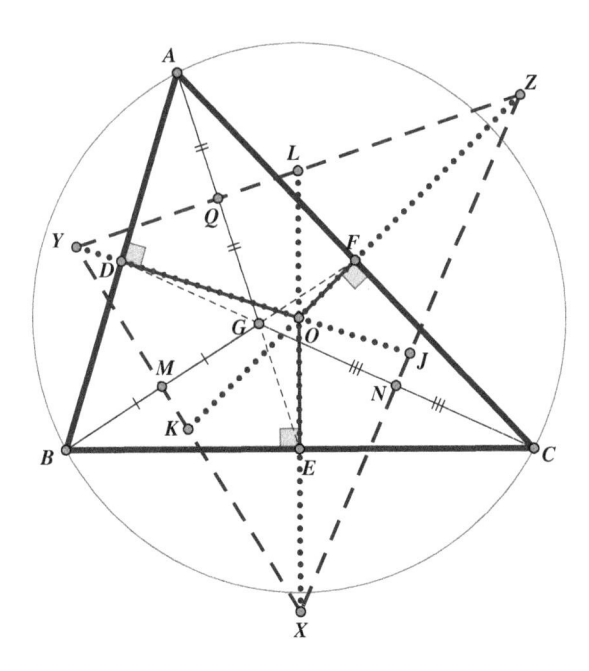

Figure 2-30

Proof

We begin in Figure 2-31 by noticing that $\angle AQZ$ and $\angle AFZ$ are right angles, therefore, A, Q, F, and Z are concyclic points and we have $\angle GAF = \angle OZQ$, as arc QF measures both angles. Since quadrilateral $CEOF$ has a pair of opposite right angles, points C, E, O, and F are concyclic. Therefore, $\angle ACE = \angle LOZ$, since both angles are supplementary to $\angle FOE$. Therefore, we have $\Delta LOZ \sim \Delta ACE$, and we get $\frac{ZL}{AE} = \frac{OL}{CE}$. Analogously, we can find that $\Delta ABE \sim \Delta YOL$, which gives us $\frac{YL}{AE} = \frac{OL}{BE}$. Since $BE = CE$, we can conclude that $ZL = YL$, which implies that XO extended meets the midpoint of ZY at point L. This can be replicated for the other two midpoints, K and J, which allows us to conclude that point O is the centroid of ΔXYZ.

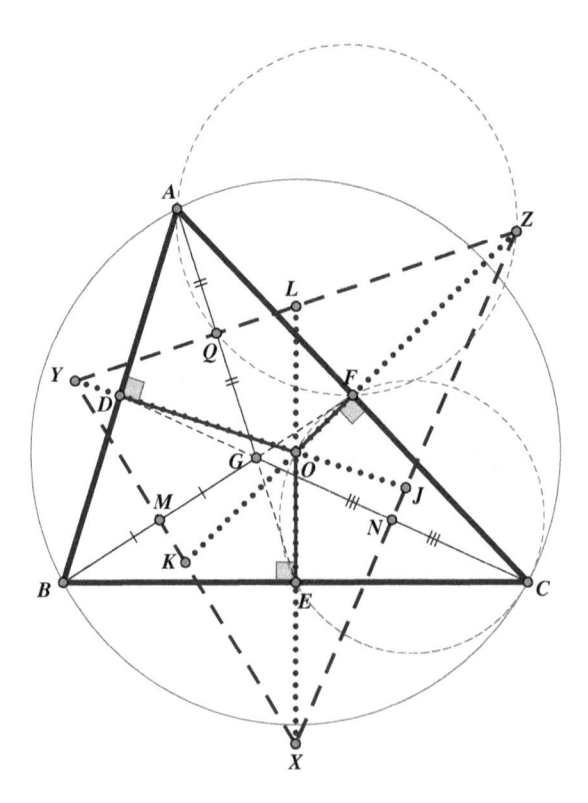

Figure 2-31

Concurrency with the Inscribed Circle

The inscribed circle can further reveal a variety of astonishments. One such is a concurrency shown in Figure 2-32, which can be found by taking the diametrically opposite points D, E, and F on the inscribed circle of triangle ABC from the points of tangency L, M, and N, and extending the lines BE, AD, and CF, which we can see are concurrent at point P. And so, another surprising concurrency has been discovered.

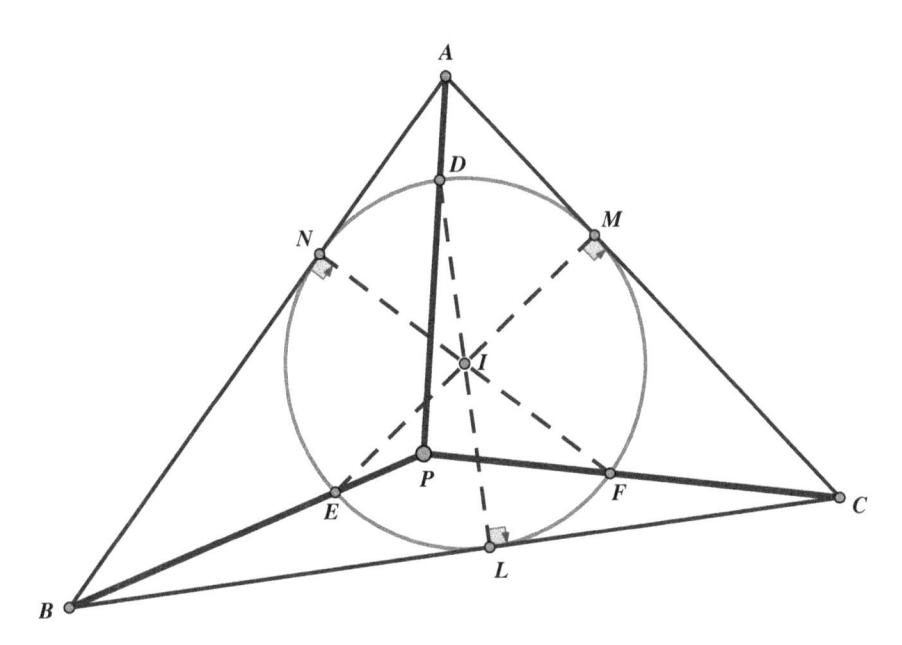

Figure 2-32

Proof

In Chapter 11, we will see that the lines AD, BE, and CF are the Nagel Cevians because they meet the opposite sides at the points of tangency of the excircles. Thus, the mentioned point P of concurrency must be the Nagel point of $\triangle ABC$.

Another Concurrency with the Inscribed Circle

Although a reader may seek and find other concurrencies, we offer one more – without proof – to show the practically limitless concurrencies in this configuration. From the previous example, in Figure 2-33, we choose any interior point P of the incircle, and from the points of tangency L, M, and N, draw lines through P that intersect the inscribed circle a second time at points D, E, and F, respectively. We then draw the lines AD, BE, and CF and observe that they are concurrent at point Q. This could be the beginning of seemingly endless concurrencies that can be discovered in this configuration.

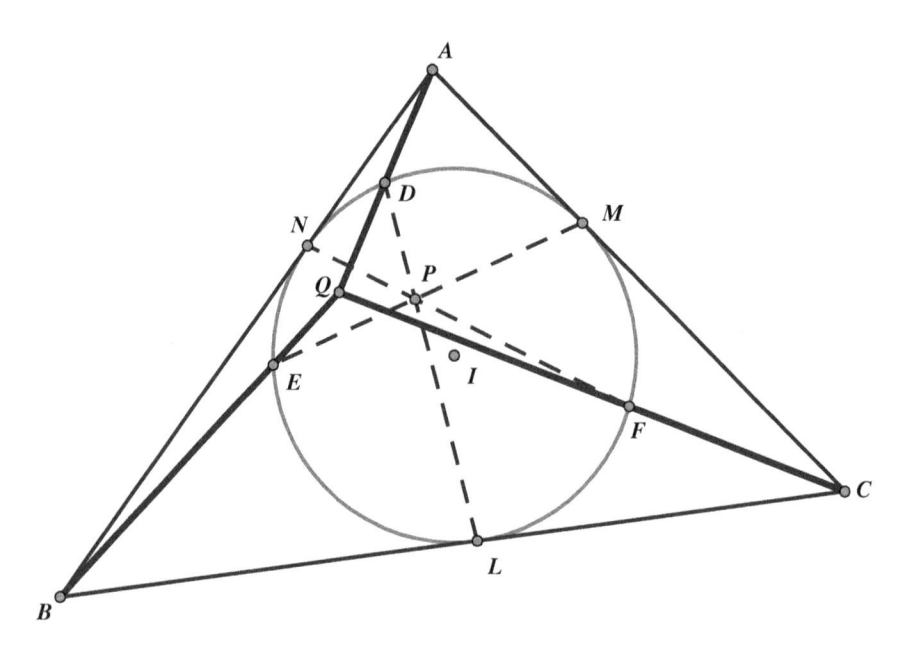

Figure 2-33

The Isogonal Conjugation: A Way of Producing a New Concurrency from a Given One

An *isogonal conjugation* works as follows. We are given a triangle $\triangle ABC$, an arbitrary interior point P, at which the concurrent Cevians AP, BP, and CP intersect, as shown in Figure 2-34. We reflect these Cevians in the corresponding angle bisectors so that at each angle the two lines reflected in the angle bisector form equal angles with the sides of the original angle. Quite unexpectedly, we find that the three reflected Cevians are concurrent at a point P', which is called the isogonal conjugate point of P. We can see that the *isogonal conjugation* produces two pairs of equal angles at each vertex of triangle ABC, which at vertex C are marked in Figure 2-34 with φ, μ.

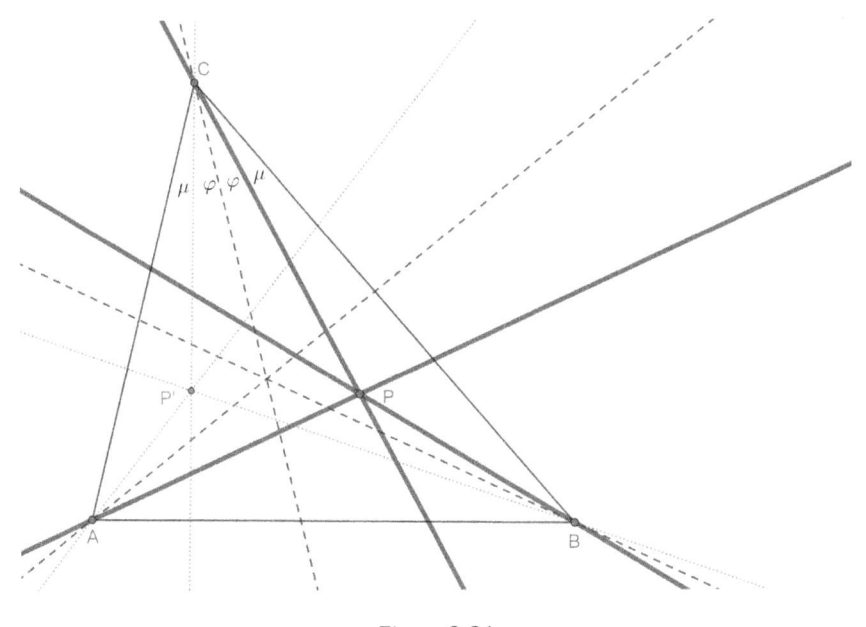

Figure 2-34

Proof

To prove that the three new reflected Cevians are actually concurrent at P', we need to apply the trigonometric version of Ceva's theorem and its converse. Regarding the original bold Cevians in Figure 2-34 and the dotted reflected ones, we see with respect to Figure 2-15 that just the angles $\alpha_1 \leftrightarrow \alpha_2$, $\beta_1 \leftrightarrow \beta_2$, and $\gamma_1 \leftrightarrow \gamma_2$ are interchanged. For example, the bold Cevians in Figure 2-34 have $\gamma_2 = \mu$, and the dotted Cevians have $\gamma_1 = \mu$. Since the bold Cevians are given to be concurrent at point P, we know by the trigonometric version of Ceva's theorem that $\frac{\sin\alpha_1}{\sin\alpha_2} \cdot \frac{\sin\beta_1}{\sin\beta_2} \cdot \frac{\sin\gamma_1}{\sin\gamma_2} = 1$. Thus, $\frac{\sin\alpha_2}{\sin\alpha_1} \cdot \frac{\sin\beta_2}{\sin\beta_1} \cdot \frac{\sin\gamma_2}{\sin\gamma_1} = 1$. The converse of the trigonometric version of Ceva's theorem lets us conclude that the dotted Cevians are also concurrent at point P'.

Here are some interesting aspects of isogonal conjugation:

1. The isogonal conjugate of the incenter I is the point I itself. This is because reflecting an angle bisector in itself does not change anything.

2. The circumcenter O and the orthocenter H of a triangle are isogonal conjugates. In order to prove this, we have to show that

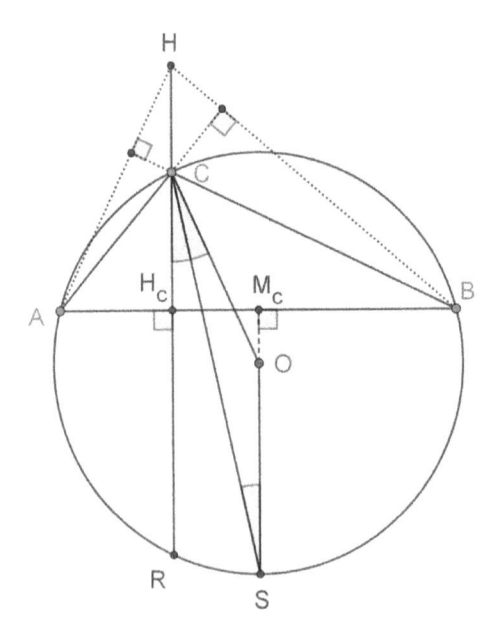

Figure 2-35

the angle bisector at C also bisects the angle between the lines $CH = CH_c = CR$ and CO (Figure 2-35) analogously at the other vertices.

Proof: In Figure 2-35, S is the intersection point of the angle bisector at C with the circumcircle. The arcs $\widehat{AS} = \widehat{BS}$ must be equal and point S must lie on the perpendicular bisector of AB. Since $\triangle CSO$ is isosceles, we have $\angle OSC = \angle SCO$. The orthocenter H lies somewhere on the line of the altitude CR, and we know that $OS \parallel CR$. Therefore, the alternate-interior angles $\angle OSC = \angle RCS$, and finally $\angle RCS = \angle SCO$, which completes the proof.

3. Whenever we know that one point is the isogonal conjugate of another point, then the concurrency of the corresponding Cevians needs not be proved. Above, we provided independent proofs for the orthocenter and circumcenter. Therefore, it would suffice to prove one of them, say, the concurrency of the perpendicular bisectors, whereupon the concurrency of the altitudes is guaranteed because the orthocenter is the isogonal conjugate of the circumcenter. In an analogous fashion, this relationship holds true for the centroid and the symmedian point of the triangle.

4. The isogonal conjugate of the centroid G is called *symmedian point* K, shown in Figure 2-36. This symmedian point is a very important point in triangle geometry, as it has many fascinating properties. The Canadian mathematician Ross Honsberger[2] (1929–2016) called it "one of the crown jewels in modern geometry".

We will encounter this point again in Chapter 3 with some unexpected properties. For the moment, we merely mention that the medians reflected in the angle bisectors are called *symmedians* and are concurrent at point K.

---

[2] R. Honsberger: *Episodes in Nineteenth and Twentieth Century Euclidean Geometry*. The Mathematical Association of America, (1995), p. 53.

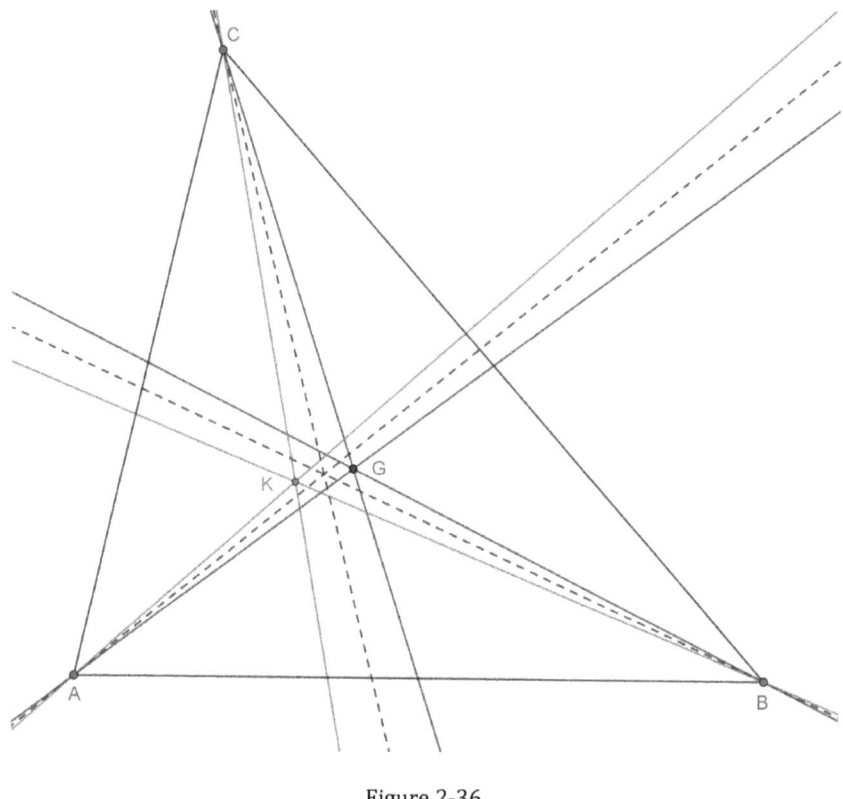

Figure 2-36

Characterization of Symmedians

In Figure 2-37, the point Q lies on the symmedian CK, if and only if, the distances of Q to the sides a and b are proportional to the sides themselves, that is $\frac{x}{y} = \frac{a}{b}$.

Isogonal conjugation interchanges angles of Cevians to the adjacent sides, which we saw in Figure 2-34: $\alpha_1 \leftrightarrow \alpha_2$, $\beta_1 \leftrightarrow \beta_2$, and $\gamma_1 \leftrightarrow \gamma_2$. Thus, it also interchanges perpendicular distances of points on the Cevians from the adjacent sides. We must show that the median m_c is the locus of all points P, where the proportion $\frac{x}{y} = \frac{b}{a}$ holds true, as shown in Figure 2-38. To prove this, we see that the perpendicular distances of M_c to a and b are $\frac{j}{2}$ and $\frac{k}{2}$, where j, k denote the lengths of the altitudes to the sides a and b, respectively. Note: $M_c E$ is a midline in $\triangle ABD$, so that, $M_c E = \frac{AD}{2} = \frac{j}{2}$; analogously, in $\triangle ABF$, we have $GM_c = \frac{1}{2}BF = \frac{k}{2}$. Furthermore, the equivalent area formulas for

Unexpected Concurrencies **87**

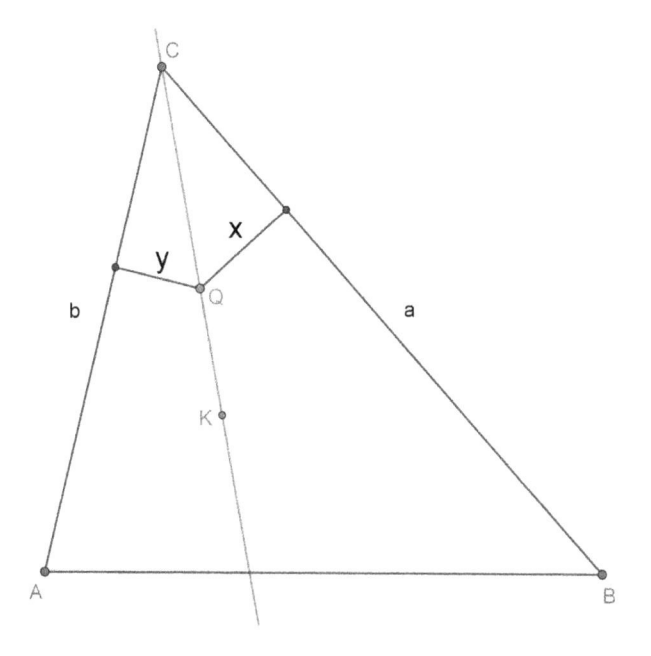

Figure 2-37

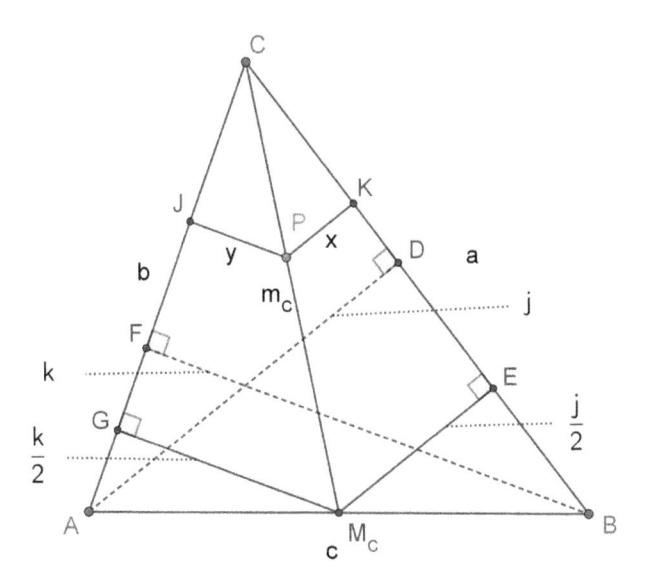

Figure 2-38

the triangle ABC are $\frac{a \cdot j}{2} = \frac{b \cdot k}{2} \Leftrightarrow \frac{j}{k} = \frac{b}{a}$. Since $\triangle CM_cE \sim \triangle CPK$ and $\triangle CM_cG \sim \triangle CPJ$, we have $\frac{j}{2} : x = CM_c : CP = \frac{k}{2} : y \Rightarrow \frac{x}{y} = \frac{j}{k}$, and this, finally, gives us $\frac{x}{y} = \frac{j}{k} = \frac{b}{a}$. Thus, we have proved $\frac{x}{y} = \frac{b}{a}$ for points on the median and established that $\frac{x}{y} = \frac{a}{b}$ for points on the symmedian, which is the median reflected in the angle bisector.

The Isotomic Conjugation: Another Way of Producing a New Concurrency from a Given Concurrency

There is an interesting relationship to the isogonal conjugation, introduced above, which is known as *isotomic conjugation*. The isogonal conjugation changed the angles of Cevians to the adjacent sides. See Figure 2-34, where we have the angle μ between the original Cevian CP and the triangle side $BC = a$ so that $\angle BCP = \mu$. For the reflected Cevian CP', we have the angle μ to the other triangle side $AC = b$ so that $\angle ACP' = \mu$. The isotomic conjugation changes distances from the intersection points of Cevians with the opposite sides to the vertices of the opposite sides. We will describe the isotomic conjugation and present a proof using a simple application of Ceva's theorem and its converse, this time using the non-trigonometric version of Ceva's theorem.

Suppose we have an arbitrary interior point P in $\triangle ABC$ and three concurrent Cevians AP, BP, and CP intersecting the triangle sides at the points A', B', and C', respectively. These "base" points on the sides are reflected in the midpoints D, E, and F of the sides, yielding the points A'', B'', and C''. The Cevians AA'', BB'', and CC'' are concurrent at point P', shown in Figure 2-39. This point P' is called the *isotomic conjugate* to P. Thus, the isotomic conjugation produces two pairs of equal line segments on each side, for example, segments x and y on side AB in Figure 2-39.

The proof is similar to that for the isogonal conjugation as it also uses Ceva's theorem and its converse. The concurrency at P yields by Ceva's theorem $\frac{AC'}{C'B} \cdot \frac{BA'}{A'C} \cdot \frac{CB'}{B'A} = 1$. But since $\frac{AC'}{C'B}$ is the reciprocal of $\frac{AC''}{C''B}$, which is analogous for the other vertices, we get $\frac{AC''}{C''B} \cdot \frac{BA''}{A''C} \cdot \frac{CB''}{B''A} = 1$, and this enables us to conclude – via the converse of Ceva's theorem – that the Cevians AA'', BB'', and CC'' are concurrent at P'.

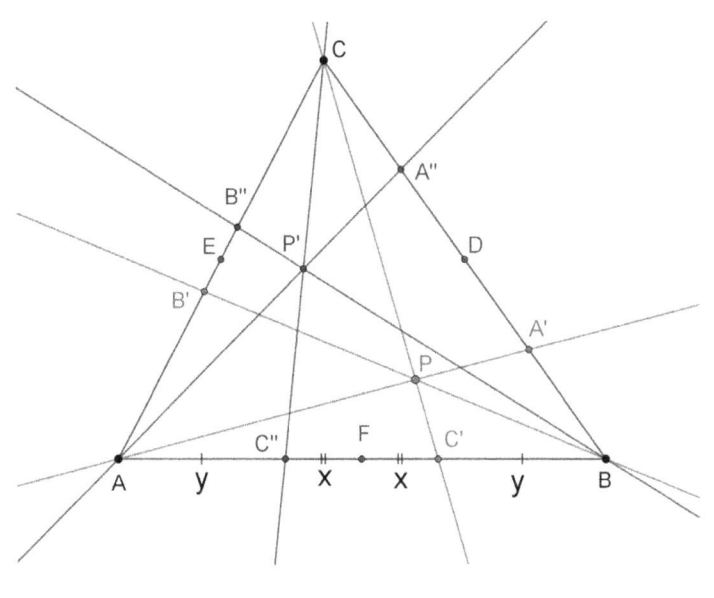

Figure 2-39

A few interesting aspects of an isotomic conjugation are as follows:

1. The isotomic conjugate of the centroid G is the point G itself.

2. Nagel point N and Gergonne point J are isotomic conjugates of each other. To prove this, one just has to know that on each triangle side the tangency points of the incircle and the corresponding excircle are symmetric to that side's midpoint. A proof of this will be provided in Chapter 11 (see Figures 11-11, 11-12, and 11-16).

3. Whenever one knows that one point is the isotomic conjugate of another point, the concurrency of the corresponding Cevians can be assumed without proof. Earlier, we gave two independent proofs for the Nagel point and the Gergonne point. It would suffice to prove one of them, say, the concurrency of the Gergonne Cevians, whereby the concurrency of the Nagel Cevians is guaranteed because the Nagel point is the isotomic conjugate of the Gergonne point.

Chapter 3
Unexpected Collinearities

As we know, any two points determine a unique line. Just as three points that are not collinear determine a unique circle, geometry is also fascinated with three points that do not determine the circle because they are collinear. Perhaps one of the most frequently used relationships that generates collinearity is known as *Menelaus' theorem*, with which we will begin this chapter.

The Famous Menelaus' Theorem

In his research, Giovanni Ceva discovered the work of the Greek mathematician Menelaus of Alexandria (70–140 CE), who in the year 100 CE discovered a relationship to which Ceva's findings are somewhat linked. Rather than concurrent lines, Menelaus developed a relationship with collinear points.

Menelaus' theorem states that we can use three collinear points, each on the side of a given triangle instead of having three concurrent lines determining points on the side of the triangle. In Figure 3-1, we have triangle ABC with three points $Z, X,$ and Y on sides AB, BC (extended), and AC, respectively, which lie on the same line; that is, they are collinear. Once again, the alternate segments determined by the three collinear points $Z, X,$ and Y will have equal products, where for triangle ABC, the following relationship holds true: $AZ \cdot BX \cdot CY = AY \cdot BZ \cdot CX$. Similarly to Ceva's theorem, Menelaus' theorem has a useful converse.

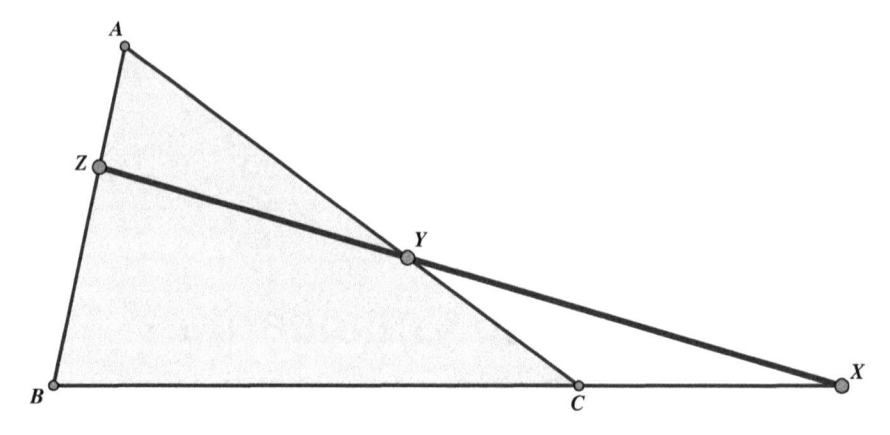

Figure 3-1

Proof of Menelaus' Theorem

According to Menelaus' theorem, we are to prove that if X, Y, and Z are collinear, then $AZ \cdot BX \cdot CY = AY \cdot BZ \cdot CX$. We begin our proof by drawing a line containing C, parallel to AB, and intersecting lines XYZ or YXZ determined by the three collinear points at point D, as we can see in Figures 3-2 and 3-3. The parallel lines determine various similar triangles as follows:

For $\triangle CDX \sim \triangle BXZ$, therefore $\dfrac{CD}{BZ} = \dfrac{CX}{BX}$, or $CD = \dfrac{BZ \cdot CX}{BX}$. (I)

For $\triangle CDY \sim \triangle AYZ$, therefore $\dfrac{CD}{AZ} = \dfrac{CY}{AY}$, or $CD = \dfrac{AZ \cdot CY}{AY}$. (II)

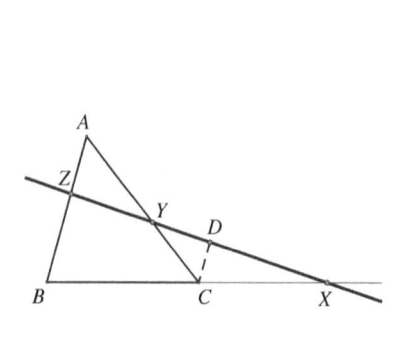

Figure 3-2 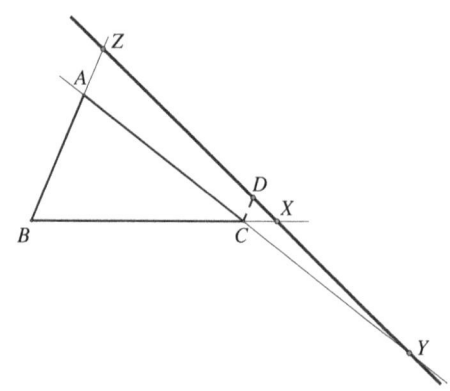   Figure 3-3

From equations (I) and (II) we get $\frac{BZ \cdot CX}{BX} = \frac{AZ \cdot CY}{AY}$, which enables us to establish that $CY \cdot AZ \cdot BX = AY \cdot BZ \cdot CX$.

Converse of Menelaus' Theorem

Now we shall prove the converse of Menelaus' theorem, which states that if the points X, Y, and Z are situated on the sides of triangle ABC (with points on the extension of the sides of the triangle) so that the equation $AZ \cdot BX \cdot CY = AY \cdot BZ \cdot CX$ is true, then the three points X, Y, and Z are collinear. This can also be stated as follows: If $\frac{AY}{CY} \cdot \frac{BZ}{AZ} \cdot \frac{CX}{BX} = 1$, then the three points X, Y, and Z are collinear.

We will let the intersection point of AB and XY be the point Z'. Then we have to prove $Z' = Z$.

Remark: There are also versions of Menalaus' theorem which work with −1 on the right side instead of 1. In both versions one has, in fact, to work with signed distances or signed segments, meaning $PQ = -QP$, in order to distinguish if a partition point of a line segment lies in its interior or exterior. Thus, especially concerning the converse of Menelaus' theorem, one has to be careful with the order of the points (again: $PQ = -QP$) so that the conclusion $Z' = Z$ is correct.

By Menelaus' theorem, we have established that $\frac{AY}{CY} \cdot \frac{BZ'}{AZ'} \cdot \frac{CX}{BX} = 1$. However, our hypothesis tells us that $\frac{AY}{CY} \cdot \frac{BZ}{AZ} \cdot \frac{CX}{BX} = 1$. Therefore, $\frac{BZ'}{AZ'} = \frac{BZ}{AZ}$. Therefore, $Z' = Z$, and the points X, Y, and Z must be collinear.

Just as we considered an external point of concurrency for Ceva's theorem, the same can be done with collinear external points of contact on the extended sides of triangle ABC, as we can see in Figure 3-3, where $CY \cdot AZ \cdot BX = AY \cdot BZ \cdot CX$ once again holds true.

Unexpected Collinear Points

In triangle ABC, shown in Figure 3-4, from a randomly selected point Q on the median AM of triangle ABC, a perpendicular is drawn to BC, intersecting it at point E. Point P is chosen on QE, whereupon perpendiculars PD and PF are drawn to sides AB and AC, intersecting the sides at points D and F, respectively, so that the points D, Q, and F are collinear. Surprisingly, we find that PA bisects $\angle BAC$.

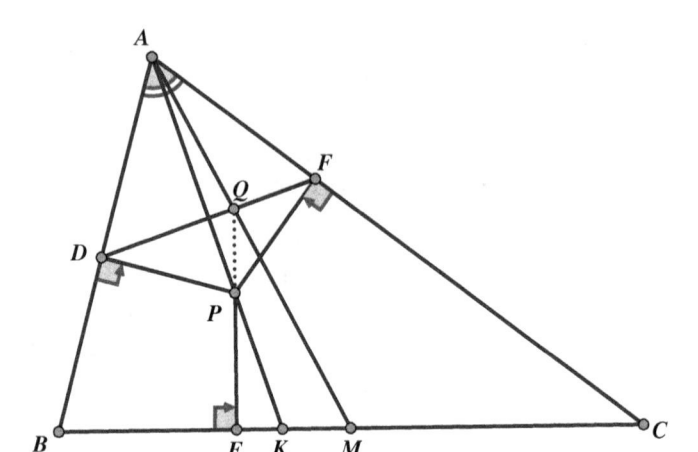

94 A Journey Through the Wonders of Plane Geometry

Figure 3-4

Proof

We should immediately notice that quadrilateral $ADPF$ is cyclic since two opposite angles are right angles. Therefore, $\angle FAP = \angle FDP = x$, and similarly, $\angle DAP = \angle DFP = z$ (see Figure 3-5). If we can show that $\angle FDP = \angle DFP$, then we will have shown that AK bisects $\angle BAC$.

The line RS is drawn parallel to BC, as shown in Figure 3-5. Since $\angle RQP$, $\angle PDR$, $\angle PQS$, and $\angle PFS$ are right angles, we know that

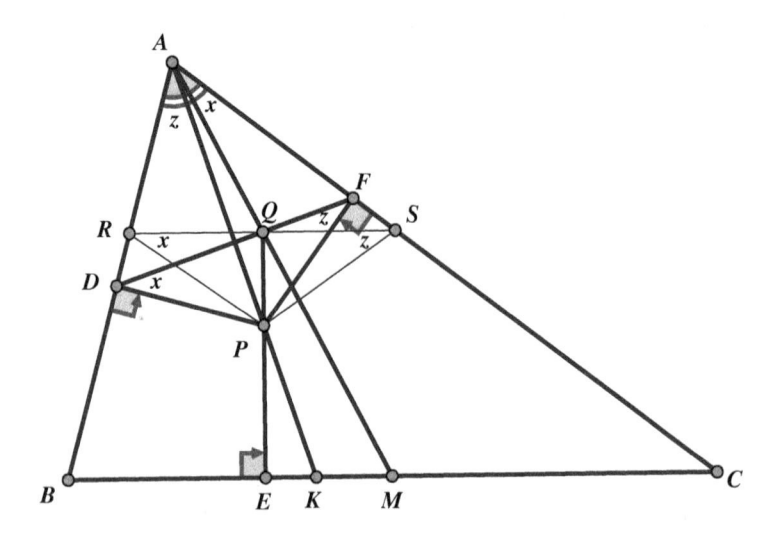

Figure 3-5

quadrilaterals *PQRD* and *PQFS* are cyclic, so that ∠*PDQ* = ∠*PRQ* and ∠*PFQ* = ∠*PSQ* (note that ∠*PRQ* = x and ∠*PSQ* = z). Therefore, *RS* is bisected at point *Q* by median *AM*. Since *QE*⊥*RS*, triangle *RPS* is isosceles and, therefore, ∠*PRS* = ∠*PSR*, which is the same as saying that $x = z$. So, we have established that ∠*BAK* = ∠*CAK*, and, therefore, *AK* bisects ∠*BAC*.

Circumscribed and Inscribed Circle Curiosities

When considering the inscribed circle and the circumscribed circle of a given triangle, unexpected wonders appear. Many concurrencies and collinearities will surprise us in this configuration. For example, the perpendicular bisectors *OS*, *OT*, and *OR* of each of the sides of triangle *ABC*, as shown in Figure 3-6, meet the circumscribed circle at

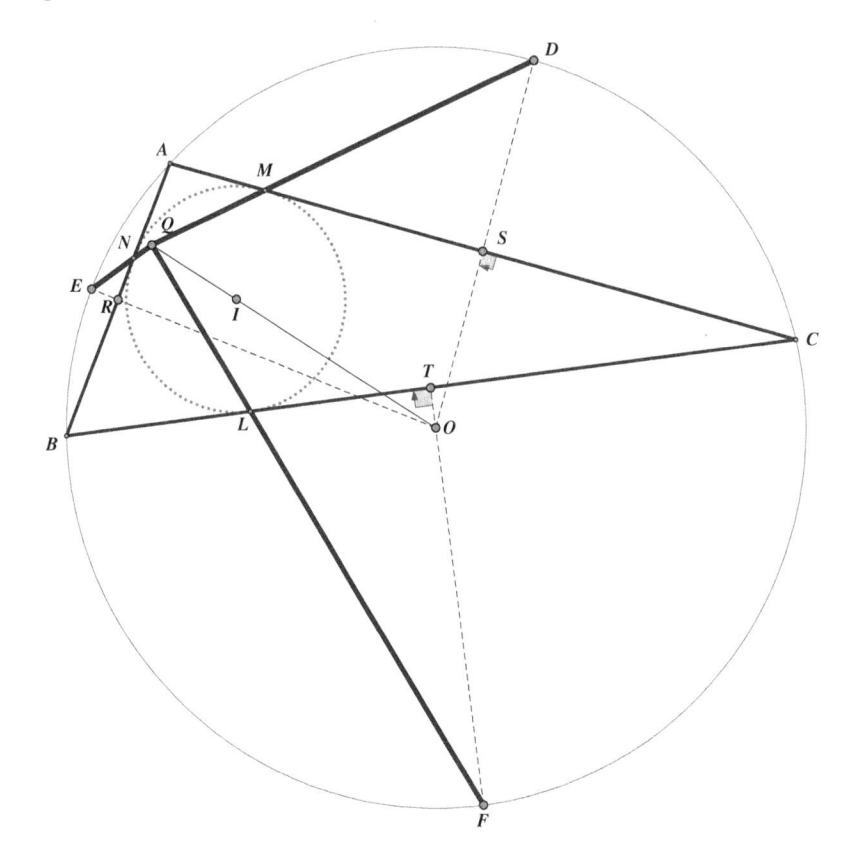

Figure 3-6

points D, F, and E, respectively, where Points D, F, and E are the midpoints of arcs $\widehat{AC}, \widehat{BC}$, and $\widehat{AB}$, respectively. The line segments EQ, DQ, and FQ connecting these latter arc-midpoints to the points of tangency of the inscribed circle of triangle ABC are concurrent, which we showed in Chapter 2 (page 77). Furthermore, another curious result is that the points O and Q are collinear with the center I of the inscribed circle of triangle ABC.

Proof

In the proof of concurrency at point Q, in Figure 3-7, which we located by extending OI, we established the collinearity of points Q, I, and O. The Figure 3-7, which we used in Chapter 2, established the similarity of triangles QIL and QOF, which further justified the collinearity.

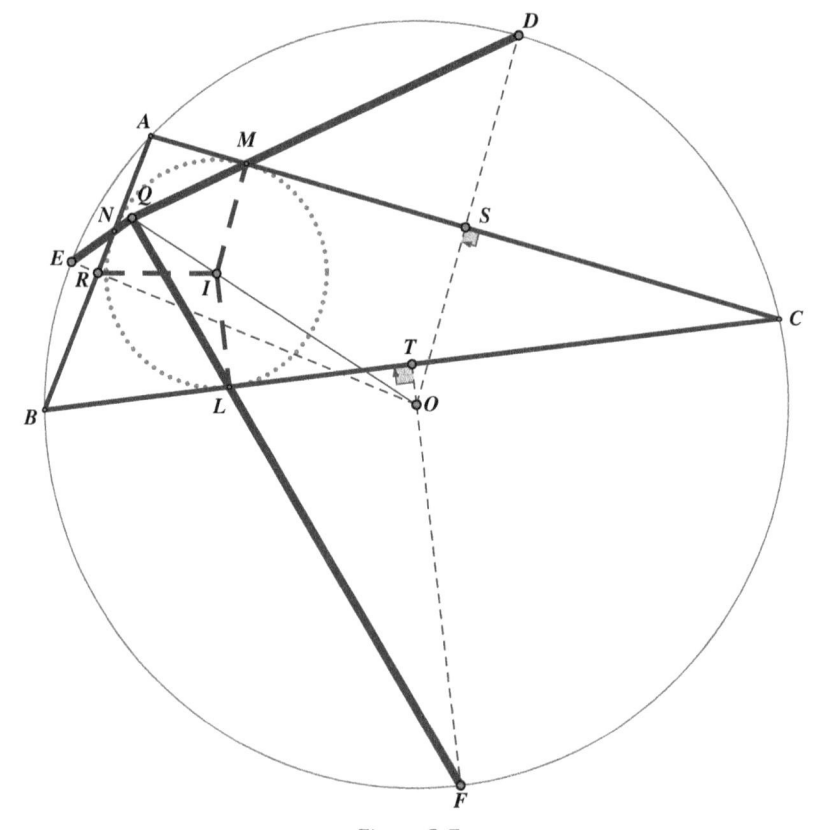

Figure 3-7

Desargues' Theorem

Let us consider a very useful geometric relationship discovered by French mathematician Gérard Desargues (1591–1661) and first presented in his book *Manière universelle de M. Desargues, pour pratiquer la perspective*. It involves two triangles placed so that the three lines joining corresponding vertices are concurrent. Remarkably, when this is the case, the pairs of corresponding sides intersect in three collinear points. In Figure 3-8, $\triangle ABC$ and $\triangle A'B'C'$ are situated in such a way that the lines joining the corresponding vertices AA', BB', and CC' are concurrent. The pairs of corresponding sides therefore intersect in three collinear points. In other words, the point Q, in which BC and $B'C'$ intersect, the point R, in which CA and $C'A'$ intersect, and the point S, in which AB and $A'B'$ intersect, are collinear.

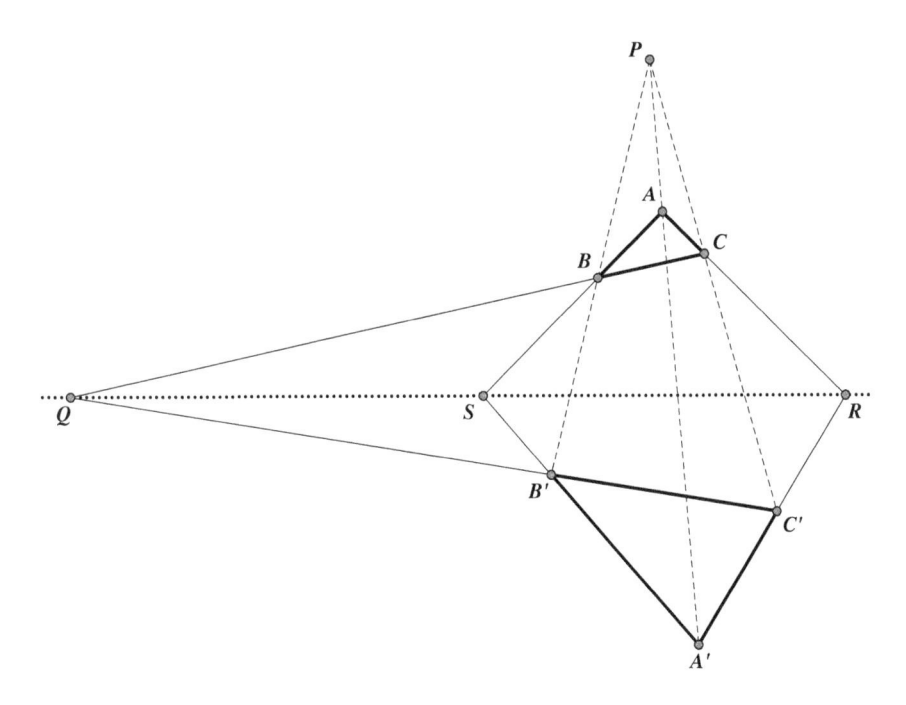

Figure 3-8

The converse is also true, which is that if the two triangles ABC and $A'B'C'$ are situated in such a way that points Q, R, and S lie on a common line, lines AA', BB', and CC' contain a common point P.

Proof: We shall prove Desargues' theorem by applying Menelaus' theorem several times. Consider line QBC as a transversal of $\triangle PB'C'$. By Menelaus' theorem, we have

$$\frac{PB}{B'B} \cdot \frac{B'Q}{C'Q} \cdot \frac{C'C}{PC} = 1. \tag{I}$$

Similarly, considering SBA as a transversal of $\triangle PB'A'$, we have

$$\frac{PA}{A'A} \cdot \frac{A'S}{B'S} \cdot \frac{B'B}{PB} = 1, \tag{II}$$

and considering RCA as a transversal of $\triangle PA'C'$, we have

$$\frac{PC}{C'C} \cdot \frac{C'R}{A'R} \cdot \frac{A'A}{PA} = 1. \tag{III}$$

By multiplying (I), (II), and (III), we get:

$$\left(\frac{PB}{B'B} \cdot \frac{B'Q}{C'Q} \cdot \frac{C'C}{PC}\right) \cdot \left(\frac{PA}{A'A} \cdot \frac{A'S}{B'S} \cdot \frac{B'B}{PB}\right) \cdot \left(\frac{PC}{C'C} \cdot \frac{C'R}{A'R} \cdot \frac{A'A}{PA}\right) = 1$$

or

$$\frac{B'Q}{C'Q} \cdot \frac{A'S}{B'S} \cdot \frac{C'R}{A'R} = 1.$$

Thus, by Menelaus' theorem, applied to $\triangle A'B'C'$, we have points A', B', and C' collinear.

It now remains for us to prove that the converse is true. We assume that ABC and $A'B'C'$ are situated such that points Q, R, and S lie on a common line, and wish to show that lines AA', BB', and CC' contain a common point P. Somewhat surprisingly, this is an immediate consequence of what we have just proved.

Consider $\triangle SB'B$ and $\triangle RCC'$. Because of our assumptions, we know that lines BC, $B'C'$, and RS have a common point Q. From the version of Desargues' theorem that we have just proved, we know that the pairs of corresponding sides of $\triangle SB'B$ and $\triangle RCC'$ intersect in three collinear points. In other words, the point P, in which BB' and CC' intersect, the point A, in which BS and CR intersect, and the point A', in which $B'S$ and $C'R$ intersect, are collinear. Lines AA', BB', and CC' therefore all pass through the common point P.

The Orthic Triangle Generates Collinear Points

Using the configuration in Figure 3-9, we will now consider an extension that leads to a collinearity instead of a concurrency. Here, the feet of the altitudes to the sides of triangle ABC are the points D, E, and F. The line FD intersects BC (extended) at point K; the line ED intersects side AC (extended) at point L; and the line FE intersects side AB (extended) at point M. We then find that the points K, L, and M are collinear.

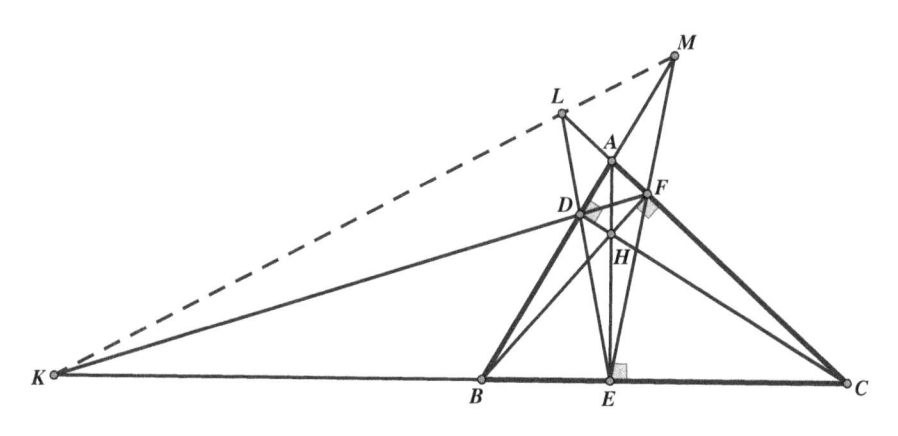

Figure 3-9

Proof

In Figure 3-9, let A, B, C and E, F, D be corresponding vertices of $\triangle ABC$ and $\triangle FED$. Since AF, CD, and BE are concurrent (they are the altitudes of △ABC), the intersections of the corresponding sides DE and BC, FE and BA, FD and CA are collinear by Desargues' theorem.

Some Median Surprises

The medians of a triangle also provide a number of other astonishing results. Consider, once again, triangle ABC with the medians AE, BF, and CD. We draw AK parallel to BF, and BK parallel to AC, as shown in Figure 3-10. The first surprise is that the three points K, D, and F are collinear. And the second surprise is that KC bisects DE, so that DN = EN.

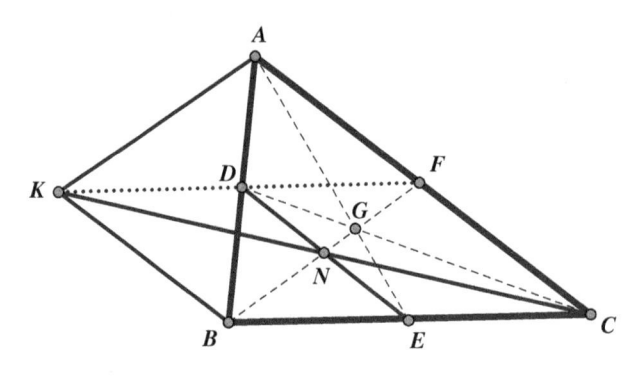

Figure 3-10

Proof

In Figure 3-10, with the parallel lines drawn as indicated, quadrilateral AFBK is a parallelogram. Therefore, since the diagonals of a parallelogram bisect each other, diagonal KF contains the midpoint of AB, which is tantamount to indicating that points K, D, and F are collinear. Since DF joins the midpoints of two sides of triangle ABC, it must be parallel to the third side, namely, BC. Therefore, quadrilateral KFCB is a parallelogram since the opposite sides are parallel. Thus, the diagonal KC bisects the line segment DE, which joins the midpoints of the opposite sides of the parallelogram, so we have DN = EN.

Median Extensions Generate Collinearity

When the medians *BN* and *CM* of triangle *ABC* are extended their own length to two points, *D* and *E*, respectively, we find that the points *D* and *E* are collinear with vertex *A*, as shown in Figure 3-11.

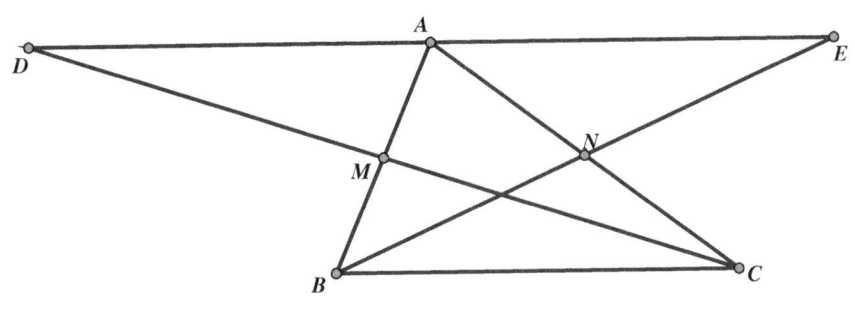

Figure 3-11

Proof

We can easily prove $\triangle DAM \cong \triangle CBM$ since $AM = BM$ and $DM = CM$, along with the vertical angles at point *M*, as shown in Figure 3-11. Therefore, $\angle DAM = \angle CBM$, which establishes that $DA \| CB$. In a similar fashion, we can prove that triangles *ANE* and *CNB* are congruent, which enables us to conclude that $AE \| CB$. Since both *DA* and *EA* are parallel to *BC*, they must be one line and points *D*, *A*, and *E* are therefore collinear.

A Truly Unexpected Collinearity

By placing an equilateral triangle inside a square and another one outside the square, as shown in Figure 3-12, we find an interesting (and unexpected) collinearity of three points. Equilateral triangles *DEC* and *BCF* are placed inside and on the outside of square *ABCD*, respectively. The result is that the points *A*, *E*, and *F* are collinear.

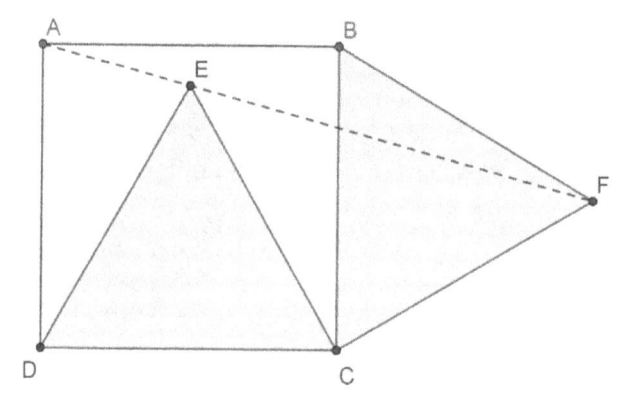

Figure 3-12

Proof

We can easily show in Figure 3-12 that △ECF is an isosceles right triangle since EC and FC are sides of congruent equilateral triangles, and both 60° angles at point C create a right angle with ∠ECB = 30°. Hence, ∠CEF = 45°. We also have isosceles △ADE, where ∠ADE = 30°. Therefore, ∠AED = 75°. The result is that ∠AED + ∠DEC + ∠CEF = 180°, which justifies that the points A, E, and F are collinear.

Tangents and Inscribed-Triangle Sides Form an Unexpected Collinearity

Collinearity can also be found by having the three tangents at the vertices of an inscribed triangle meet the opposite sides. We can see this configuration in Figure 3-13, where tangents AE, BD, and CF to the circumscribed circle O of triangle ABC meet the opposite sides at points D, E, and F, which turn out to be collinear.

Proof

In Figure 3-13, triangles ADB and BDC are similar because ∠BCD = ∠ABD, since both are measured by one-half arc AB, and therefore

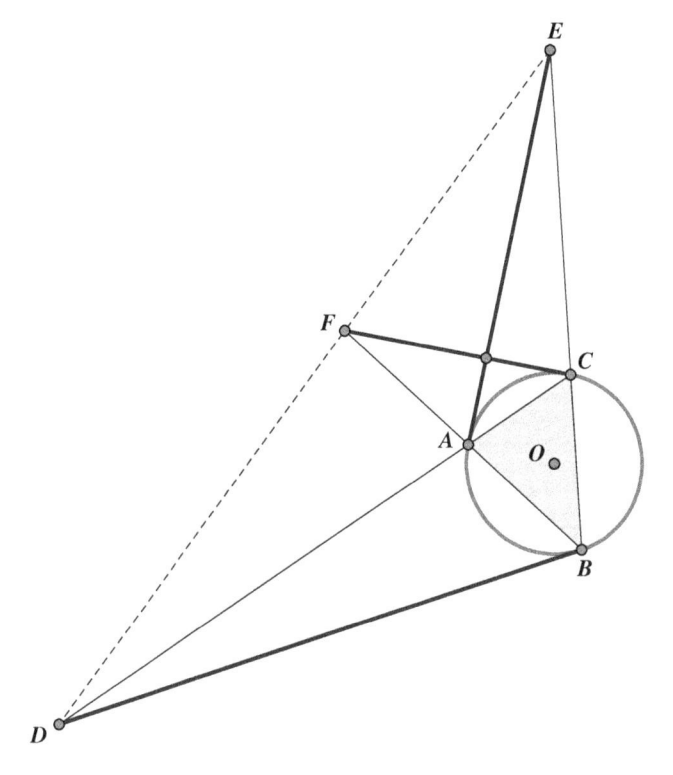

Figure 3-13

$\frac{AD}{BD} = \frac{BD}{CD} = \frac{AB}{CB} = \frac{x}{y}$. This leads to $\frac{AD}{CD} = \frac{AD}{BD} \cdot \frac{BD}{CD} = \frac{x^2}{y^2}$. Just as point D partitioned line DC proportionally, the same is true for points E and F, so that $\frac{CE}{BE} = \frac{z^2}{x^2}$ and $\frac{BF}{AF} = \frac{y^2}{z^2}$. Therefore, the product of these ratios into which the points E, F, and D partitioned the sides of triangle ABC gives us $\frac{AD}{CD} \cdot \frac{CE}{BE} \cdot \frac{BF}{AF} = \frac{x^2}{y^2} \cdot \frac{z^2}{x^2} \cdot \frac{y^2}{z^2} = 1$, which, according to Menelaus' theorem, indicates that the points D, E, and F are collinear.

Four Angle Bisectors and Perpendiculars Determine Four Collinear Points

An unexpected collinearity of four points can be seen when we draw angle bisectors to both interior angles and exterior angles, as we show in Figure 3-14, where triangle *ABC* has interior angle bisectors *CG* and *BD*, and exterior angle bisectors *BE* and *CJ*. When we draw perpendiculars from vertex *A* to each of these four angle bisectors meeting them at points *E*, *G*, *D*, and *J*, we find that these four points lie on the same straight line, which is to say they are collinear.

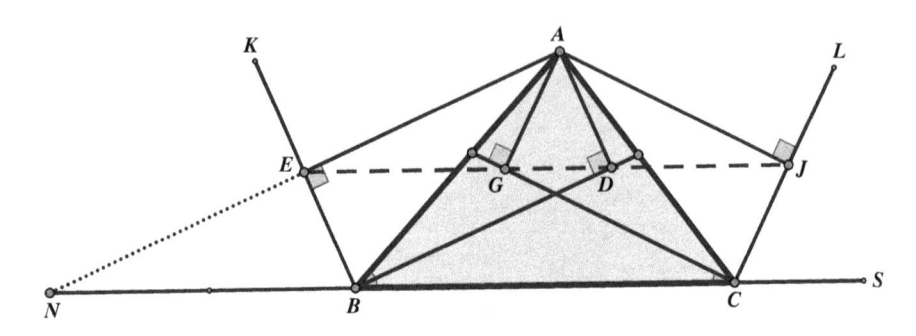

Figure 3-14

Proof

In Figure 3-15, let *BD* and *BE* represent the interior and exterior angle bisectors of $\angle B$ in $\triangle ABC$, with *D* and *E* being the feet of the perpendiculars to *BD* and *BE* from *A*. Extend *AD* and *AE* to meet *BC* at *M* and *N*, respectively. We see that $\triangle BDA \cong \triangle BDM$ (ASA), so that $AD = DM$. Also, $\triangle BEA \cong \triangle BEN$ (ASA), so that $AE = EN$. Therefore, *DE* is the midline of $\triangle AMN$. The midline of a triangle is parallel to the third side (in

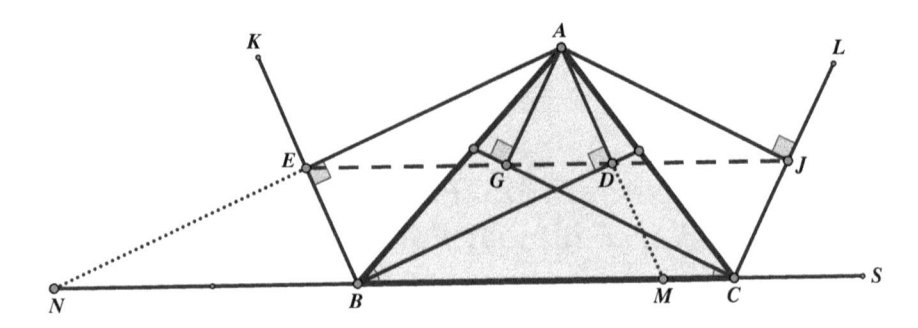

Figure 3-15

this case, *DE*∥*NBMC*) and lies halfway between that third side and the opposite vertex.

Analogously, the feet of the perpendiculars from *A* to the interior and exterior bisectors of ∠*C* are also on a line that is parallel to *BC* and lies halfway between *BC* and *A*. Therefore, all four feet of the perpendiculars from vertex *A* to the interior and exterior bisectors of the other two angles of △*ABC* are collinear.

An Unexpected Collinearity

In Figure 3-16, we find that point *D* is randomly placed on side *AB* of isosceles triangle *ABC*, where *AB* = *AC*. It turns out that the center *I* of the inscribed circle of triangle *ADC* and the center *O* of the circumscribed circle of triangle *BCD* lie on the median (or angle bisector, or altitude) of triangle *ABC*. This implies that the points *A*, *I*, *O*, and *M* (the midpoint of side *BC*) are all collinear.

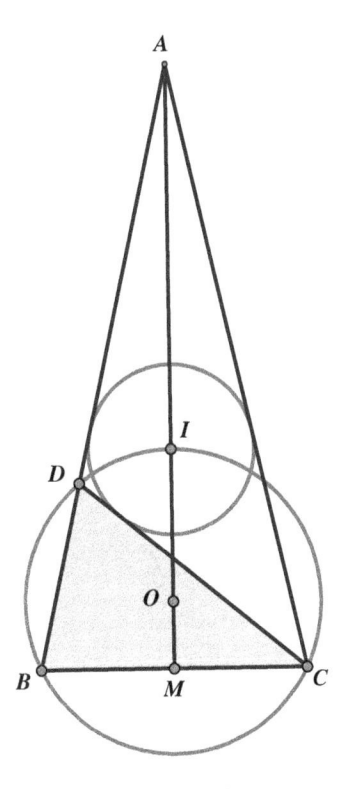

Figure 3-16

Proof

Let M be the midpoint of BC in Figure 3-16. Since point O is the circumcenter of triangle BCD, line OM is the perpendicular bisector of BC. Furthermore, since I is the incenter of triangle ACD, line AI is the angle bisector of $\angle A$ and passes through M, since $AB = AC$. Therefore A, I, and O lie on the perpendicular bisector of BC.

Another Unexpected Collinearity

In Figure 3-17, we show triangle ABC with its inscribed circle center at I and its circumscribed circle center at point O. The line parallel to side AB through point I meets the line tangent to the circumscribed

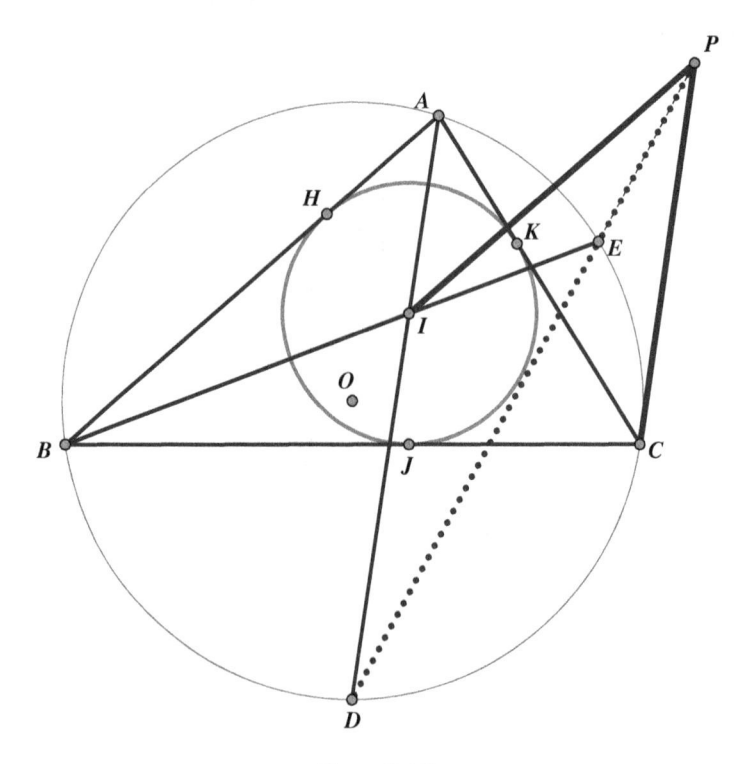

Figure 3-17

circle at point C at point P. Line BI (extended) intersects the circumscribed circle at point E, and line AI intersects the circumscribed circle at point D. Unanticipatedly, we find that the points D, E, and P are collinear.

Proof

In Figure 3-18, let α, β, and γ be the measures of the interior angles of $\triangle ABC$ at the vertices A, B, and C. First, we will show that $DI = DC$. Since ABDC is a cyclic quadrilateral, we have $\angle BCD = \frac{\alpha}{2}$, which gives us $\angle ICD = \frac{\alpha}{2} + \frac{\gamma}{2}$. We see that $\angle DIC$ is an exterior angle of $\triangle AIC$, which shows $\angle DIC = \frac{\alpha}{2} + \frac{\gamma}{2}$. This proves that $\triangle IDC$ is isosceles. Analogously, $EI = EC$ and, thus, both points E and D lie on the perpendicular bisector of IC. It remains to be shown that P also lies on this perpendicular bisector, or that $\angle PCI = \angle PIC$. Since $\angle PCA$ is an angle between the chord AC and the tangent at C, we know $\angle PCA = \beta$, thus, $\angle PCI = \beta + \frac{\gamma}{2}$.

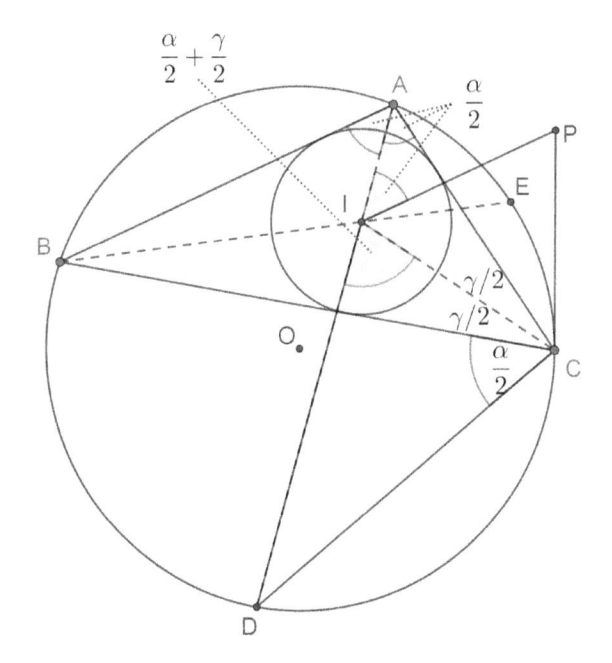

Figure 3-18

For $\angle PIC$, we get via the angle sum in $\triangle AIC$ of 180° the relation $\angle PCI = \left(180° - \frac{\alpha}{2} - \frac{\gamma}{2}\right) - \frac{\alpha}{2}$ (note that $PI \parallel AB$ yields $\angle PIA = \frac{\alpha}{2}$). This equals $\beta + \frac{\gamma}{2}$, which equals the angle $\angle PCI$.

Another Unexpected Collinearity

In Figure 3-19, the circle inscribed in triangle ABC has points of tangency D, E, and F with the sides AB, AC, and BC, respectively. The lines joining points of tangency intersect the extensions of the triangle's opposite sides, so that FE intersects BA at point M, FD intersects CA at point L, and ED intersects CB at point K. Surprisingly, the points M, L, and K are collinear.

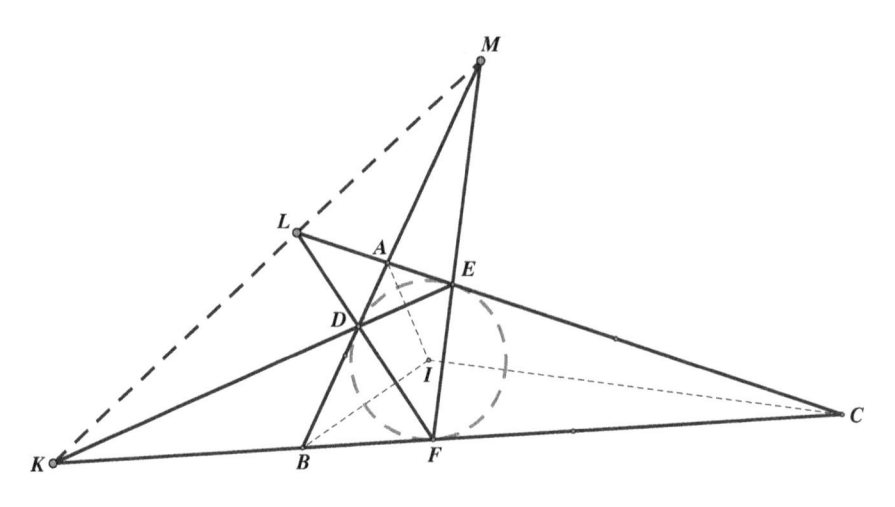

Figure 3-19

Proof

Since tangents to a circle from an external point are equal, we have the following equalities: $CE = CF$, $AD = AE$, and $BD = BF$. Therefore, we have $\frac{AD}{BD} \cdot \frac{BF}{CF} \cdot \frac{CE}{AE} = 1$. Thus, by Ceva's theorem, the lines AF, BE, and CD are concurrent. Because these are the lines joining the corresponding vertices of triangle *ABC* and triangle *FED*, by Desargues' theorem the intersections of the corresponding sides are collinear, namely, the points *K*, *L*, and *M*.

The Meeting of the Three Famous Triangle Centers

Within a triangle, there is a special significance about the orthocenter (the point of intersection of the altitudes of the triangle), the center of the circumscribed circle, and the centroid (the point of intersection of the medians of a triangle), which in Figure 3-20 are designated by points *H*, *O*, and *G*, respectively. Not only are these three points always collinear, creating a famous line in triangle geometry called the *Euler line* of the triangle, but they produce the following relationship: $OG = \frac{1}{2}GH$.

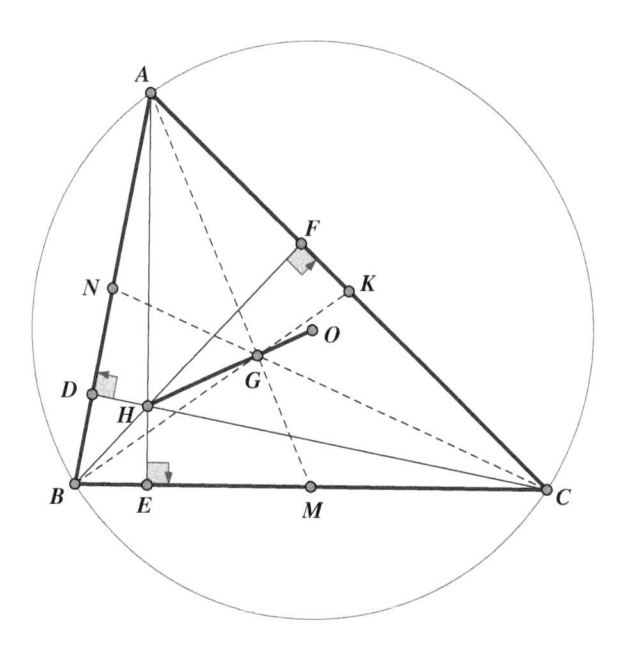

Figure 3-20

Proof

Using Figure 3-21, we will prove that the point P on the extension of line segment OG (beyond G) with $PG = 2GO$ is actually the orthocenter H. We know that $\triangle GAP$ and $\triangle GMO$ are similar triangles since $GA = 2GM$, $GP = 2GO$, and $\angle AGP = \angle MGO$ (vertex angles). Thus, $AP \parallel OM$, and since $OM \perp BC$, we also have $AP \perp BC$ intersecting BC at point E. Hence, P must lie on the altitude AE. Similarly, P must lie on the other altitudes, which tells us that P must actually be the orthocenter H.

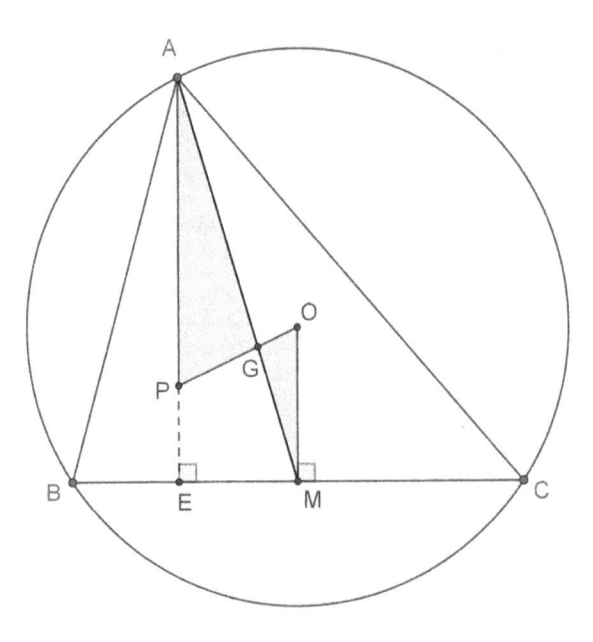

Figure 3-21

Another proof of this phenomenon using homothety is given in Chapter 11. There, one can also find a proof for another famous line, the *Nagel line*. It contains the Nagel point N (see Chapter 2), the incenter I, and the centroid G of a triangle. And, analogously to the *Euler line*, we have on the *Nagel line*: $IG = \frac{1}{2}NG$.

Simson's Theorem

When considering collinear points involving triangles, the famous Simson theorem must be acknowledged. One of the great injustices in the history of mathematics involves this theorem. It was originally published by William Wallace (1768–1843) in Thomas Leybourn's *Mathematical Repository* (1799–1800), but the theorem, through careless misquotes, has been attributed to Robert Simson (1687–1768), the famous English interpreter of Euclid's *Elements*, whose book has been the basis for the study of geometry in the English-speaking world and greatly influenced the American high school geometry course. To conform to the norm, we shall use the popular term *Simson's theorem*.

Simson's theorem states that the feet of the perpendiculars drawn from *any* point on the circumscribed circle of a triangle to the sides of the triangle are collinear. This is shown in Figure 3-22, where point P is any point on the circumscribed circle of triangle ABC. We draw $PY \perp AC$ at Y, $PZ \perp AB$ at Z, and $PX \perp BC$ at X. According to Simson's theorem, points X, Y, and Z are collinear, regardless of where point P is positioned on the circumscribed circle of the triangle. This line is usually referred to as the *Simson line*. It should be noted that the converse of this theorem is also true.

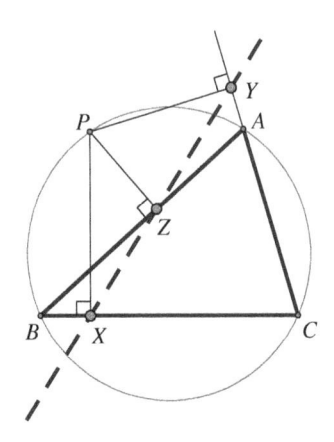

Figure 3-22

Proof

In Figure 3-23, $\angle PYA$ is supplementary to $\angle PZA$ so that quadrilateral $PZAY$ is cyclic. We draw lines PA, PB, and PC. Therefore, $\angle PYZ = \angle PAZ$. (I)

Similarly, since $\angle PYC$ is supplementary to $\angle PXC$, quadrilateral $PXCY$ is cyclic, and it follows that $\angle PYX = \angle PCB$. (II)

However, quadrilateral $PACB$ is also cyclic, since it is inscribed in the given circumcircle, and therefore, $\angle PAZ = \angle PCB$. (III)

From (I), (II), and (III) we can conclude by transitivity $\angle PYZ = \angle PYX$, and thus, points X, Y, and Z are collinear.

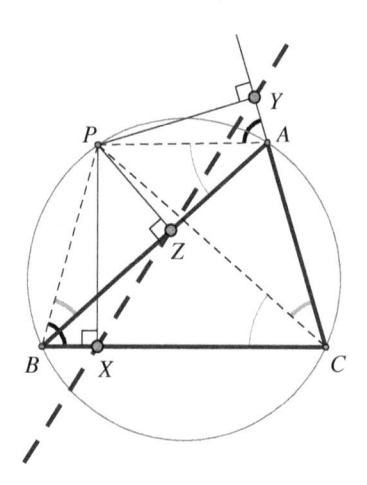

Figure 3-23

Simson's Theorem (Proved Using Menelaus' Theorem)

We begin by drawing PA, PB, and PC as shown in Figure 3-23. We have $\angle PBA = \frac{1}{2}\widehat{AP}$, and $\angle PCA = \frac{1}{2}\widehat{AP}$. Therefore, $\angle PBA = \angle PCA = \alpha$. Thus, $\frac{BZ}{PZ} = \cot \alpha = \frac{CY}{PY}$ (in $\triangle BPZ$ and $\triangle CPY$) or $\frac{BZ}{PZ} = \frac{CY}{PY}$, which implies $\frac{BZ}{CY} = \frac{PZ}{PY}$. (I)

Similarly, $\angle PAB = \angle PCB = \beta$ (as both are equal in measure to $\frac{1}{2}\widehat{BP}$).

Therefore, $\frac{AZ}{PZ} = \cot\beta = \frac{CX}{PX}$ (in $\triangle APZ$ and $\triangle CPX$) or $\frac{AZ}{PZ} = \frac{CX}{PX}$, which implies $\frac{CX}{AZ} = \frac{PX}{PZ}$. (II)

Since $\angle PBC$ and $\angle PAC$ are opposite angles of an inscribed (cyclic) quadrilateral, they are supplementary. However, $\angle PAY$ is also supplementary to $\angle PAC$.

Therefore, $\angle PBC = \angle PAY = \gamma$, and so $\frac{BX}{PX} = \cot\gamma = \frac{AY}{PY}$ (in $\triangle BPX$ and $\triangle APY$) or $\frac{BX}{PX} = \frac{AY}{PY}$, which implies $\frac{AY}{BX} = \frac{PY}{PX}$. (III)

By multiplying (I), (II), and (III), we obtain: $\frac{BZ}{CY} \cdot \frac{CX}{AZ} \cdot \frac{AY}{BX} = \frac{PZ}{PY} \cdot \frac{PX}{PZ} \cdot \frac{PY}{PX} = 1$.

Thus, by Menelaus' theorem, the points X, Y, and Z are collinear. These three points determine the *Simson line* of $\triangle ABC$ with respect to point P.

An Extension of Simson's Theorem

Generating the Simson line from the intersection of the circumscribed circle with the extension of one of the triangle's altitudes produces an intriguing curiosity. This Simson line is parallel to the tangent at the vertex from which this altitude emanates. For example, in Figure 3-24, the altitude $BD = h_b$ of triangle ABC meets the circumscribed circle at P (and at B), and the Simson line of triangle ABC with respect to P is parallel to the line tangent to the circle at B.

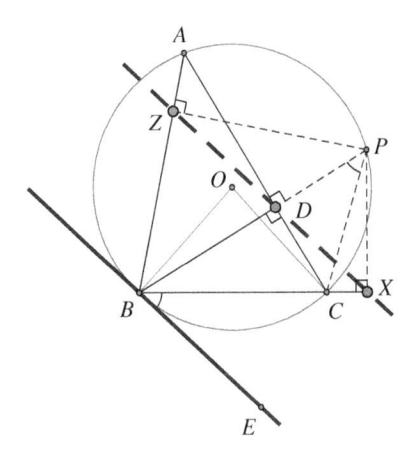

Figure 3-24

Proof

This can be easily justified. We know that in Figure 3-24, point D is one of the points on the Simson line. We also have PX and PZ perpendicular, respectively, to sides BC and AB of triangle ABC. Therefore, points X, D, and Z determine the Simson line of P with respect to triangle ABC. Next, we will draw PC. Quadrilateral $PDCX$ is cyclic because $\angle PDC = \angle PXC = 90°$. The two inscribed angles subtending the same arc $\widehat{DC}$ of the circumscribed circle of quadrilateral $PDCX$ are equal.

Therefore, $\angle DXC = \angle DPC$. (I)

However, in the circumscribed circle (with circumcenter O) of triangle ABC, $\angle EBC = \frac{1}{2}\widehat{BC}$, and $\angle DPC\ (=\angle BPC) = \frac{1}{2}\widehat{BC}$.

Therefore, $\angle EBC = \angle DPC$. (II)

From (I) and (II), by transitivity, $\angle DXC = \angle EBC$ are alternate-interior angles, thus making the Simson line XDZ parallel to tangent line EB.

A Parallel to the Simson Line

When we extend one of the perpendiculars from point P, say, perpendicular PX, to meet the circumscribed circle at point Q, lo and behold, we find that AQ is parallel to the Simson line XZY, as shown in Figure 3-25.

Proof

In Figure 3-26, since P and Q both lie on the circumcircle of ABC, we have $\angle AQP = \angle ABP$. Noting that $\angle PXB = \angle PZB = 90°$, and points P, B, X, and Z also lie on a common circle, we have $\angle ZXP = \angle ZBP$. Thus, we have $\angle ZXP = \angle ZBP = \angle ABP = \angle AQP$, and hence, $AQ \| ZX$.

Unexpected Collinearities **115**

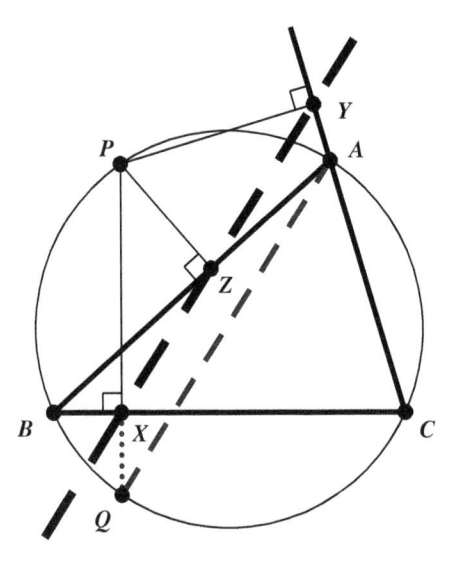

Figure 3-25

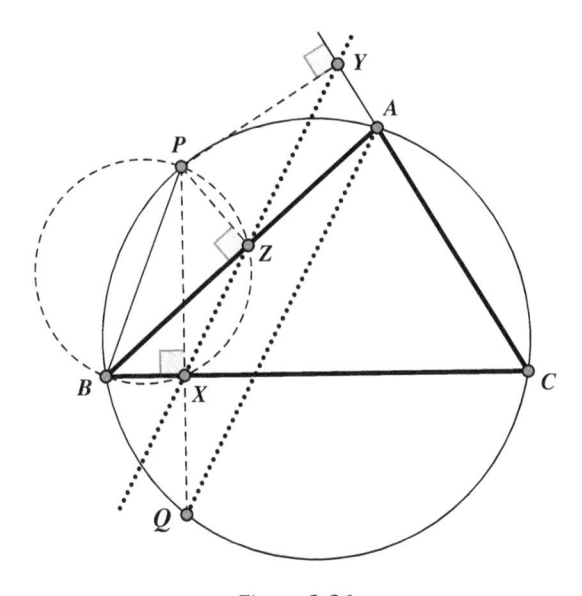

Figure 3-26

A Characteristic of Simson's Theorem

We can admire the Simson line *NML* for another unusual characteristic, as it bisects the line joining the Simson point *P* on the circle with the orthocenter *H* of the triangle *ABC*. That is, *PR* = *HR* (see Figure 3-27).

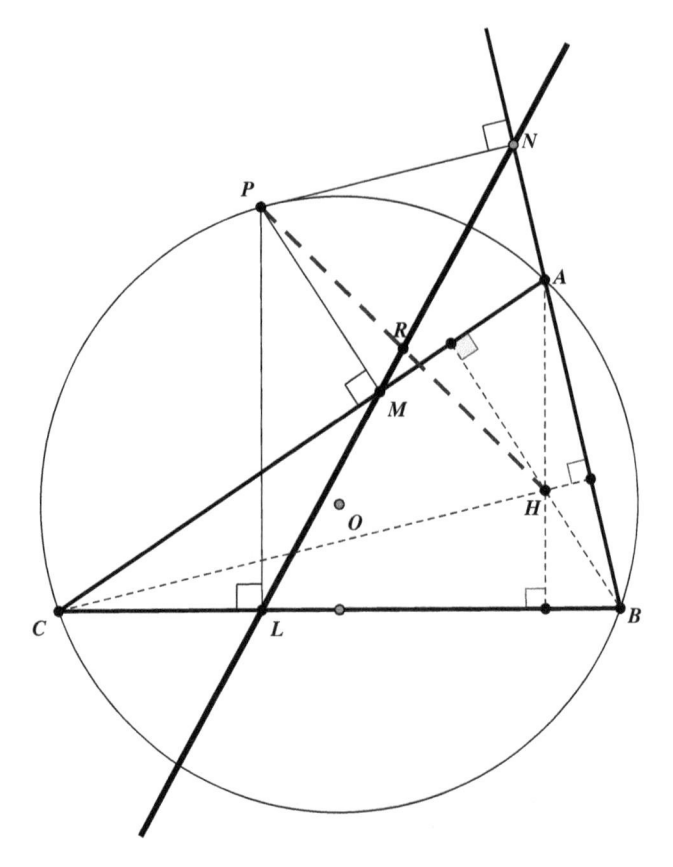

Figure 3-27

Proof

As shown in Figure 3-28, we extend altitude *BE* to meet circle *O* at point *T* and extend *PM* to meet circle *O* at point *K*. After constructing *BK*, a line from point *H* parallel to *BK* will meet *PK* at point *U*. We then have parallelogram *BKUH*, whereby *HU* = *BK*. We also have isosceles

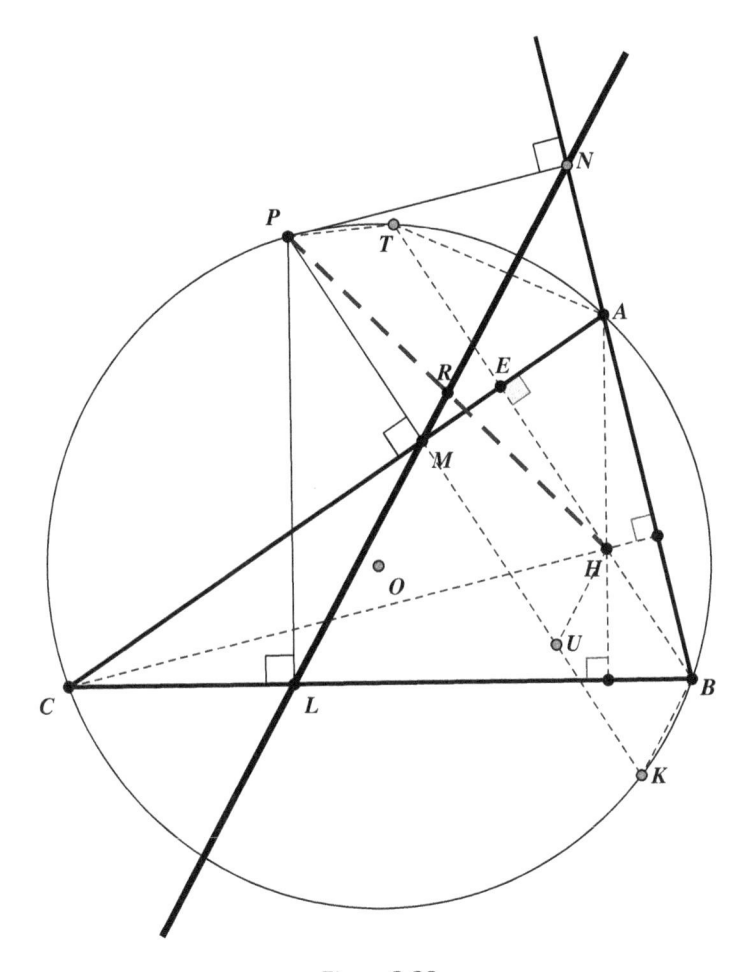

Figure 3-28

trapezoid *PTBK*, so that $PT = BK$. Therefore, *TPUH* is an isosceles trapezoid.

We then have $\angle ATB = \angle ACB$. Also, $\angle AHE = \angle ACB$, since their respective sides are perpendicular. Therefore, triangle *TAH* is isosceles, whereupon *E* is the midpoint of *TH*. Since *AE* extended passes through point *M*, point *M* is the midpoint of side *PU* of triangle *PHU*. Due to the above-mentioned phenomenon ("A Parallel to the Simson Line"), *MR* is parallel to side *HU* of triangle *PHU*; hence, it must also intersect the other side of triangle *PHU* at its midpoint, *R*. Therefore, *PR = HR*.

Another Simson Line as Bisector

There seems to be no end to the wonders produced by the Simson line. In Figure 3-29, the Simson line of equilateral triangle *ABC* bisects the radius of the circumscribed circle whose endpoint is *P*.

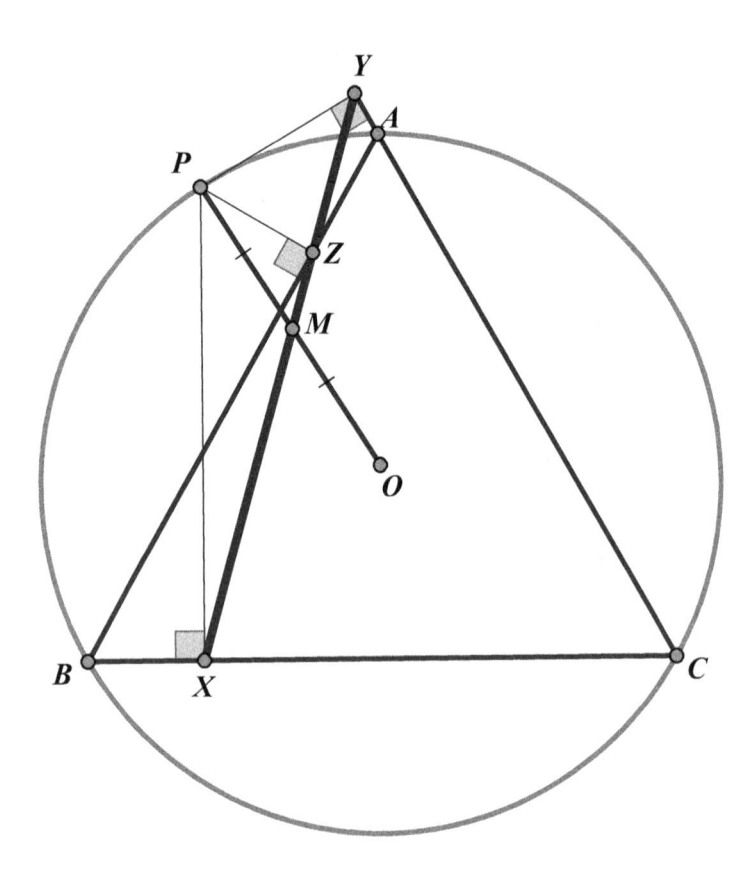

Figure 3-29

Proof

For the equilateral triangle in the given situation, the center *O* of the circumscribed circle is also the orthocenter *H*. Therefore, we know immediately that the Simson line bisects *OP* at *M* since we have already proved this for the general triangle in the previous example, shown in Figure 3-28.

Yet Another Characteristic of Simson's Theorem

When the three perpendiculars from point *P*, namely, *PX*, *PZ*, and *PY*, to the sides of triangle *ABC* are extended their own length to points *K*, *L*, and *N*, we find that these three points are collinear and parallel to the Simson line, as shown in Figure 3-30.

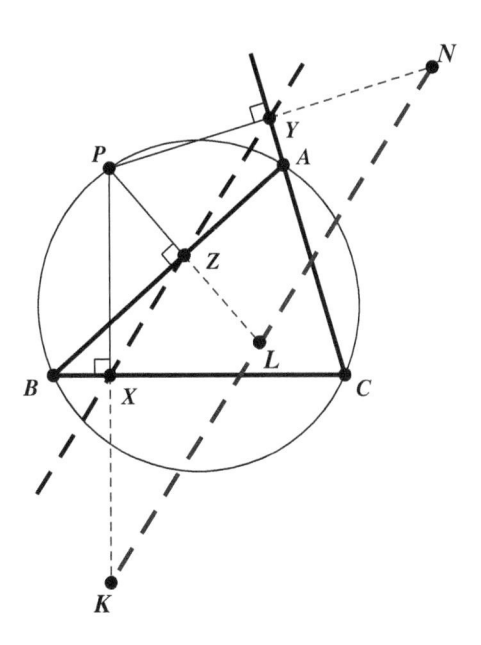

Figure 3-30

Proof

We know that a line joining the midpoints of two sides of a triangle is parallel to the third side. Therefore, in triangle *NPL* in Figure 3-30, we have *ZY* parallel to *NL*. Similarly, in triangle *LPK*, we have *ZX* parallel to *KL*. Therefore, *NLK* is parallel to *YZX*, which once again gives us another parallel to the Simson line.

The Newton Line Revisited

In Figure 3-31, *ABCD* is a convex quadrilateral that is not a parallelogram,[1] and *I* an interior point. Joining *I* with the four vertices *A, B, C,* and *D* yields four triangles, two of which are shaded (△*AIB* and △*DIC*) and the other two unshaded (△*AID* and △*BIC*). One can experiment with dynamic geometry programs such as Geometer's Sketchpad or GeoGebra to determine the location of point *I* that will yield an equality between the sum of the two shaded triangles and the sum of the non-shaded triangles.

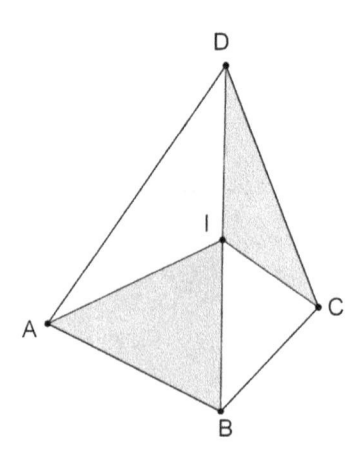

Figure 3-31

---

[1] In the case of parallelograms, all interior points have the property described.

There is a remarkable finding when we consider convex quadrilaterals and the area balance property. (In the following, we restrict to *interior* points, but this would not be necessary.) The line $g = MN$, shown in Figure 3-32, which joins the midpoints of the diagonals of the quadrilateral (*Newton line*), is the locus of all points I with the area balance property, establishing a truly unexpected collinearity of infinitely many points. This theorem is also often called *Anne's Theorem*.[2]

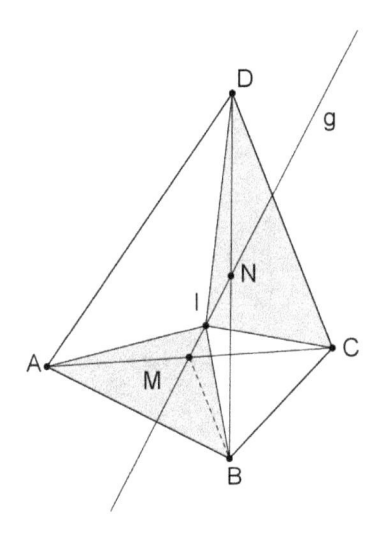

Figure 3-32

One can see immediately that M and N have the area balance property because in this case we have two pairs of adjacent and differently shaded triangles with equal areas (note that, e.g., triangles AMB and CMB have equal areas because they have equal bases, $AM = MC$, and the same altitude from point B).

Proof

First, we prove the following statement: When moving the point I on g, the area sums of the gray and white triangles do not change.

---
[2] Pierre-Leon Anne, French mathematician (1806–1850).

122 A Journey Through the Wonders of Plane Geometry

Because the points *M*, *N* have the area balance property, then *all* points of *g* have this property, too.

Let *I*, *K* be two different points of *g*; we will show first that the two arrow-shaped quadrilaterals *AIBK* and *CIDK*, shown in Figure 3-33 with dotted and dashed lines, have the same area.

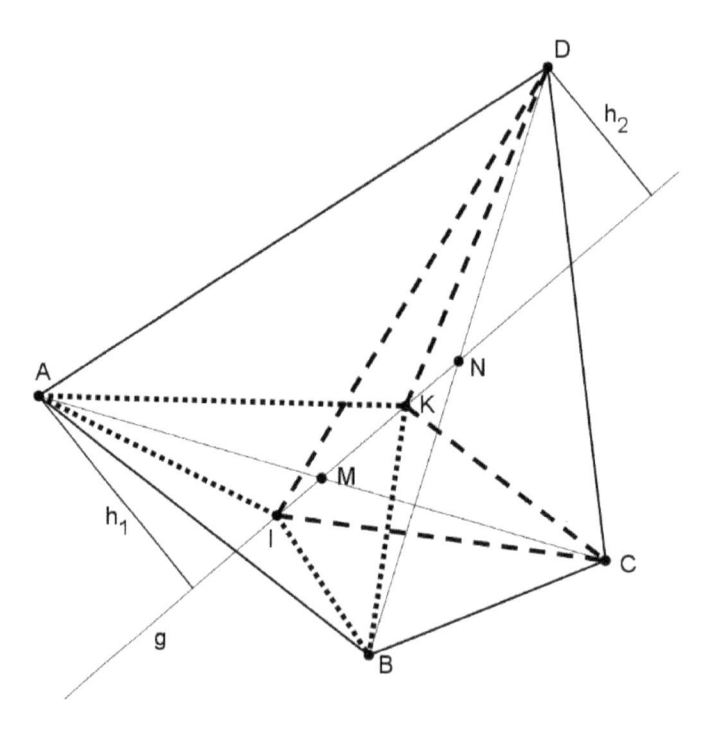

Figure 3-33

Because *M* lies on *g*, the points *A* and *C* are the same distance h_1 from *g*; analogously, *B*, *D* are the same distance h_2 from *g* because *N* also lies on *g*. Both arrows consist of two triangles with the same side *IK*; one of the triangles has the altitude h_1, and the other one the altitude h_2. Therefore, both arrows have the same area = $\frac{1}{2} \cdot IK \cdot (h_1 + h_2)$.

When we move point *I* to *K*, we can establish the following considerations of "increasing and decreasing" for the gray and white areas,

Unexpected Collinearities   **123**

Figure 3-34

Figure 3-35

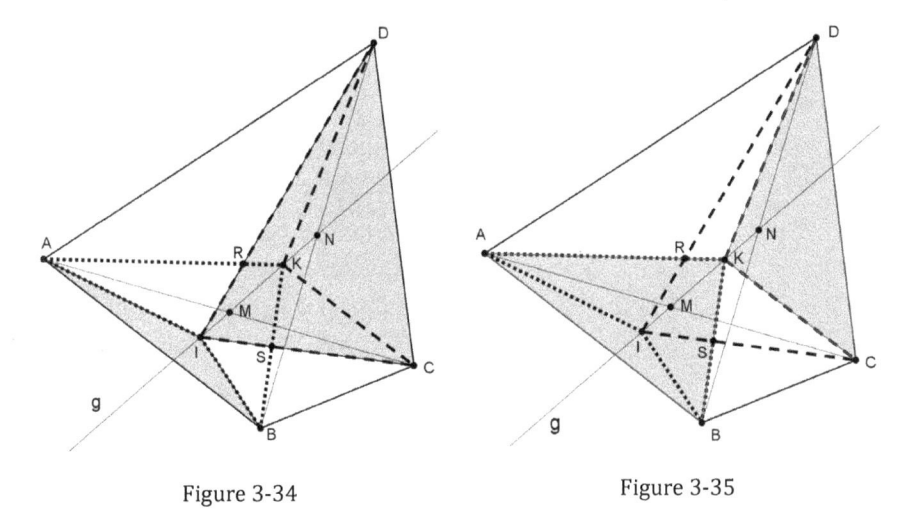

as we see in Figure 3-34 before, and Figure 3-35 afterwards; the color of triangles, such as △ARD or △AIB, does not change.

1. The quadrilateral *IRKS* (overlapping of the arrows) was shaded and stays shaded.
2. The triangles *RKD* and *SKC* change their color from shaded to white.
3. The triangles *RIA* and *SIB* change their color from white to shaded.

Because the arrows have equal areas, the area sums of the remaining triangles in Figure 3-34 and in Figure 3-35, respectively, are equal.

We now need to see what happens when point *I* does *not* lie on line *g*.

For *I* = *M* , the area sums are equal; moving *I* away from *M* on the diagonal *AC* will usually change the area sums (see Figure 3-36). Because the triangle *MID* changes to shaded, the triangle *MIB* changes to unshaded; the triangles usually have different areas because the altitudes from *B* and *D* to *MI* and *AC* are different (except when *N* happens to lie on *AC*; in this case, we would take the other diagonal *BD* for the described operation).

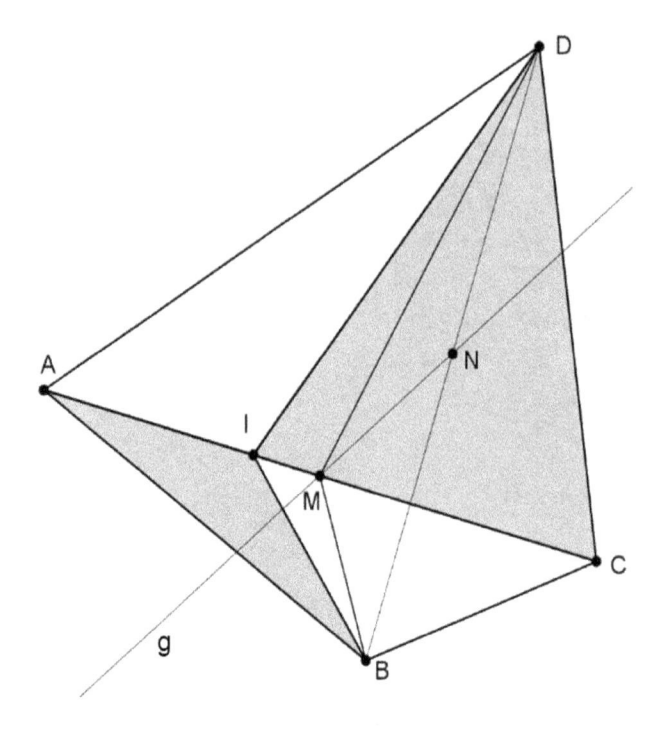

Figure 3-36

The remaining part of the argumentation relies on an observation that is based on a more general question: For which points I is the area sum of equally colored triangles *constant* (not necessarily the half of the quadrilateral area)? The following conjecture seems likely (which can easily be confirmed with dynamic geometry): Consider when I moves on a line p that is parallel to g. The above proof for I on g also holds true also in this case if one considers the following (see Figure 3-33; one has to interpret the figure dynamically): When translating the diagonal IK of an arrow ("base side" of "its" triangles) to a line which is parallel to g, the altitude of the one triangle will increase and the altitude of the other triangle will decrease by the same amount. Thus, the sum of the altitudes and the area of the arrow will not change. Both arrows (the not moved and the moved one) have the same area $= \frac{1}{2} \cdot IK \cdot (h_1 + h_2)$.

As a consequence, we have: If I does not lie on g, then move I parallel to g onto a diagonal (the area sums do not change); for I on a diagonal (say *AC*), we know from Figure 3-36 that the area sums for *I* = *M* and *I* ≠ *M* are different. This completes the proof.

In the case of a tangential quadrilateral (one having an incircle), we know as a bonus by the more general considerations above that the incenter *I* lies on the Newton line (Figure 3-37) because it has the area balance property. (Note that the *inradius r* is the equal altitude of all four triangles; also note, since the two tangential segments from an exterior point to a circle are equal, that *BF* = *BE*, *CF* = *CG*, *AH* = *AE*, and *DH* = *DG*, so that *BF* + *CF* + *AH* + *DH* = *BE* + *CG* + *AE* + *DG*, or stated another way, *BC* + *AD* = *AB* + *CD*.)

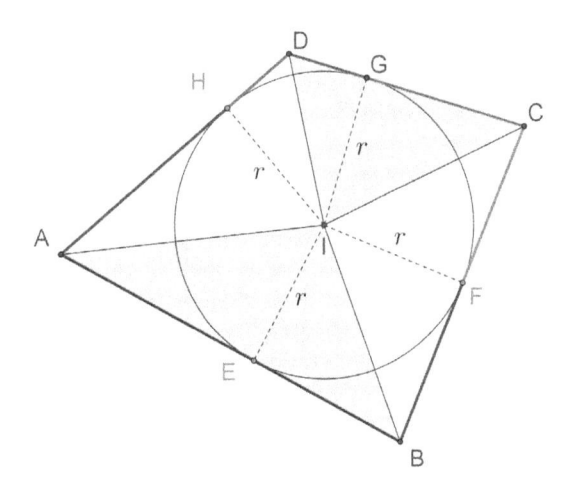

Figure 3-37

Chapter 4
Squares on Triangle Sides

The relationship between squares and triangles has always been a fascinating aspect of the plane geometry world. Many unusual and challenging relationships evolve when squares and triangles are involved in a geometric configuration. Perhaps the best-known example of squares placed on the sides of a triangle is seen when we exhibit the Pythagorean theorem geometrically. Rather than say that the sum of the squares of the legs of a right triangle is equal to the square of the hypotenuse, typically written as $a^2 + b^2 = c^2$, we can state the same idea geometrically by saying that the sum of the areas of the squares on the legs of a right triangle is equal to the area of the square on the hypotenuse, which is shown in Figure 4-1.

The Pythagorean Theorem

We begin by considering the squares drawn on the sides of a right triangle, as shown in Figure 4-1. There are currently over 400 proofs of the Pythagorean theorem, 367 of which can be found in *The Pythagorean Proposition* (National Council of Teachers of Mathematics, 1968) by the American mathematician Elisha S. Loomis (1852–1940). One of these proofs was developed by the 20th president of the United States of America, James A. Garfield (1831–1881), while he was a member of Congress.

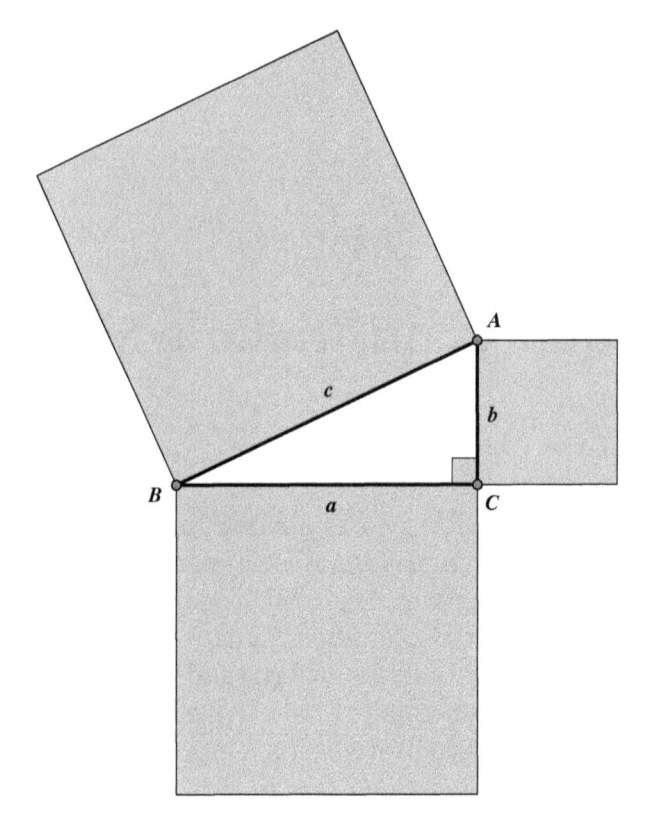

Figure 4-1

Proof

Garfield's proof begins by extending CB to point D so that $AC = BD$, as shown in Figure 4-2. We draw $DE \perp CBD$ so that $DE = BC$, and then $\triangle ABC \cong \triangle BDE$. Therefore, $AB = BE$. Consider trapezoid $ACDE$, whose area is $\frac{1}{2}(CD)(AC+DE)=\frac{1}{2}(a+b)(a+b)=\frac{1}{2}(a+b)^2$. And since $\angle DBE + \angle ABC = 90°$, we have $\angle ABE = 90°$. The area of right triangle ABE is $\frac{1}{2}c^2$. Furthermore, the area of triangle ABC is $\frac{1}{2}ab$. The area of trapezoid $ACDE$ $\left[\frac{1}{2}(a+b)^2\right]$ is equal to the sum of the areas: $\frac{1}{2}(a+b)^2 = \triangle ABC + \triangle BDE + \triangle ABE = \frac{1}{2}ab+\frac{1}{2}ab+\frac{1}{2}c^2$. We can simplify that to read $a^2 + 2ab + b^2 = c^2 + 2ab$, which produces the Pythagorean theorem, namely, $a^2 + b^2 = c^2$.

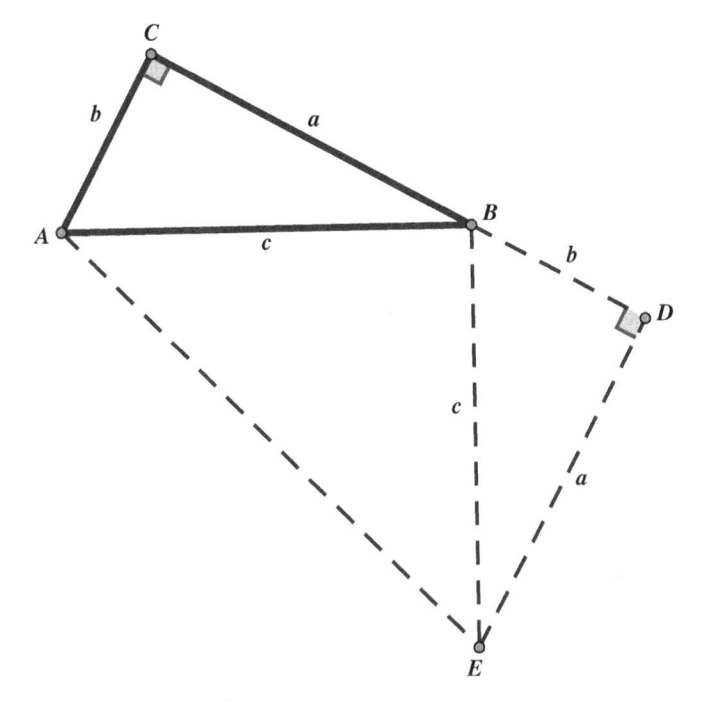

Figure 4-2

Squares on the Legs of a Right Triangle

As we continue our journey through geometry that relates squares and triangles, we place a square on each of the legs of a right triangle to generate an unusual result. In Figure 4-3, right triangle ABC has square $AEDB$ and square $AFGC$ on its sides AB and AC, respectively. When we draw BG and DC, we have created equal segments along the legs, where $AK = AJ$.

Proof

To justify this unexpected result, we will use triangle similarity. Let the right triangle sides opposite angles A, B, and C be a, b, and c, respectively, as seen in Figure 4-3. Since AB is parallel to ED, we have $\triangle CKA \sim \triangle CDE$, therefore, $\frac{AK}{c} = \frac{b}{b+c}$, and $AK = \frac{b \cdot c}{b+c}$.

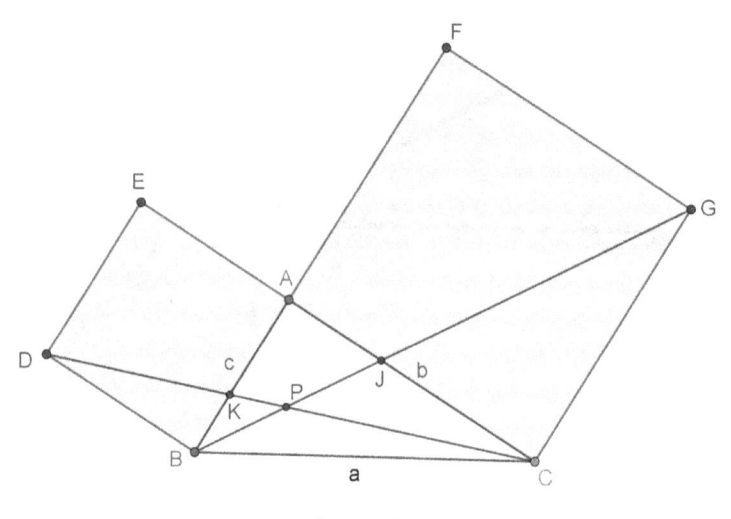

Figure 4-3

Analogously, $\triangle BAJ \sim \triangle BFG$, therefore, $\frac{AJ}{b} = \frac{c}{b+c}$, and $AJ = \frac{b \cdot c}{b+c}$. Therefore, $AK = AJ$, which is what we set out to prove.

A Square on the Hypotenuse of a Right Triangle

The configuration in Figure 4-4 demonstrates two more surprising phenomena. The first one is that there arises another square $AKLJ$ with L on the hypotenuse BC. In Figure 4-4, we assume that point L is

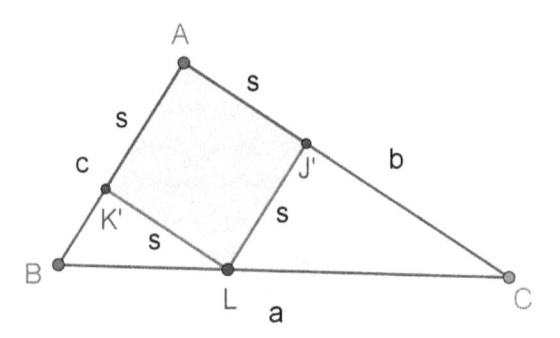

Figure 4-4

on hypotenuse *BC*. We label the side length of the square with *s* and the points on the legs of the right triangle *ABC* with *K'*, *J'*. We need to show that these points are actually the points labeled in Figure 4-3 as *K* and *L*. From the similar triangles *ABC* and *J'LC*, we have $\frac{s}{b-s} = \frac{c}{b}$, which yields $s = \frac{bc}{b+c}$, and this is the same value as in Figure 4-3 for *AK* and *AJ*.

A Surprising Altitude to a Hypotenuse of a Right Triangle

Another interesting item in Figure 4-3 is that $AP \perp BC$; in other words, *P* lies on the altitude through *A*. To prove that, we draw the perpendicular from *A* to intersect *BC* at point *Q* and calculate all the segments on the perimeter of the triangle *ABC*, which is shown in Figure 4-5. For the segments of the hypotenuse, we use the similarity of the triangles $\triangle ABQ \sim \triangle ABC \sim \triangle AQC$ to get $BQ = \frac{c^2}{a}$ and $QC = \frac{b^2}{a}$. We use the converse of Ceva's theorem to prove that *BJ*, *CK*, and *AQ* are concurrent, and this completes the proof since *BJ* and *GK* intersect at point *P* (see Figure 4-3). We take the products of alternate segments along the sides, as shown in Figure 4-5. This generates the following equality:
$\frac{bc}{b+c} \cdot \frac{c^2}{a} \cdot \frac{b^2}{b+c} = \frac{c^2}{b+c} \cdot \frac{b^2}{a} \cdot \frac{bc}{b+c}$.

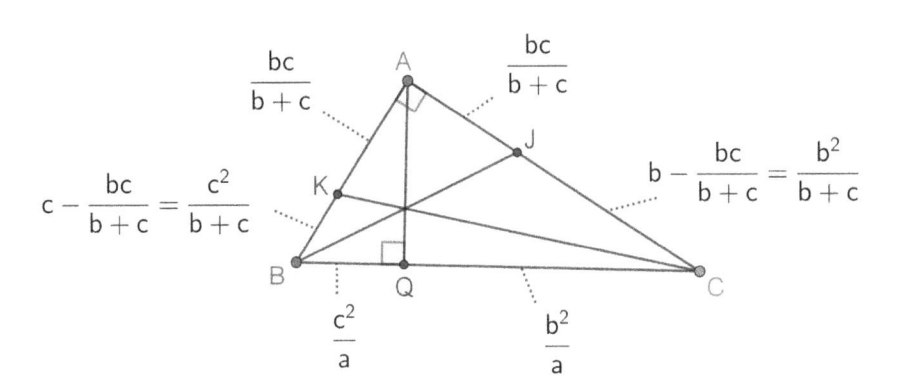

Figure 4-5

Another Surprise When Squares are Placed on the Legs of a Right Triangle

An unusual result can be found by once again placing squares on the legs of a right triangle, as we show in Figure 4-6, where squares *ACST* and *BCRP* are placed on the sides *AC* and *BC*, respectively, of the right triangle *ABC*. From vertex *P*, the perpendicular is drawn to intersect the extension of *AB* at point *D*, and from vertex *T* a perpendicular is drawn to intersect the extension of *AB* at point *E*. What results is quite unexpected, namely, that *DP* + *ET* = *AB*.

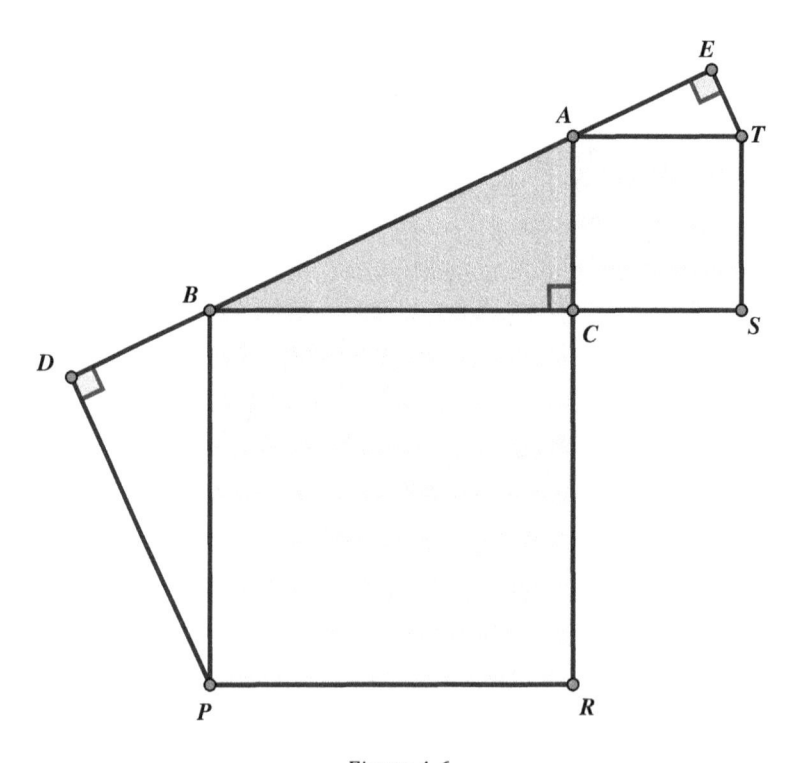

Figure 4-6

Proof

To prove that $DP + ET = AB$, we begin by constructing altitude HC to hypotenuse AB of right triangle ABC, as shown in Figure 4-7. Since $\angle DBP$ is complementary to $\angle HBC$, and $\angle DPB$ is complementary to $\angle DBP$, we have $\angle DBP = \angle HCB$. We also have $BP = BC$, so that $\triangle PDB \cong \triangle BHC$ and $DP = BH$. In an analogous fashion, we can prove that $\triangle AHC \cong \triangle TEA$, so that $ET = AH$. We can then see that $BH = DP$, so we can conclude that $DP + ET = AB$.

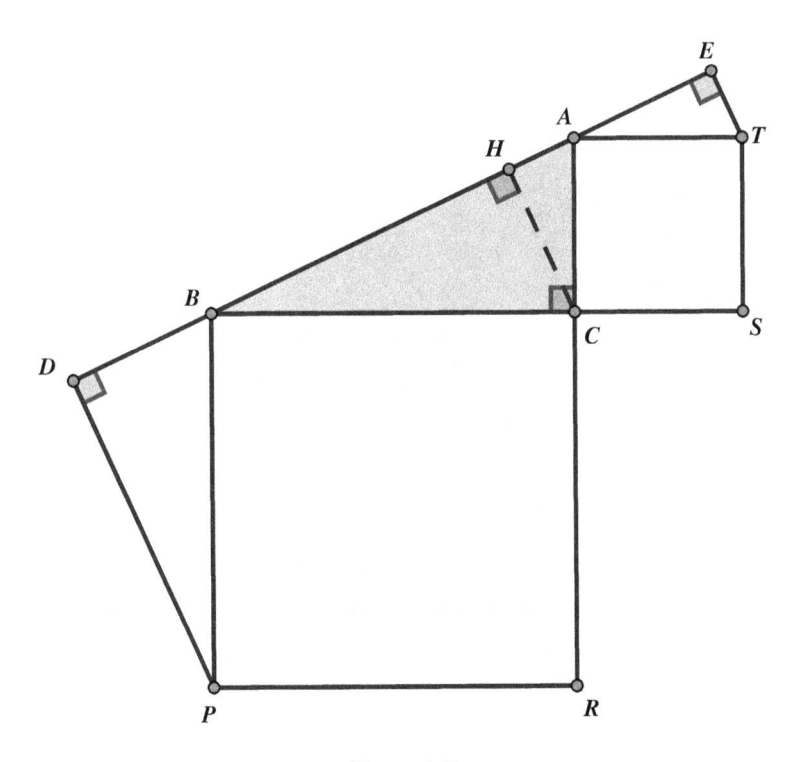

Figure 4-7

More Surprises When Squares are Placed on the Legs of a Right Triangle

The previous configuration (Figure 4-7) leads to another highly unexpected result. Once again, squares are drawn on each leg of the right triangle *ABC*, whose vertex *A* has a right angle, as shown in Figure 4-8. From point *D*, a perpendicular to *BC* (extended) intersects it at point *K*, and from point *G*, the perpendicular to *BC* (extended) intersects it at point *L*. The surprising result is that area△*DKB* + area△*GLC* = area△*ABC*. When comparing this result to the previous one we can see a distinct similarity, yet it shows another point of view.

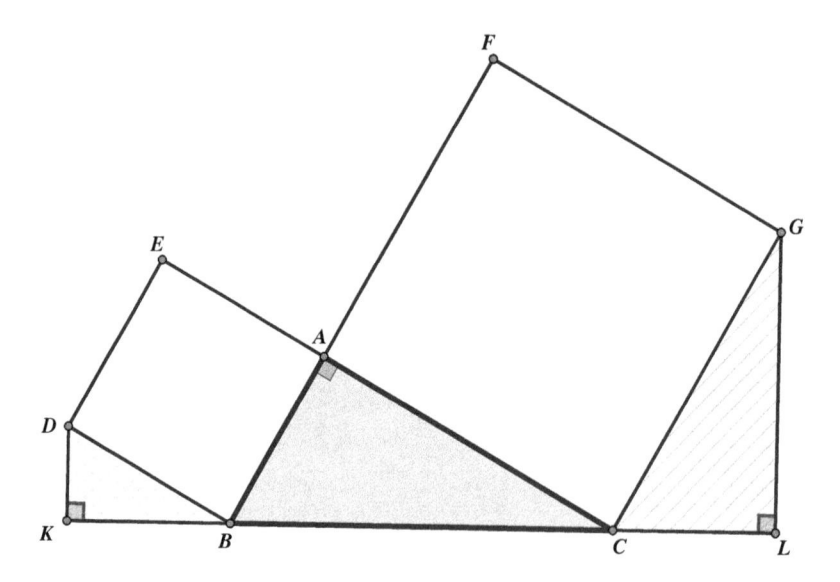

Figure 4-8

Proof

In Figure 4-9, we begin by drawing altitude AM in triangle ABC. We then have the following: $\angle ABM$ is complementary to $\angle DBK$, and $\angle KDB$ is complementary to $\angle DBK$. Therefore, $\angle ABM = \angle KDB$. Since $AB = DB$, we have triangle $DKB \cong \triangle BMA$. Using the same technique, we can show that $\triangle AMC \cong \triangle CLG$. Thus, we conclude that area$\triangle DKB$ + area$\triangle GLC$ = area$\triangle ABC$.

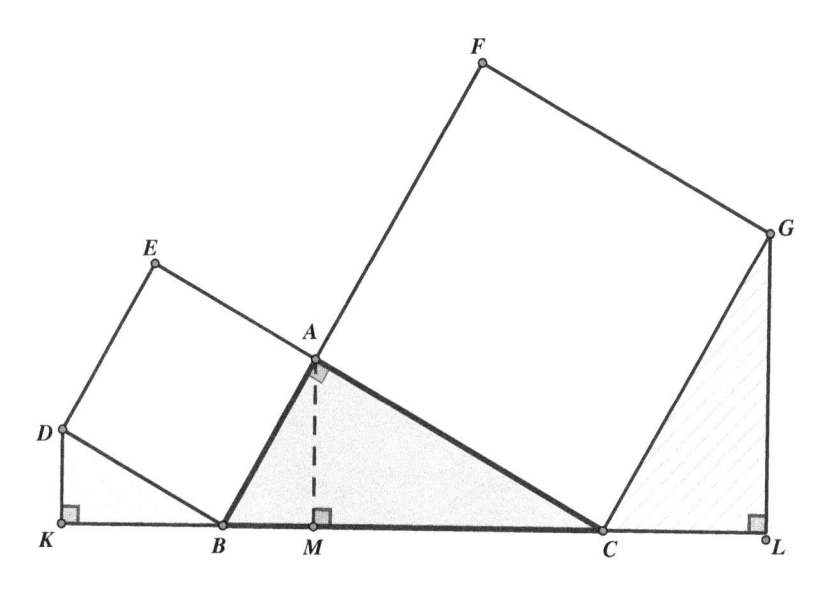

Figure 4-9

More Placements of Squares on Right Triangle

This time, we will be placing a square on a leg and a square on the hypotenuse of a right triangle to produce more unexpected results. In Figure 4-10, where right triangle ABC has square $ACFG$ placed on side AC and square $BCDE$ placed on hypotenuse BC, we select a point L anywhere on AC. Counterintuitively, we can conclude that area$\triangle LCF$ = area$\triangle LCD$. Remarkable of this aspect rests on the fact that the triangle equality will hold true regardless of where point L is situated on line AC.

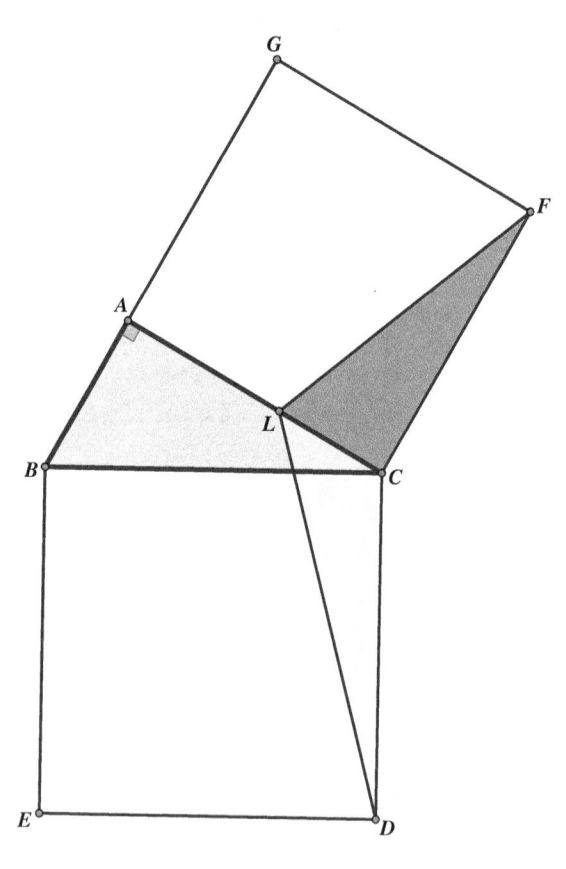

Figure 4-10

Proof

As we set out to prove in Figure 4-11 that triangle *LCD* and triangle *LCF* are equal in area, we note that they share a common base *LC*; thus, we must show that the altitudes *PD* and *FC* are equal. Since ∠*BCD* is a right angle, we know that ∠*PCD* and ∠*ACB* are complementary. However, in triangle *ABC*, ∠*ABC* is complementary to ∠*ACB*. Therefore, ∠*PCD* = ∠*ABC*. Furthermore, *BC* = *DC*, which enables us to conclude that ∠*ABC* ≅ ∠*PCD*. Thus, *PD* = *AC* = *CF*, and the two triangles Δ*LCD* and Δ*LCF* are therefore equal in area because they share a common base and have equal altitudes.

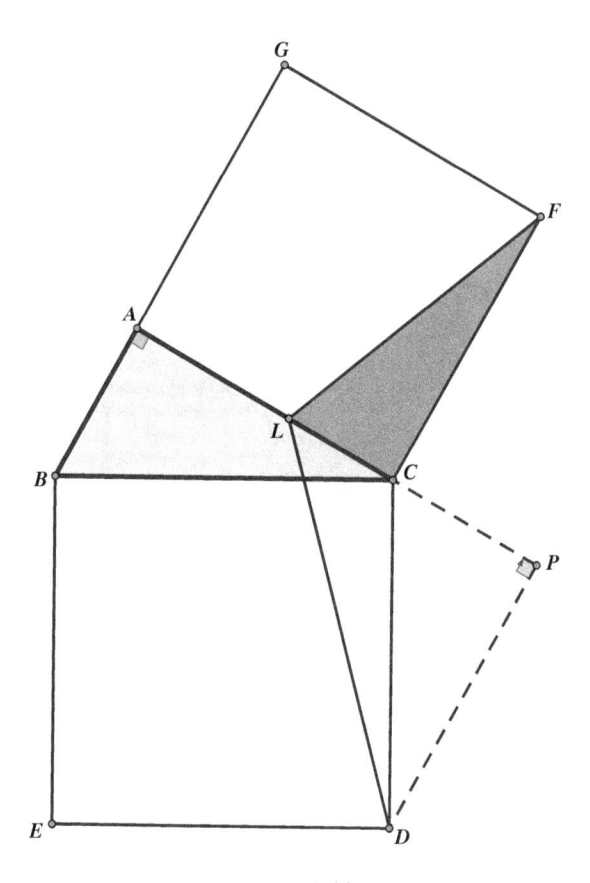

Figure 4-11

One Square on the Hypotenuse of a Right Triangle

It is rather easy to divide a square into two equal parts. Two simple ways are by drawing a diagonal of the square or by joining the midpoints of a pair of opposite sides of the square. Once again, the right triangle placed on the square will provide a most unanticipated procedure to divide the square into two equal areas. In Figure 4-12, we begin by drawing any right triangle *AEB* with its hypotenuse on the side *AB* of square *ABCD*. The bisector of the right angle *AEB* intersects sides *AB* and *DC* at points *K* and *L*, respectively. It is this angle bisector that partitions the square into two equal-area regions. In other words, the area of quadrilateral *ADLK* is equal to the area of quadrilateral *BKLC*.

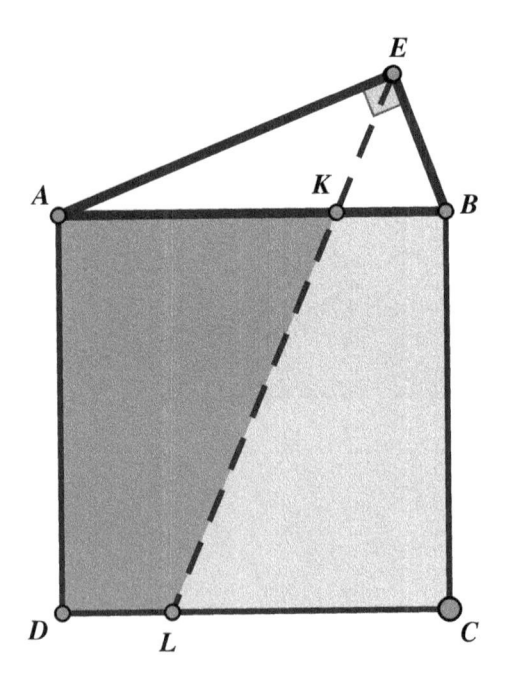

Figure 4-12

Proof

As we show in Figure 4-13, we encase the square *ABCD* in another square *EGFH*, where *EF* must be the diagonal of the square since it bisects ∠*E*. We find that we have replicated triangle *ABE* on each side of the square, and we can easily show that △*FLC* ≅ △*EKA* and △*KEB* ≅ △*LFD*. The diagonal *EF* of square *EGFH* also divides the square *ABCD* into equal parts, as we can see when we subtract the various pairs of congruent triangles, namely, △*KEB* ≅ △*LFD*, △*FLC* ≅ △*EKA*, and △*BDG* ≅ △*DHA*, from each side of the diagonal *EF* of square *EGFH*. The two quadrilaterals that remain are equal in area, namely, area*AKLD* = area*BKLC*, which is what we set out to prove.

Remark: If one knows that any line through the center of a square divides it into two congruent parts it suffices to show that the line *KL* passes through the center of *ABCD*. And this immediately follows by the congruence of the triangles *EKA* and *FLC*.

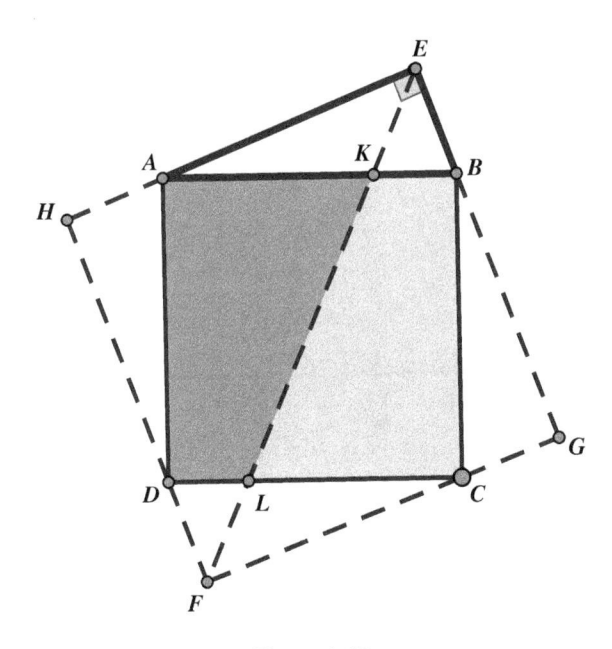

Figure 4-13

Another Square on the Hypotenuse of a Right Triangle

Once again, a right triangle situated on a side of the square, as shown in Figure 4-14, where the hypotenuse of right triangle ABC is placed on square $BCDE$, will produce a completely unexpected result:

$$\text{area}\triangle ABE + \text{area}\triangle ACD = \frac{1}{2}\text{area}BCDE.$$

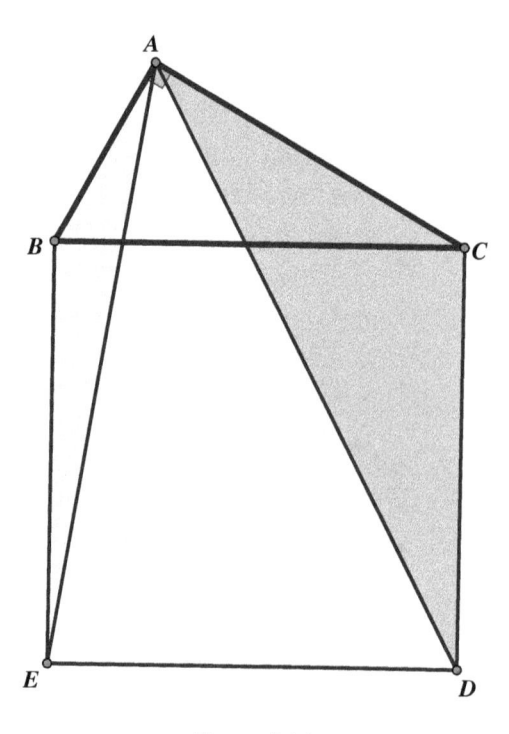

Figure 4-14

Proof

To prove this area equality, we will simply apply the popular formula for the area of a triangle, which is one-half the product of the altitude and its base. In Figure 4-15, the altitude of triangle ABE is BF with a base of BE. Therefore, area$\triangle ABE = \frac{1}{2}(BF)(BE)$. Similarly, area$\triangle ACD = \frac{1}{2}(CF)(CD)$. The sum of these two triangle areas yields area$\triangle ABE +$ area$\triangle ACD = \frac{1}{2}(BF)(BE) + \frac{1}{2}(CF)(CD)$. However, since $BE = CD$, this sum is $\frac{1}{2}(BE)(BF+CF) = \frac{1}{2}(BE)(BC)$, which is one-half the area of square $BCDE$.

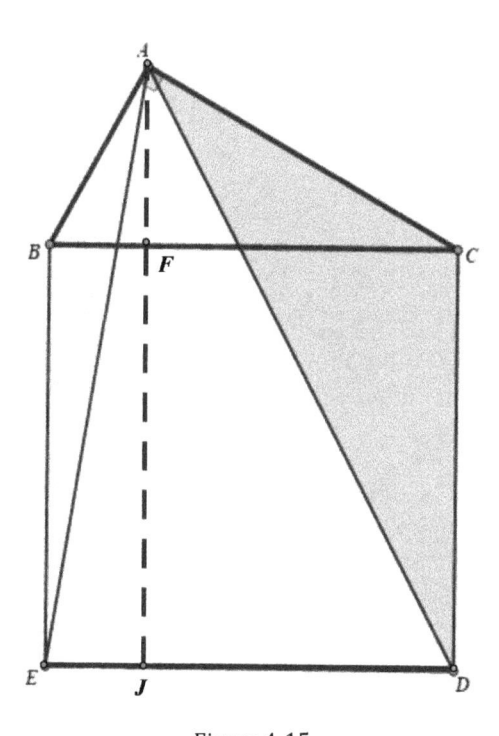

Figure 4-15

A Square on the Hypotenuse of a Right Triangle Produces Another Surprise

Once again, we draw a right triangle so that its hypotenuse is congruent to the side of a square, as we show in Figure 4-16, where right triangle *ABC* is placed on side *AB* of square *ABSR*. With the intersection of the

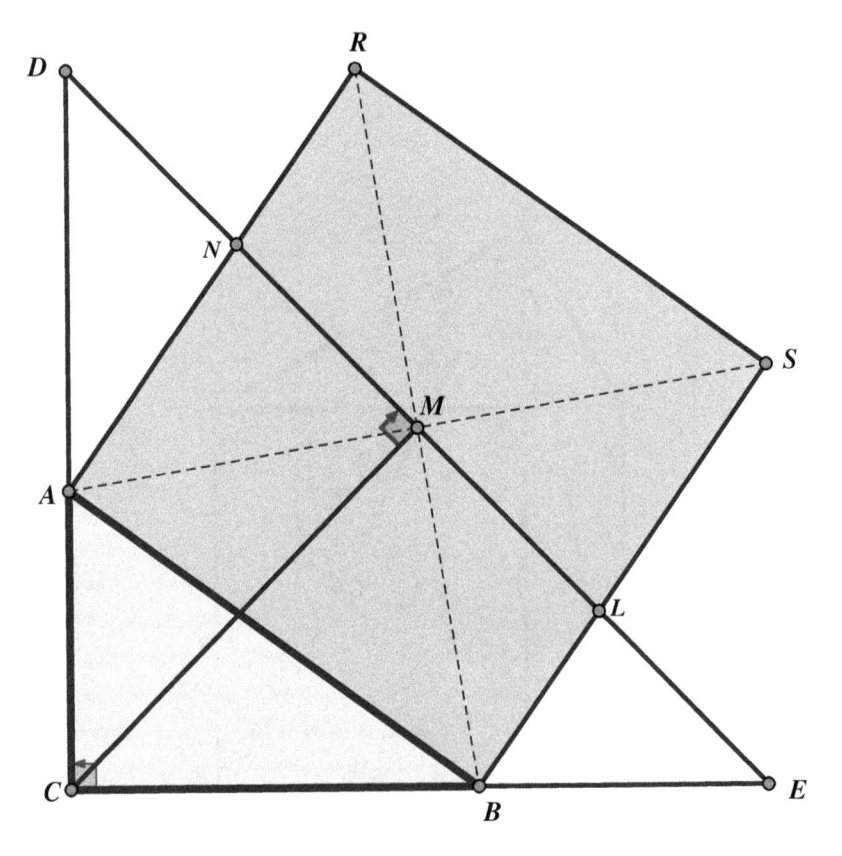

Figure 4-16

diagonals at point *M*, we construct a line perpendicular to *CM* intersecting the extensions of sides *CA* and *CB* at points *D* and *E*, respectively. The unexpected result is that *BE* = *AC* and *AD* = *BC*. Another surprising feature here is that *CM* bisects ∠*ACB*, further demonstrating the ongoing amazements that can be found in geometric configurations.

Proof

We will first seek to prove two triangles congruent, namely, △*MBE* ≅ △*MAC*, shown in Figure 4-16. Since the right ∠*ABL* lies on the straight-line *CBE*, we can conclude that ∠*LBE* is complementary to ∠*ABC*. In right triangle *ABC*, ∠*CAB* is complementary to ∠*ABC*. Therefore, ∠*LBE* = ∠*CAB*. Also, since ∠*MAB* = ∠*MBA* = 45°, by addition we have ∠*MBE* = ∠*MAC*. We also have ∠*AMC* = ∠*BME* since we removed ∠*CMB* from two right angles. Thus, we have △*MBE* ≅ △*MAC* and *BE* = *AC*. Analogously, we can prove △*MBC* ≅ △*MAD*, which, in turn, enables us to say that *AD* = *BC*. By adding these two equalities, we get *AD* + *AC* = *BE* + *BC*. This is equivalent to *DC* = *EC*, making △*DCE* an isosceles right triangle, where the altitude *CM* is the angle bisector of ∠*DCE*, which was the second thing we wanted to prove.

More Surprises of Squares and Triangles

A square on a triangle with a 45° vertex angle enables us to produce two squares where one has double the area of the other. In Figure 4-17, we have triangle *ABC*, where ∠*A* = 45° and perpendiculars from vertices *B* and *C* intersect the opposite sides at points *D* and *E*, respectively. Unexpectedly, we find that the square on *BC*, *BCFG*, has twice the area of the square *DEIH* on *DE* – even though it may not appear that way.

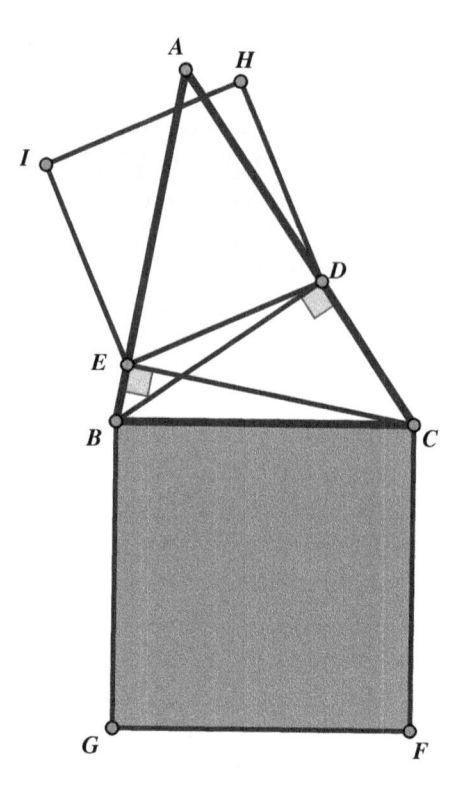

144 *A Journey Through the Wonders of Plane Geometry*

Figure 4-17

Proof

We begin by noticing in Figure 4-17 the special feature of triangle *ABC*, namely, that ∠*A* = 45°. Triangle *ADB* is an isosceles right triangle with *AD* = *BD*. We also know that quadrilateral *BEDC* is cyclic with diameter *BC* since right triangles *BEC* and *BDC* share a common hypotenuse. Therefore, ∠*DEB* + ∠*DCB* = 180°. We also can see that ∠*DEB* + ∠*AED* = 180°. Thus, we have ∠*DCB* = ∠*AED*. This enables us to establish △*ADE* ~ △*ABC* so that $\frac{DE}{BC} = \frac{AD}{AB}$. Furthermore, $\frac{AD}{AB} = \frac{1}{\sqrt{2}}$, which is a common property of an isosceles right triangle. We then have $\frac{DE}{BC} = \frac{1}{\sqrt{2}}$, which we can write as $BC = DE\sqrt{2}$, or $BC^2 = 2DE^2$. Hence, the area of square *BCFG* equals BC^2 and the area of square *DEIH* is DE^2, so that the area of *BCFG* is twice the area of square *DEIH*.

Squares on All Sides of a Triangle

Placing squares on the sides of a randomly drawn triangle can surprisingly create a perpendicularity, as we can see in Figure 4-18, where squares are placed on the sides of triangle *ABC* and we find that *AM*⊥*CT*. Furthermore, *AM* and *CT* are equal!

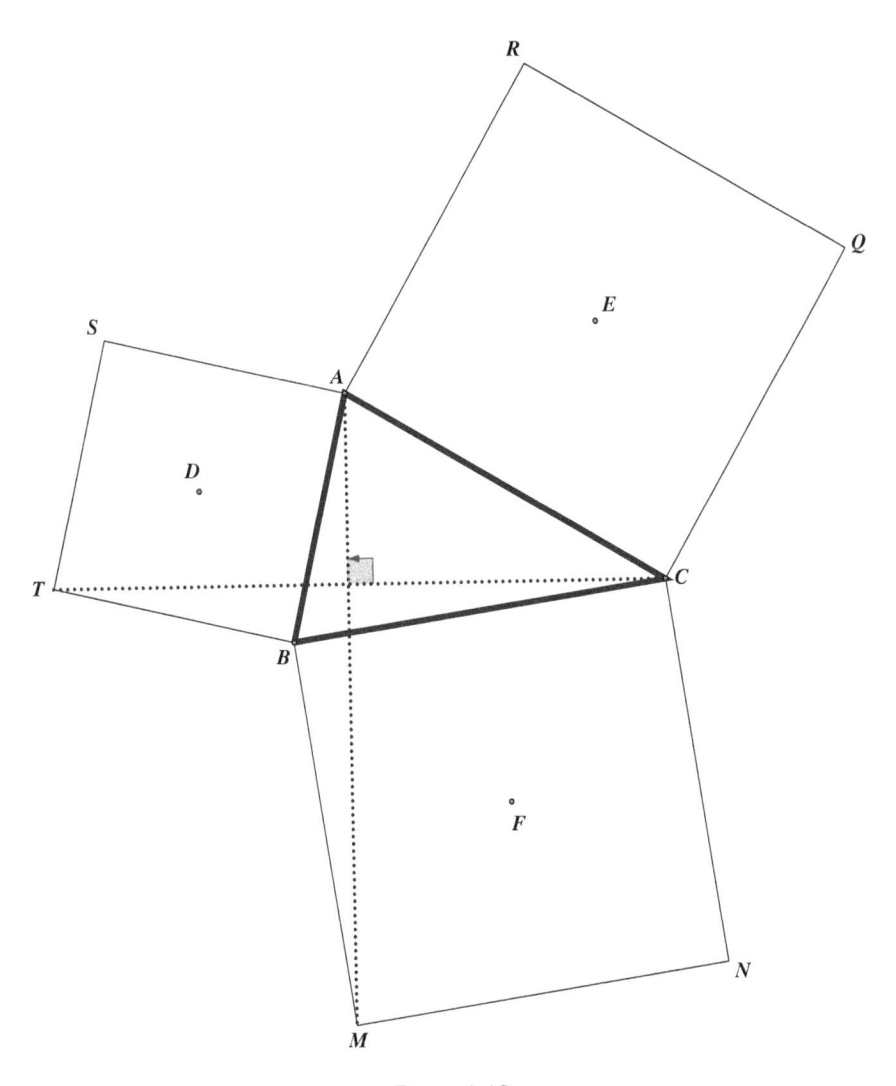

Figure 4-18

Proof

To prove the perpendicularity and equality, we must prove that $\triangle ABM \cong \triangle TBC$, shown in Figure 4-19. From the sides of the squares, we have $TB = AB$ and $BC = BM$. Also, $\angle TBC = \angle ABC + 90°$, and $\angle ABM = \angle ABC + 90°$. Therefore, $\angle TBC = \angle ABM$ and we consequently have $\triangle ABM \cong \triangle TBC$. A rotation by 90° clockwise about point B will take triangle TBC to triangle ABM. We therefore conclude that $AM \perp CT$ and $AM = CT$.

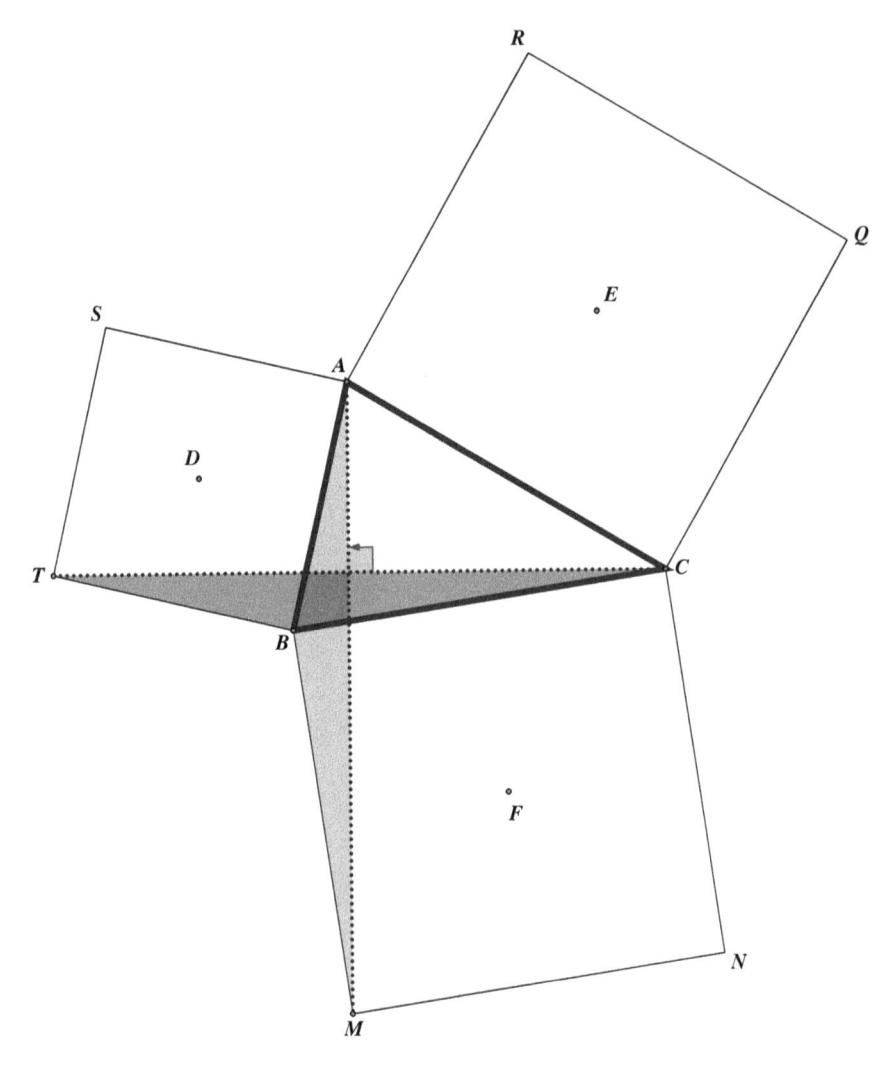

Figure 4-19

More Surprises of Squares on the Sides of a Random Triangle

When we place squares on each side of a randomly constructed triangle, such as triangle *ABC* in Figure 4-20, we can use this scheme to create a right triangle. We begin by constructing two parallelograms, *BSXQ* and *CRZN*. By connecting the points *A*, *X*, and *Z*, we establish the right triangle *XAZ*, where $\angle XAZ = 90°$. Furthermore, this right triangle is an isosceles right triangle since *AX* = *AZ*.

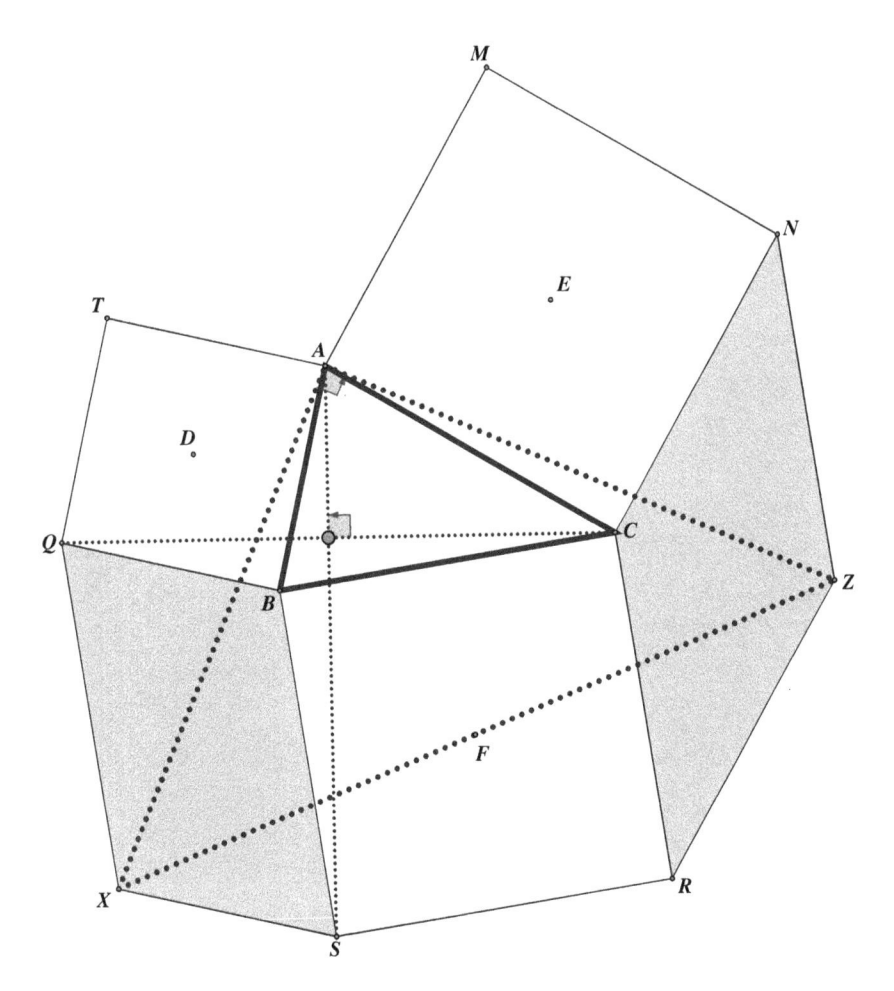

Figure 4-20

148 *A Journey Through the Wonders of Plane Geometry*

Proof

To prove this relationship, we consider auxiliary lines shown in Figure 4-21. Since $AC = NC$, as they are the sides of a square, when we extend BC to intersect the sides of the parallelograms at points D and E, we find $\angle ACA' + \angle NCD = 90°$, and $\angle CND + \angle NCD = 90°$ therefore, $\angle ACA' = \angle CND$. We then have $\triangle AA'C \cong \triangle CDN$, so that $DN = A'C = n$ and $CD = AA' = h$. Also, $NZ = CR = BC = n + m = a$, and $A'Z' = DZ = NZ - DN = a - n = m$. It follows that $ZZ' = DA' = n + h$. Therefore, $AZ' = m + h$.

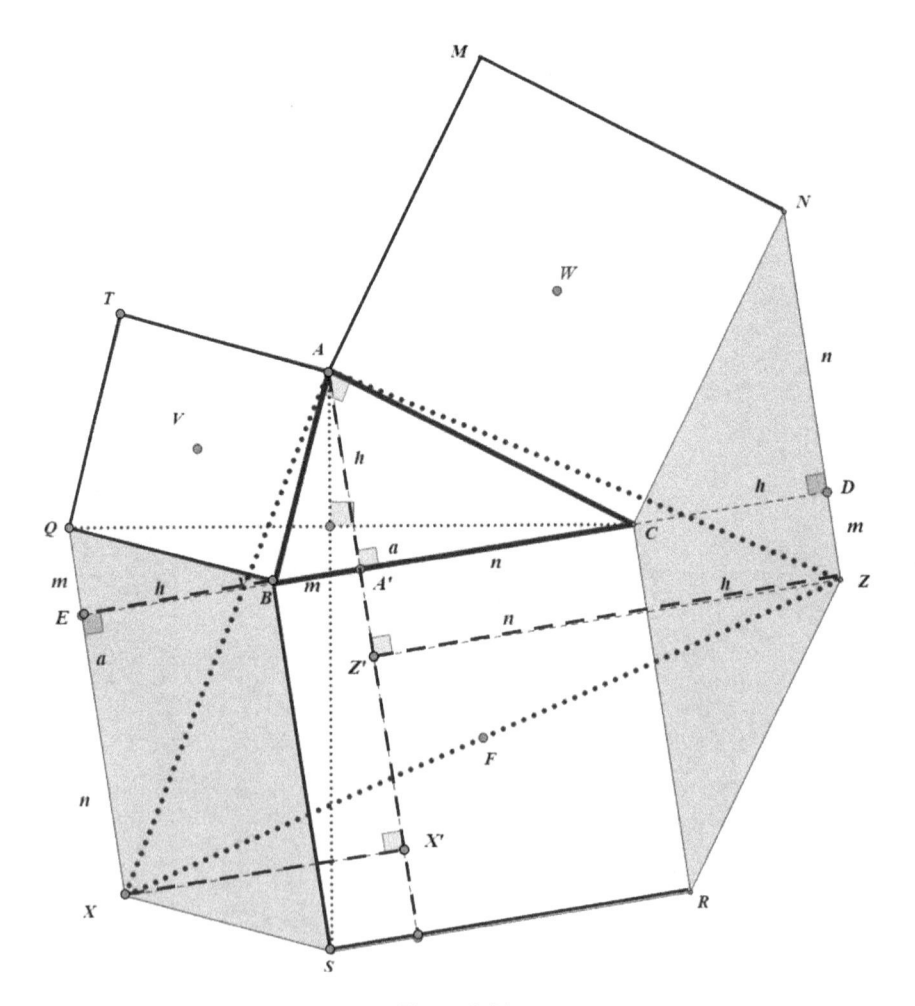

Figure 4-21

Furthermore, $\triangle AA'B \cong \triangle BEQ$, and $BE = AA' = h$, and $QE = BA' = m$. Then, $XX' = A'E = A'B + BE = m + h$. Also, $QX = BS = BC = a$, and $A'X' = EX = QX - QE = a - m = n$. Then, $XX' = m + h = AZ'$. Thus, we have $\triangle AX'X \cong \triangle ZZ'A$, and $AX = AZ$ so that $\angle XAZ = \angle XAX' + \angle ZAZ' = \angle XAX' + \angle AXX' = 90°$, which is what we set out to prove.

Unusual Concurrency by Squares on the Sides of a Triangle

Placing squares on the sides of a triangle continues to lead to a variety of other concurrencies. This time, we will extend the external side of each square placed on triangle ABC until they meet at points X, Y, and Z, thus creating triangle XYZ, shown in Figure 4-22. By joining these points to the nearest vertices of triangle ABC, we find a concurrency emerges at point P.

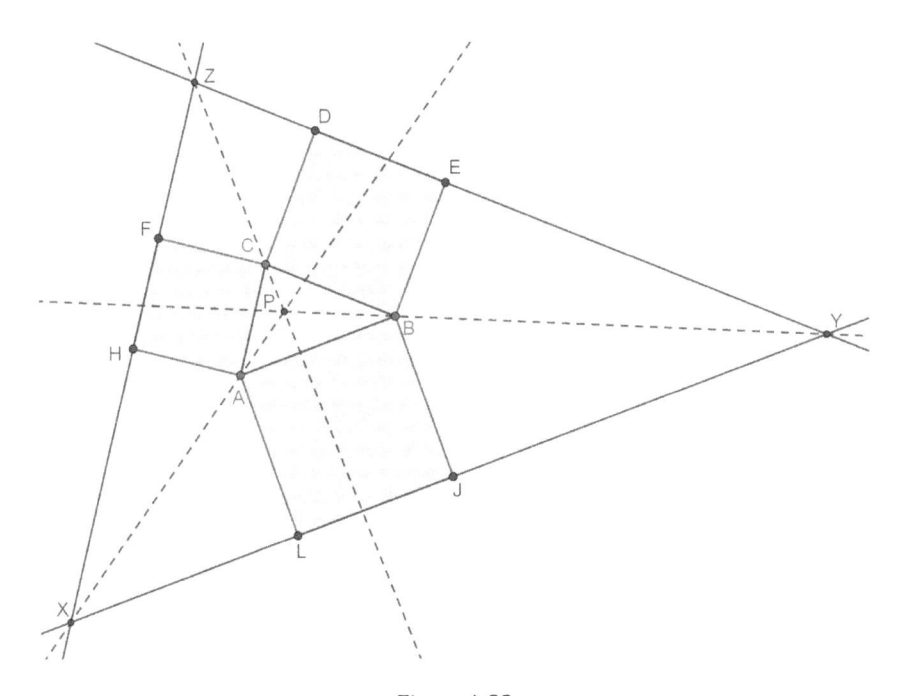

Figure 4-22

150 A Journey Through the Wonders of Plane Geometry

Proof

Explaining this phenomenon thoroughly will open up new pathways to consider. First, we must recall our encounter with the *symmedian point* of a triangle. The symmedian point is the *isogonal conjugate* of the triangle's centroid G, as mentioned in Chapter 2.

In Figure 4-23, the cevian CQ is a symmedian (which contains the symmedian point K) if and only if the distances x and y to the sides a and b are proportional to the sides themselves so that $\frac{x}{y} = \frac{a}{b}$. Using $\frac{x}{y} = \frac{a}{b}$, we can see that the symmedians of a triangle and the symmedian point K can be established in another way – that is, aside from reflecting the medians in the angle bisectors. We now consider squares placed outwardly on the triangle sides, as shown in Figure 4-23, where the square's sides are extended so that they are parallel to the triangle's sides and intersect each other. One such intersection point is Z, shown in Figure 4-24. The line ZC is a *symmedian* of $\triangle ABC$ at vertex C, as will be explained below. When this is done analogously at the other two vertices, we have another way to construct the

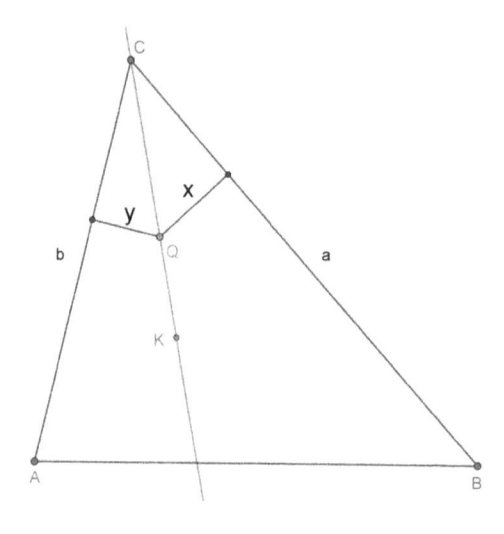

Figure 4-23

Squares on Triangle Sides **151**

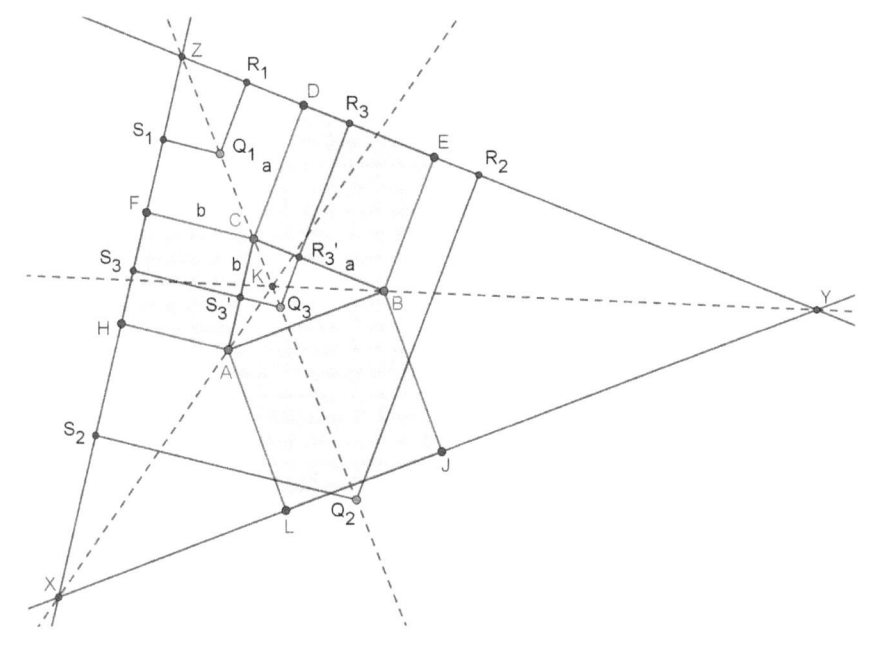

Figure 4-24

symmedian point K using outwardly erected squares to find the intersection point of XA, YB, and ZC.[1] Now the explanation: Let Q be an arbitrary point on ZC (Figure 4-24). For any point Q on CZ, the ratio of the distances to YZ and XZ is $a:b$, which can be seen in the case where $Q = C$. For the general case, note the similar triangles: $\Delta ZCD \sim \Delta ZQ_1R_1 \sim \Delta ZQ_2R_2 \ldots$ and $\Delta ZCF \sim \Delta ZQ_1S_1 \sim \Delta ZQ_2S_2\ldots$, respectively. From the proportion $\frac{x}{y} = \frac{a}{b}$, line ZC must be the symmedian of ΔXYZ at the vertex Z. The triangles ΔABC and ΔXYZ are homothetic[2] with center K. Since the triangles ΔZQ_3R_3 and $\Delta CQ_3R_3'$ are similar (analogously, $\Delta ZQ_3S_3 \sim \Delta CQ_3S_3'$,

---

[1] This was first detected and proved by Ernst Wilhelm Grebe (1804–1874), a German mathematician. This is the reason why the symmedian point is, especially in older German publications, sometimes called the *Grebe point*. This point is also called the Lemonie point, named after the French mathematician Émile Lemoine (1840–1912).
[2] Homothety will be discussed in Chapter 11.

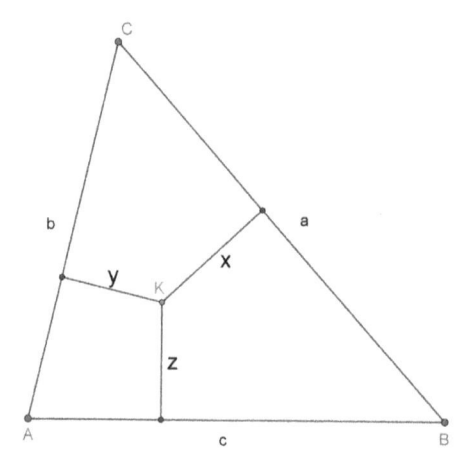

Figure 4-25

as seen in Figure 4-24), we have $\frac{Q_3 R_3'}{Q_3 S_3'} = \frac{Q_3 R_3}{Q_3 S_3} = \frac{a}{b}$. This, in turn, means that the point Q_3 on ZC in the interior of $\triangle ABC$ lies on the symmedian of $\triangle ABC$ at the vertex C. Altogether, line ZC is both the symmedian of $\triangle XYZ$ (at vertex Z) and the symmedian of $\triangle ABC$ (at vertex C).

From $\frac{x}{y} = \frac{a}{b}$, it follows immediately that the symmedian point K is uniquely defined by (Figure 4-25) $x:y:z = a:b:c$, or equivalently, $\frac{x}{a} = \frac{y}{b} = \frac{z}{c}$.

More on the Symmedian Point

The symmedian point has many other famous properties and is thus a truly special and important point in triangles. One of these properties is highlighted in Figure 4-26, where we have squares on the sides of a triangle.

The following is a striking minimum property of the symmedian point K: Of all the points in a triangle, the sum of the squares of the distances to the outer sides of the squares is least when taken from point K. In other words, if we were to take any other point for finding

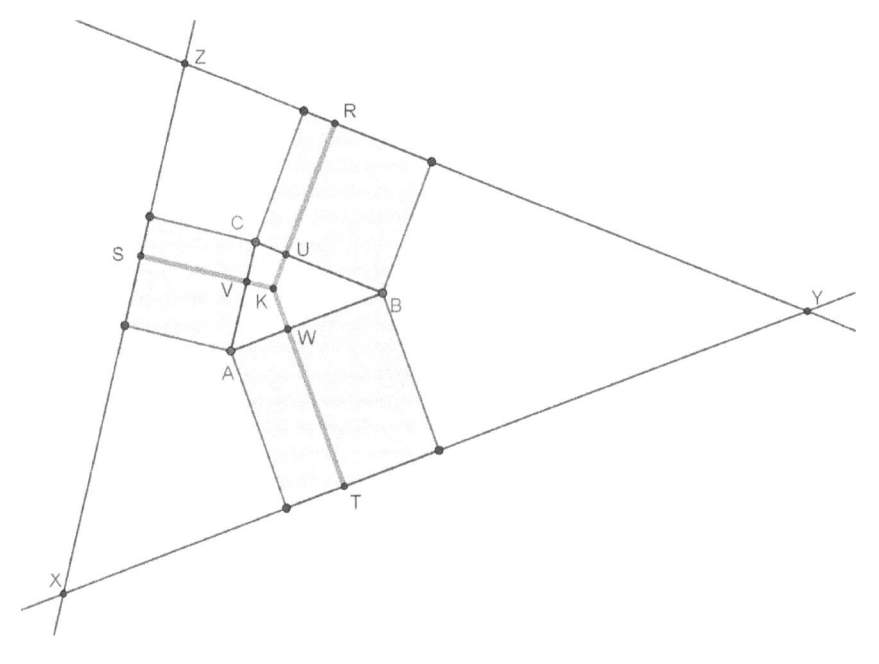

Figure 4-26

the sum of the squares of the distances to the remote sides of each square, it would be greater than those from point K. Mathematically, $KR^2 + KS^2 + KT^2$ is the least for any other point in $\triangle ABC$, as shown in Figure 4-26. Furthermore, K minimizes the sum of squared distances to the sides of $\triangle ABC$; that is, $KU^2 + KV^2 + KW^2$ is smaller than that for any other interior point in the triangle.

Proof

Let us first prove that $KU^2 + KV^2 + KW^2$ is a minimum, analogous to what is shown in Figure 4-26, where K minimizes $x^2 + y^2 + z^2$. For this purpose, one can use the following identity: $(a^2 + b^2 + c^2)(x^2 + y^2 + z^2) = (ax + by + cz)^2 + (ay - bx)^2 + (bz - cy)^2 + (cx - az)^2$.

Now, let a, b, and c denote the side lengths of the triangle $\triangle ABC$ and $x, y,$ and z the distances of an interior point P to the triangle sides,

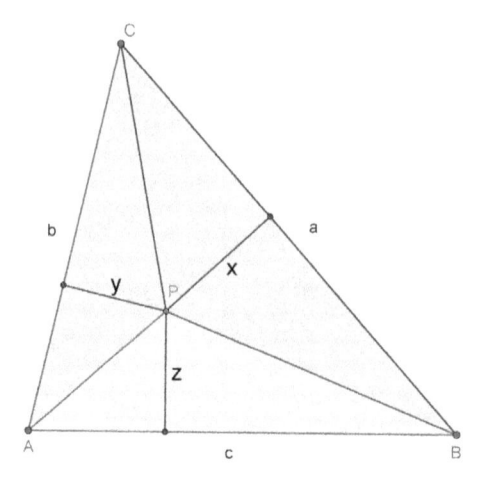

154 A Journey Through the Wonders of Plane Geometry

Figure 4-27

which is shown in Figure 4-27. Note that for any interior point P, $\triangle ABC$ can be partitioned by the line segments PA, PB, and PC into three triangles with sides a, b, and c and altitudes x, y, and z. Then, $ax + by + cz$ is twice the area of $\triangle ABC$ and thus a constant.

The first parenthesis on the left side of $(a^2 + b^2 + c^2)(x^2 + y^2 + z^2) = (ax + by + cz)^2 + (ay - bx)^2 + (bz - cy)^2 + (cx - az)^2$, which is $(a^2 + b^2 + c^2)$, is a constant as well. Thus, we can conclude that $x^2 + y^2 + z^2$ becomes a minimum if and only if $(ay - bx)^2 + (bz - cy)^2 + (cx - az)^2$ does the same. But this sum of squares clearly is not negative and takes its minimum 0 for $x:y:z = a:b:c$. According to $\frac{x}{a} = \frac{y}{b} = \frac{z}{c}$, this defines the symmedian point K, which completes the proof that the symmedian point K minimizes $KU^2 + KV^2 + KW^2$, as in Figure 4-26, which in Figure 4-25 is $x^2 + y^2 + z^2$.

Now, we are to prove that $KR^2 + KS^2 + KT^2$ is also a minimum in Figure 4-26. First, note that $KR = a + x$, $KS = b + y$, $KT = c + z$; hence, we must minimize

$$(a+x)^2 + (b+y)^2 + (c+z)^2 = \underbrace{(a^2 + b^2 + c^2)}_{\text{const.}} + (x^2 + y^2 + z^2) + \underbrace{2(ax + by + cz)}_{\text{const.}}^{\overbrace{}^{4 \times \text{area}(\triangle ABC)}}.$$

We see that $KR^2 + KS^2 + KT^2$ is a minimum if and only if $x^2 + y^2 + z^2$ is a minimum, and from above we know that this is the case for the symmedian point K. Thus, we have proved our second claim: the symmedian point K minimizes $KR^2 + KS^2 + KT^2$.

More Unusual Concurrencies by Squares on the Sides of a Triangle

Placing squares on each of the sides of a randomly drawn triangle *ABC*, as we show in Figure 4-28, produces yet more interesting surprises. We obtain concurrent lines by joining the center of each square with the remote vertex of the original triangle *ABC*. Furthermore, we find that these three lines, *AD*, *BE*, and *CF*, are perpendicular to the lines joining the midpoints of adjacent squares. In other words, *AD*⊥*EF*, *BE*⊥*DF*, and *CF*⊥*DE*. There is even more to admire in this configuration, namely, that these lines are not only perpendicular but also equal to each other. In other words, we have *AD* = *EF*, *BE* = *DF*, and *CF* = *DE*. Truly unexpected occurrences!

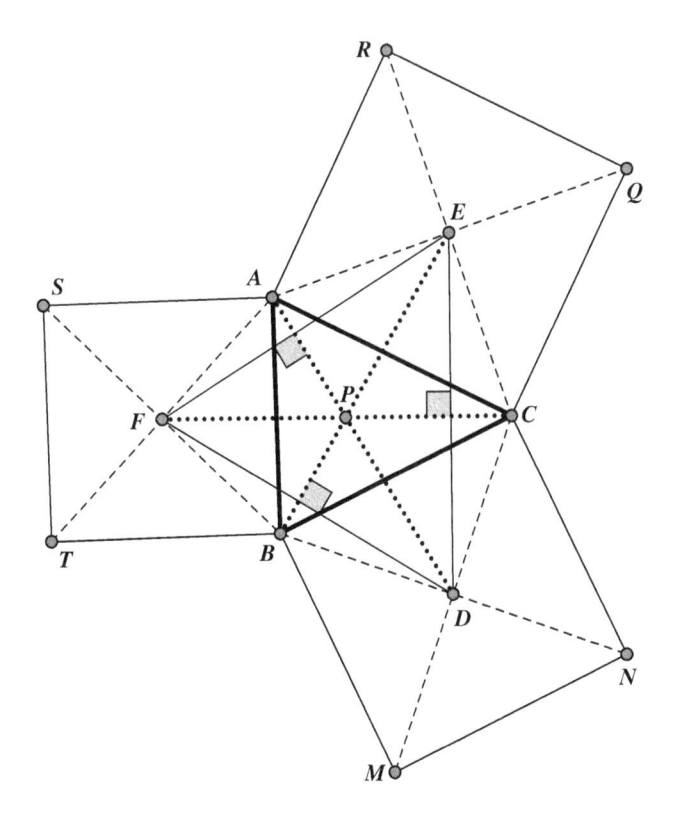

Figure 4-28

Proof

This phenomenon is named after a French geometer whom we know only by his surname, Vecten. He published under the name M. Vecten, but "M." means "Monsieur" (French for mister), and his given name is unknown. We will come back to proving this relationship with homothety in Chapter 11 when we consider Vecten's theorem and the Vecten point.

Another Concurrency by Squares on the Sides of a Triangle

In Figure 4-29, if we join midpoints *G*, *H*, and *J* of the line segments *SR*, *QN*, and *MT* with the midpoints of the remote sides of the original triangle *ABC* to generate the lines *GU*, *JV*, and *HW*, we find that these lines are concurrent at point *L*.

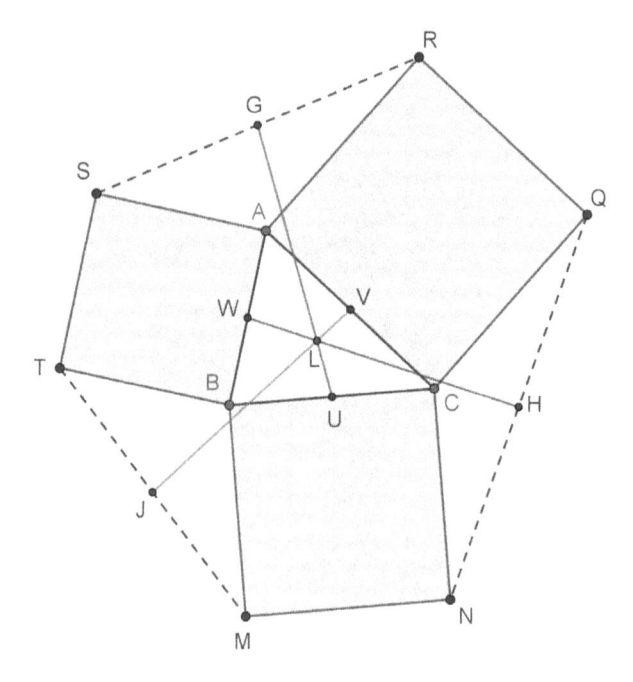

Figure 4-29

To prove the concurrency shown in Figure 4-29, we must first discuss the Finsler–Hadwiger theorem,[3] which describes a third square derived from any two squares that share a vertex.

Finsler–Hadwiger Theorem

In Figure 4-30, we have two squares $ABCD$ and $AB'C'D'$ with common vertex A. Let E and F be the midpoints of BB' and DD', respectively, and let M and M' be the centers of the two squares. Unexpectedly, we find that $EMFM'$ is a square as well.

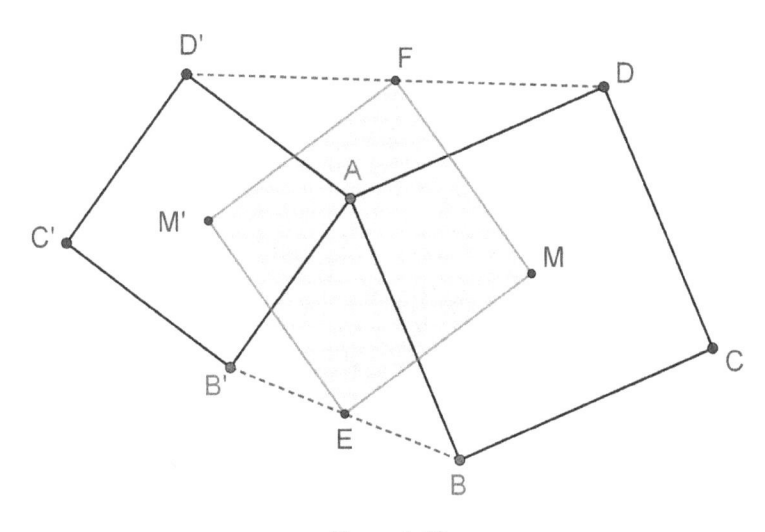

Figure 4-30

---

[3] This theorem is named after the Swiss-German mathematicians Paul Finsler and Hugo Hadwiger, who published it in 1937.

Proof

To prove this theorem, we consider Figure 4-31, where we use the triangles $\triangle ADD'$ and $\triangle AB'B$ to construct parallelograms $ADPD'$ and $AB'QB$, respectively. Because of the two squares and vertex A, $\angle D'AD$ is supplementary to $\angle B'AB$.

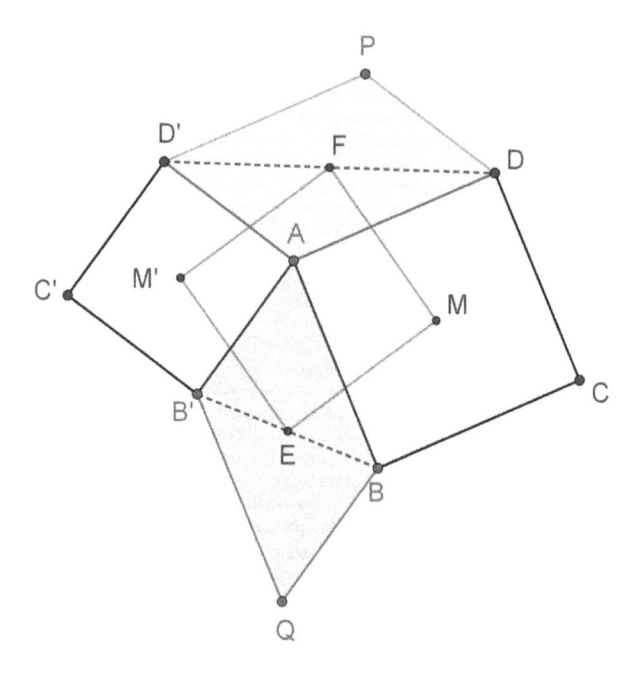

Figure 4-31

Next, we see that a 90° clockwise rotation of the parallelogram $ADPD'$ around M' yields the parallelogram $AB'QB$ (note $D' \mapsto A, A \mapsto B'$). Thus, we know $M'F$ and $M'E$ are equal and perpendicular. Analogously, by the 90° counterclockwise rotation around M, we see that MF and ME are equal and perpendicular. Since $\triangle FM'E$ and $\triangle FME$ are right isosceles triangles with the same hypotenuse *EF*, we find that the quadrilateral *EMFM'* is a square.

Now we are prepared to prove the concurrency in Figure 4-29. Due to the Finsler–Hadwiger theorem, *GU* is a diagonal of the square *UFGD*, as seen in Figure 4-32. Thus, *GU* is the perpendicular bisector of *DF*. Analogously, *JV* is the perpendicular bisector of *DE*, and *HW* the perpendicular bisector of *EF*. Hence, the three lines *GU*, *JV*, *HW* meet at the circumcenter *L* of △*DEF*.

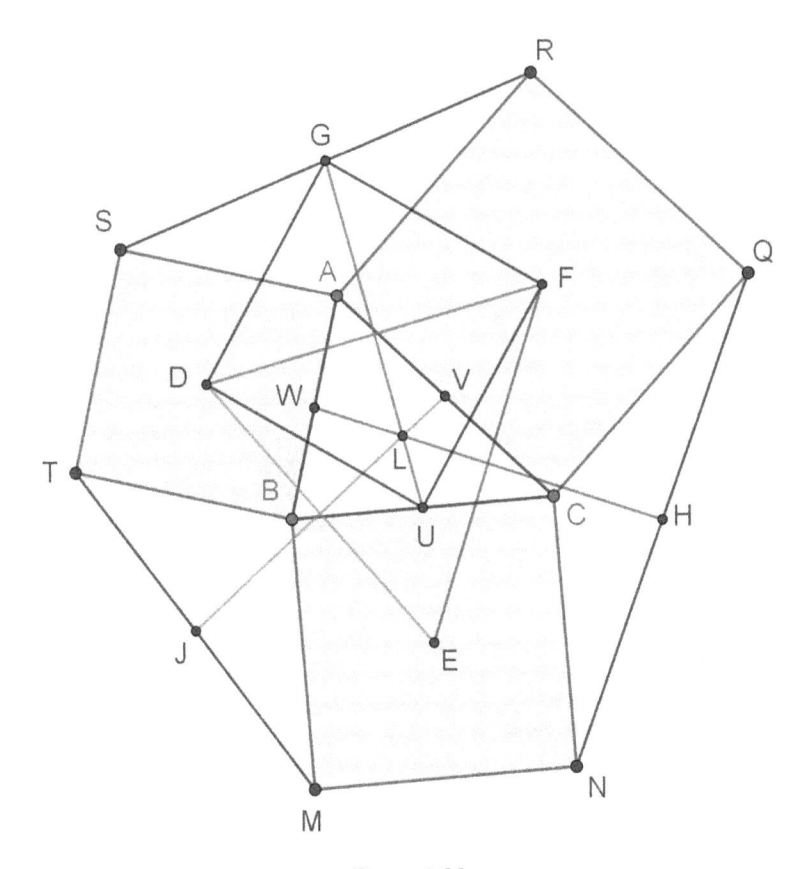

Figure 4-32

Yet More Unusual Concurrencies by Squares on the Sides of a Triangle – A Challenge!

Once again, we begin by constructing a square on each side of a random triangle. Connecting the midpoints *G*, *H*, and *J* of the line segments *SR*, *QN*, and *MT* with the midpoints of the remote squares generates the lines *GE*, *HD*, and *JF*. We find that these lines are concurrent at point *P*, as we see in Figure 4-33.

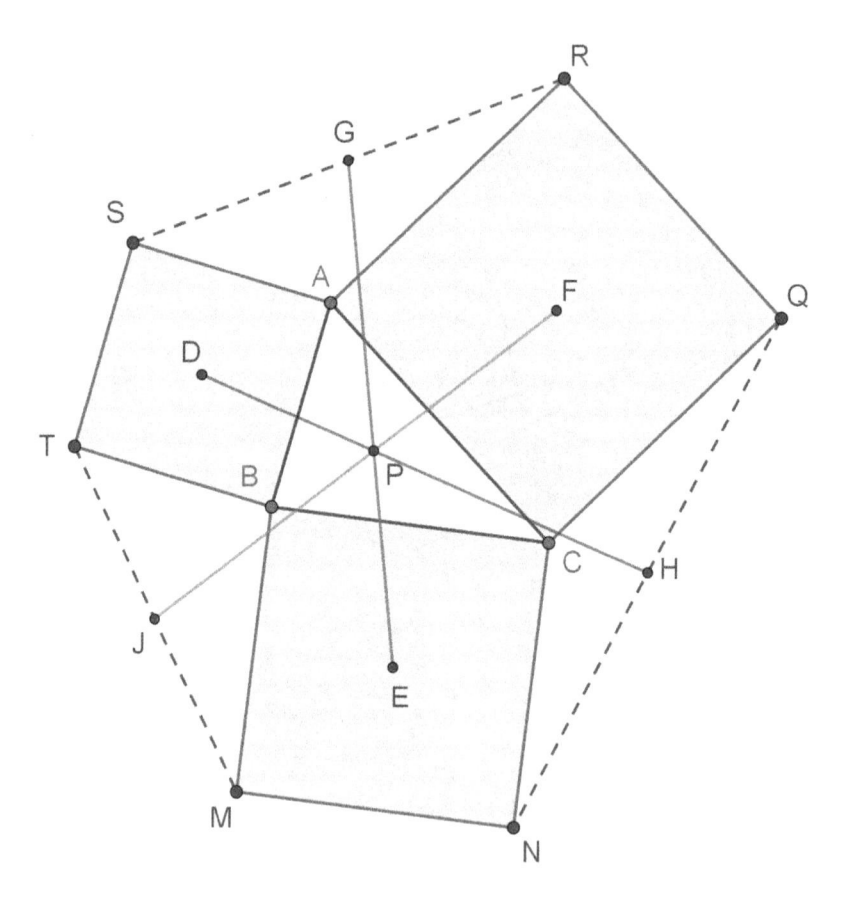

Figure 4-33

Furthermore, as we can see in Figure 4-34, the line segments *GE*, *HD*, and *JF* are equal and perpendicular to *JH*, *JG*, and *GH*, respectively, which allows us to conclude that *P* is the orthocenter of Δ*JHG*.

We leave the proof of this interesting relationship as an entertaining challenge for the reader.

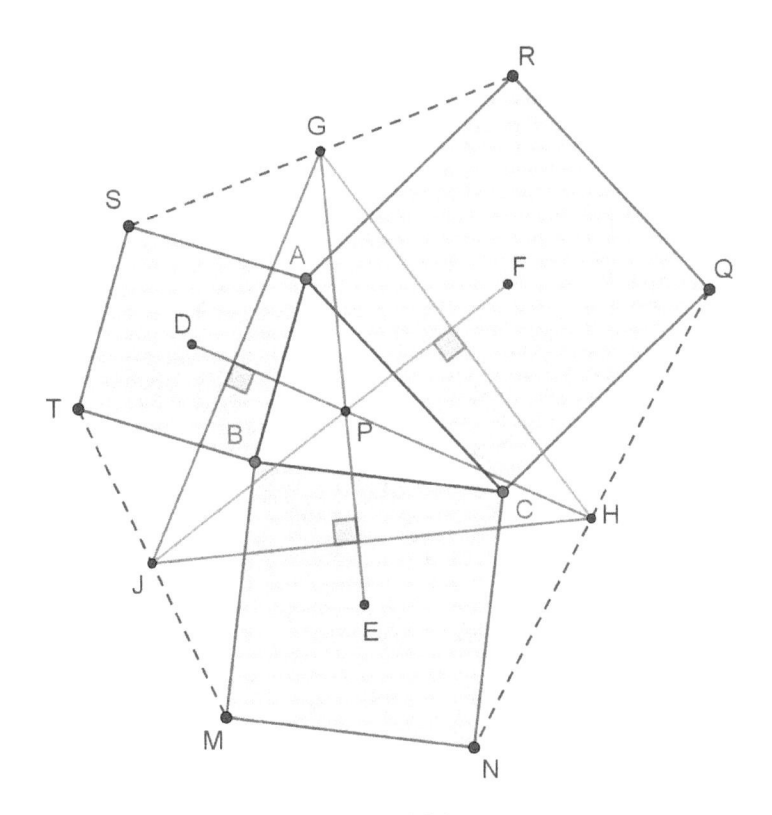

Figure 4-34

Squares on Quadrilateral Sides

Although this chapter has focused on squares on triangles, there is much to be said about squares on the sides of a quadrilateral, such as those shown in Figure 4-35. We find that the lines joining the centers of the opposite squares are equal, that is, $PN = OM$. Furthermore, we find that PN and OM are also perpendicular. This was detected and first proven by the Belgian mathematician Henri van Aubel (1830–1906), who published this result in 1878.

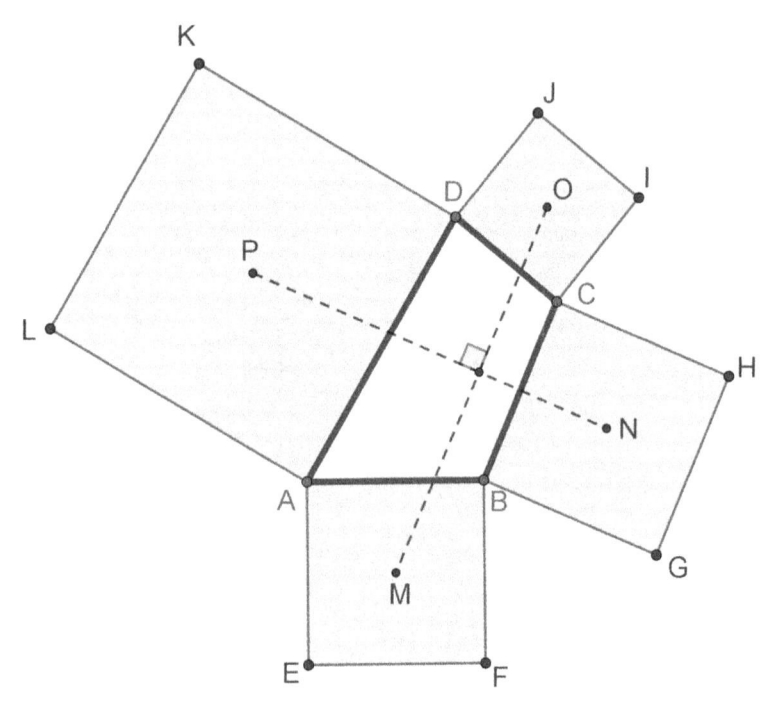

Figure 4-35

Proof

Let R be the midpoint of diagonal AC. According to the Finsler–Hadwiger theorem from above, we know $PR = RO$ and $PR \perp RO$ (Figure 4-36), and analogously, $MR = RN$ and $MR \perp RN$. Thus, a 90° counterclockwise rotation about R maps M onto N and O onto P. This, finally, gives us $\triangle RMO \cong \triangle RNP$, and the claimed result that $PN = MO$ and $PN \perp MO$.

The theorem of van Aubel is closely related to Vecten's theorem, which we will prove in Chapter 11 independent of van Aubel's theorem. On the other hand, Vecten's theorem is an immediate consequence of van Aubel's theorem by letting two points of the quadrilateral coincide (e.g., C and D, as well as the points I, J, O, as

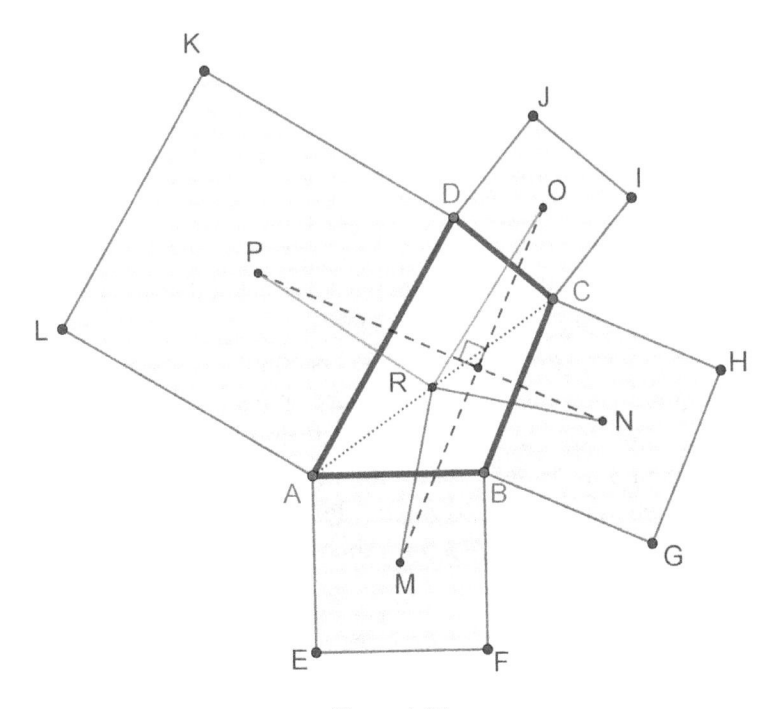

Figure 4-36

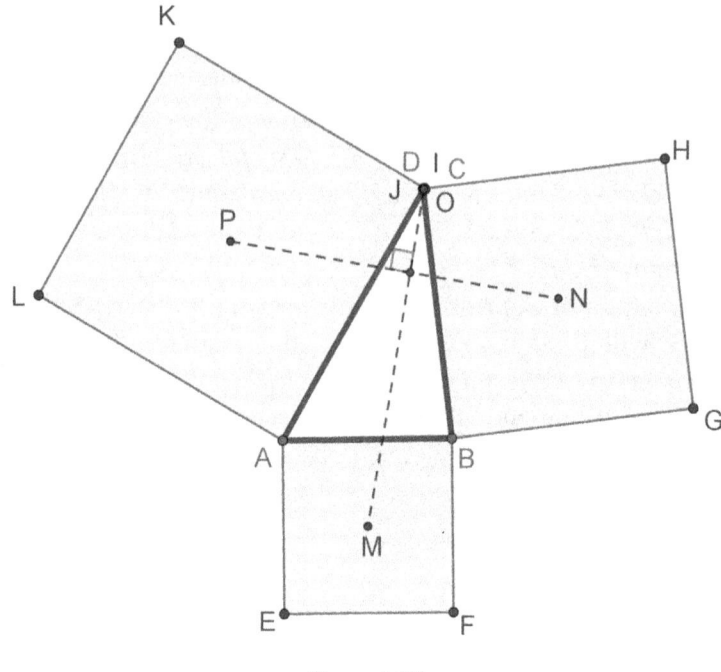

Figure 4-37

seen in Figure 4-37). Vecten's theorem states the following: If squares are erected on triangle sides outwardly, then the line segment joining two square centers (*PN* in Figure 4-37) is equal and perpendicular to the line segment joining the third square center with the remote triangle vertex (*MC* in Figure 4-37). Considering Vecten's theorem as a special case of van Aubel's theorem returns us to the topic of this chapter, namely, squares on *triangle* sides. For more details and a further discussion concerning Vecten's theorem, see Chapter 11.

Chapter 5

Similar Triangles on Triangle Sides

The Fermat–Torricelli Point

Recall some of the significant points in a triangle. The concurrency point of the angle bisectors of a triangle is the center of the triangle's inscribed circle. The point of intersection of the perpendicular bisectors of the sides of a triangle is the center of the triangle's circumscribed circle. The point of concurrency of the altitudes of a triangle is the triangle's orthocenter. The point of intersection of the medians of a triangle is the triangle's centroid, or center of gravity.

Another significant point of a triangle is the point where the sides subtend equal angles, that is, the angles opposite specific sides. For example, in Figure 5-1 we have a triangle with all interior angles less than 120°, and the point P is so situated that ∠APB = ∠BPC = ∠CPA. This, unexpectedly, is also the very same point in a triangle from which the sum of the distances to the vertices is the smallest. That is, where AP + BP + CP is smaller than the sum of the distances from any other point in the triangle to the vertices. These are just two properties of a very special point in a triangle that will provide us with some surprising results. Follow along!

We begin our exploration of this particularly significant point in a triangle by constructing equilateral triangles on all sides of a

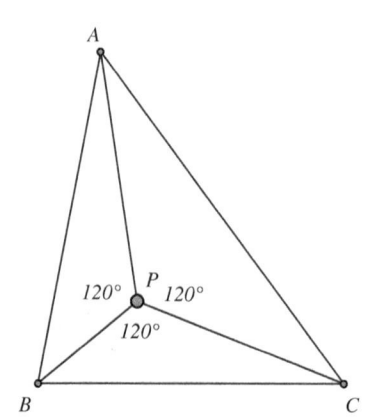

166 A Journey Through the Wonders of Plane Geometry

Figure 5-1

randomly drawn triangle *ABC*, as shown in Figure 5-2. We draw the lines *AA′*, *BB′*, and *CC′*. Using some elementary geometry, we can easily show that these three lines are equal by proving pairs of triangles congruent. Now we say "easily," yet most high school students who attempt to prove this equality – usually given as an exercise with congruence proofs early in the course – struggle doing it, simply because they have difficulty identifying the triangles that have to be proved congruent to establish the line segment equality. Once the triangles have been identified, the proof is quite simple. That is, to show *AA′* = *CC′*, we merely prove that △*AA′B* ≅ △*BC′C*, as shown in Figure 5-3 (by the side-angle-side congruence relationship). Similarly, *AA′* = *BB′* by showing, in similar fashion, that △*AA′C* ≅ △*AB′BC*.

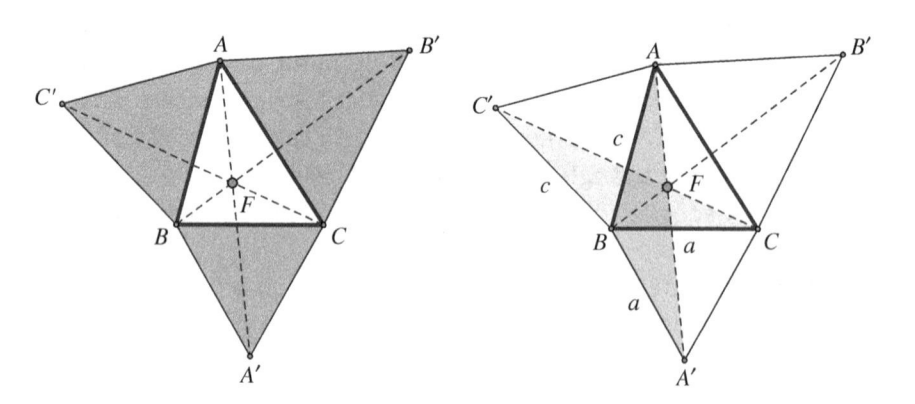

Figure 5-2

Figure 5-3

Another way to see that $AA' = CC'$ is to rotate AA' by 60° counterclockwise about B. Since $A \mapsto C'$ and $A' \mapsto C$, and since rotations do not change lengths, we have $AA' = CC'$.

Now having justified that the lines AA', BB', and CC' are equal, note that the lines are concurrent at point F. This point is sometimes called the *Fermat point*, named after the French mathematician Pierre de Fermat (1607–1665). It is also called the *Fermat–Torricelli* point, since the Italian mathematician Evangelista Torricelli (1608–1647) also dealt with this problem of finding the point of a triangle ABC that minimizes the sum of the distances to A, B, and C.

To prove this concurrency, we draw the circumscribed circles of the three equilateral triangles and demonstrate that they contain a common point F, as shown in Figure 5-4 for the case that each interior

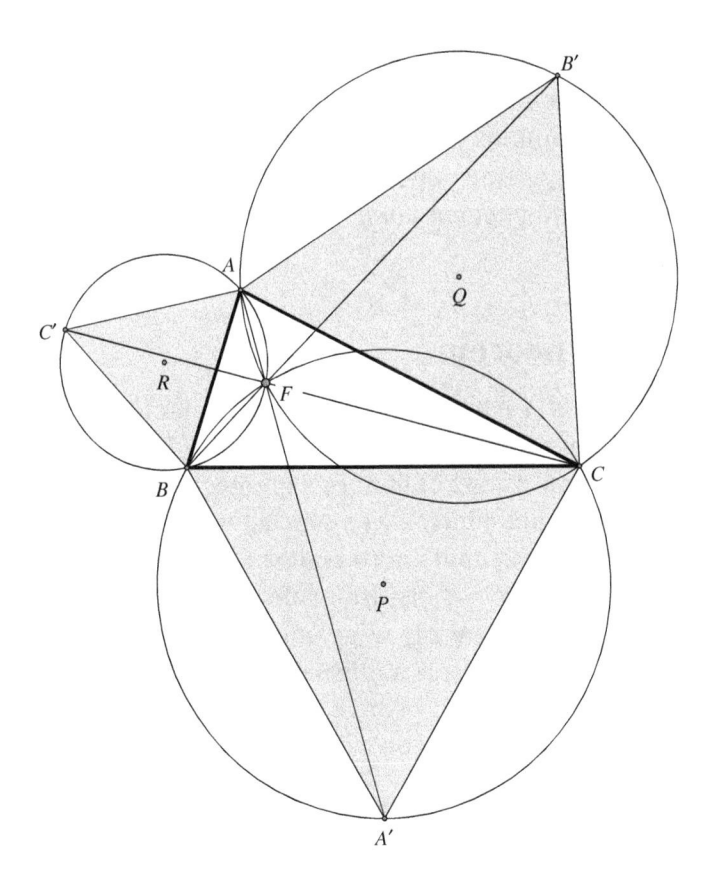

Figure 5-4

angle of △ABC is less than 120°. The cases where one interior angle is exactly 120° or even larger work analogously.

Let's consider the circumscribed circles of the three equilateral triangles △BCA′, △ACB′, and △ABC′ with respective centers P, Q, and R, as shown in Figure 5-4. Circles Q and R intersect at points F and A. We want to show that the point F is also on circle P. Since $\overarc{AC'B} = 240°$, we know that the inscribed angle $\angle AFB = \frac{1}{2}(\overarc{AC'B}) = 120°$. Similarly, $\angle AFC = \frac{1}{2}(\overarc{AB'C}) = 120°$. Therefore, $\angle BFC = 120°$, since a complete revolution is 360°.

Since $\overarc{BA'C} = 240°$, $\angle BFC = 120°$ is an inscribed angle and point F must, therefore, lie on circle P. Thus, we can see that the three circles are concurrent, intersecting at point F.

We join point F with points A, B, C, A′, B′, and C′ to find that $\angle B'FA = \angle AFC' = \angle C'FB = 60°$; therefore, B′FB is a straight line. Similarly, C′FC and A′FA are also straight lines, which establishes the concurrency of the lines AA′, BB′, and CC′ at F. In this way, we can determine the point in △ABC at which the three sides subtend (i.e., determined by being opposite) congruent angles. The point F is also called the *equiangular point* of △ABC since $\angle AFB = \angle AFC = \angle BFC = 120°$.

Napoleon's Theorem

At this point, we are ready to embark on a famous theorem in geometry attributed to Napoleon Bonaparte (1769–1821), who, aside from his fame as one of history's greatest military commanders, also distinguished himself as a top-flight mathematics student at the Paris Military School and member of the *Institut de France*, a prestigious scientific society. He prided himself on his talent in mathematics, and particularly in geometry. However, the theorem that bears his name was first published by the English mathematician William Rutherford (1798–1871) in "The Ladies' Diary" in 1825, four years after Napoleon's death. To this day, it is uncertain

Similar Triangles on Triangle Sides **169**

if Napoleon ever knew of the relationship we are about to investigate.[1]

To show Napoleon's theorem, we consider the centers of the three circumscribed circles of the three equilateral triangles drawn on the sides of triangle *ABC*, shown in Figure 5-5. These centers determine an equilateral triangle *PQR*, known as the *Napoleon triangle*. The Napoleon triangle is equilateral because its sides make angles of 60°, which can be easily proved. Due to symmetry, we have $BF \perp RP$ and $CF \perp QP$, and because $\angle YFZ = 120° = \angle BFC$, we can conclude that $\angle ZPY = \angle QPR = 60°$ since it is the remaining angle of quadrilateral *YFZP*, whose angle sum is 360°. Similarly, we can show that $\angle PRQ = 60° = \angle PQR$, which makes $\triangle PQR$ equilateral.

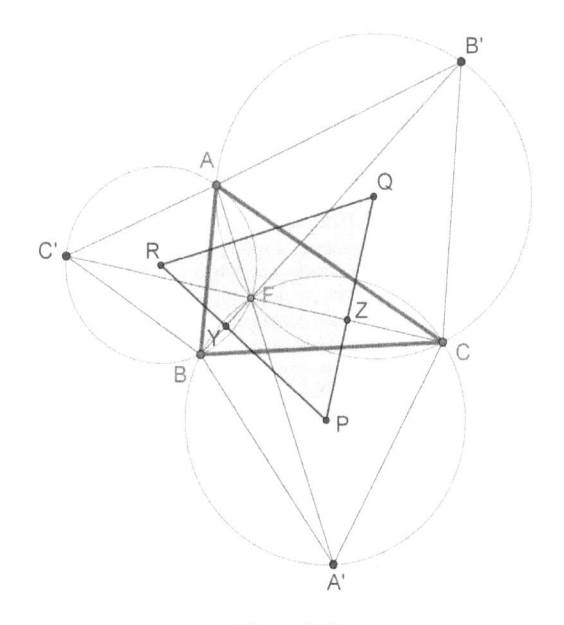

Figure 5-5

---

[1] See B. Grünbaum, "Is Napoleon's Theorem *Really* Napoleon's Theorem?" *American Mathematical Monthly* 119, 6, 495–501 (2012).

170 A Journey Through the Wonders of Plane Geometry

There are many unusual relationships tying the Napoleon triangle to the original triangle. For one, the Napoleon triangle and the original triangle share a common centroid. As we establish this interesting relationship, we will encounter some other curiosities along the way. We begin this quest in Figure 5-6, where we have point G as the centroid of triangle ABC, and point P as the centroid of triangle BCA'. We denote point M_a as the midpoint of BC. Because the centroid of a triangle trisects each of the medians, we have $AM_a = 3GM_a$ and $A'M_a = 3PM_a$. Since GP partitions AM_a and $A'M_a$ proportionally, we can conclude that $\triangle M_a GP \sim \triangle M_a AA'$, and $AA' = 3GP$. In other words, the

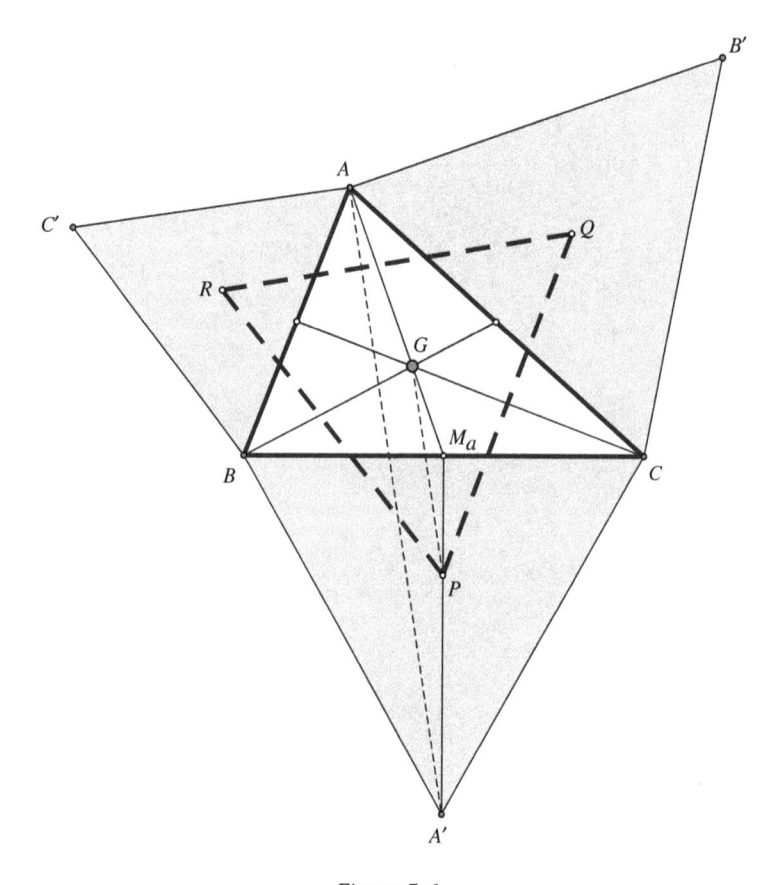

Figure 5-6

distance between the centroids is one-third the length of the line segment joining a vertex of the original triangle with the remote vertex of the relevant equilateral triangle.

In a similar fashion, we can show in Figure 5-7 that $CC' = 3GR$ and $BB' = 3GQ$. We have shown above that $AA' = BB' = CC'$. Therefore, $GP = GQ = GR$. Since triangle PQR is equilateral and the distances from point G to its vertices are equal, we can conclude that G is also the

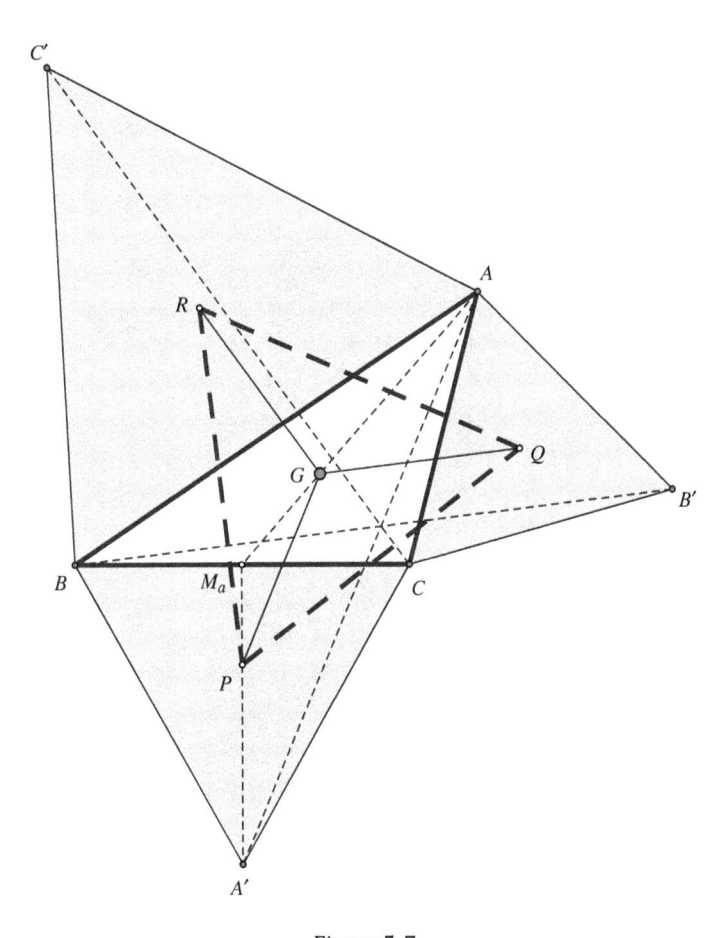

Figure 5-7

172 *A Journey Through the Wonders of Plane Geometry*

centroid of triangle *PQR*. We have, therefore, shown that point *G* is the centroid of both the (outer) Napoleon triangle *PQR* and the original triangle *ABC*.

All that we have said so far about the three equilateral triangles drawn on the sides of a randomly selected triangle was based on their being drawn *externally* to the original triangle. Yet, as you might have expected, we can make all the same arguments about a configuration that has the three equilateral triangles drawn *internally* to the given triangle, or shall we say, overlapping the original triangle, as shown in Figure 5-8. Triangle *UVW* is equilateral and shares its centroid point *G* with the centroid of triangle *PQR* and triangle *ABC*.

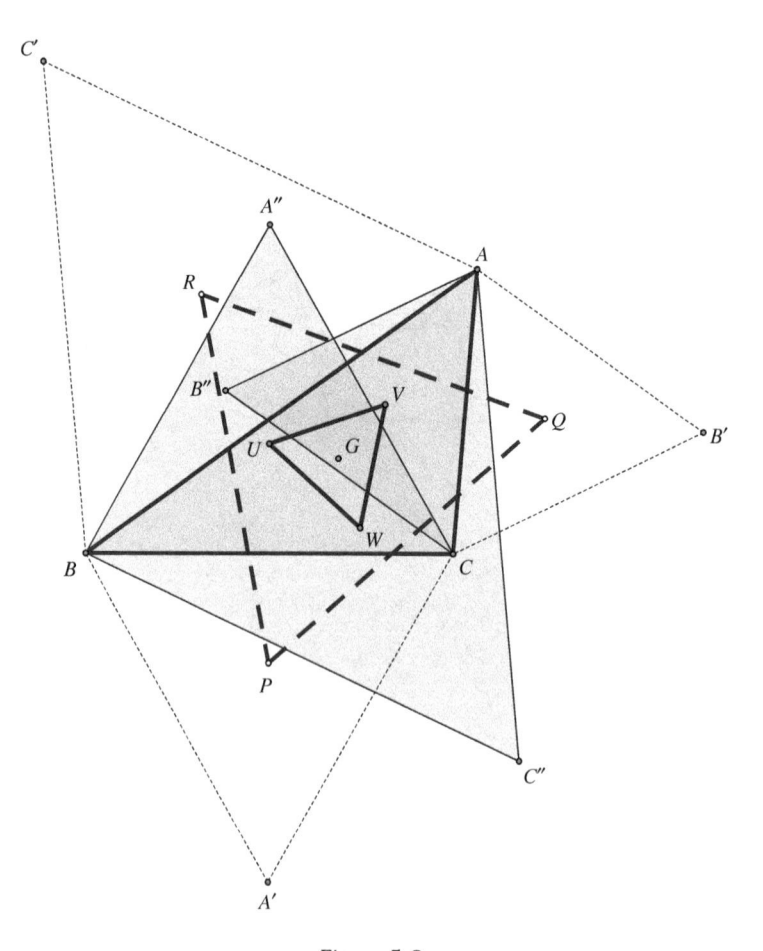

Figure 5-8

Similar Triangles on Triangle Sides **173**

Referring back to any of the previous few figures, say Figure 5-7, suppose we now leave triangle *BCA'* fixed and move point *A* to various positions (even to the other side of *BC*). As long as the point *A* does not land on points *B* or *C*, where triangle *ABC* would then have a zero area, all that we have established above will still hold true (note that in some cases triangles will degenerate to line segments with area 0 or even to points!). This is truly an amazing relationship! It can be easily shown with dynamic geometry software, such as Geometer's Sketchpad or GeoGebra.

As if the configuration shown in Figure 5-2 did not already produce enough unexpected equilateral triangles, we can find yet another one. All we need to do is to construct a parallelogram *AC'CD* as shown in Figure 5-9, and we can identify an equilateral triangle, namely, *AA'D*. The same size equilateral triangle can be produced at all sides of this configuration, since each of these equilateral triangles will have for a side one of the equal lengths: *AA'*, *BB'*, and *CC'*. To justify that triangle *AA'D* is in fact equilateral, we can show that *AD = AA'*, since both are equal to *CC'*, and $\angle DAA' = 60°$ since it is an alternate-interior angle to $\angle AFC' = 60°$, where *F* is the Fermat point. That is,

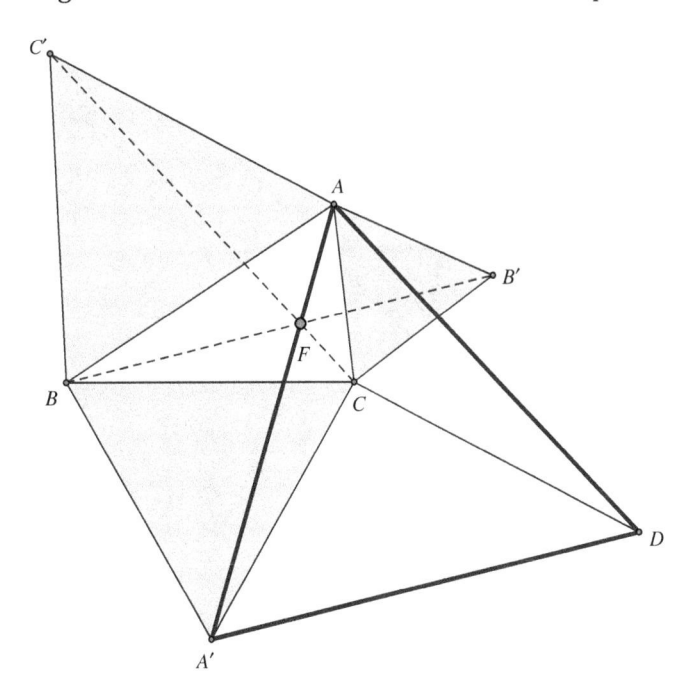

Figure 5-9

174 *A Journey Through the Wonders of Plane Geometry*

AA'D is an isosceles triangle with a 60° vertex angle, which makes it equilateral.

By connecting the outer vertices *A'*, *B'*, *C'* of the outwardly erected equilateral triangles, respectively, with the corresponding vertices *P*, *Q*, *R* of the Napoleon triangle, we find the lines concurrent at the center *O* of the circumscribed circle of triangle *ABC*, as shown in Figure 5-10. Furthermore, we have another unexpected feature: these concurrent lines are perpendicular bisectors of the sides of the initial △*ABC*.

We are not yet finished with this rich equilateral triangle configuration. We need to focus again on the Fermat point *F*. Not only is point *F* the equiangular point, but, as we indicated earlier, it is also the *minimum-distance point* from the three vertices of triangle *ABC* – that

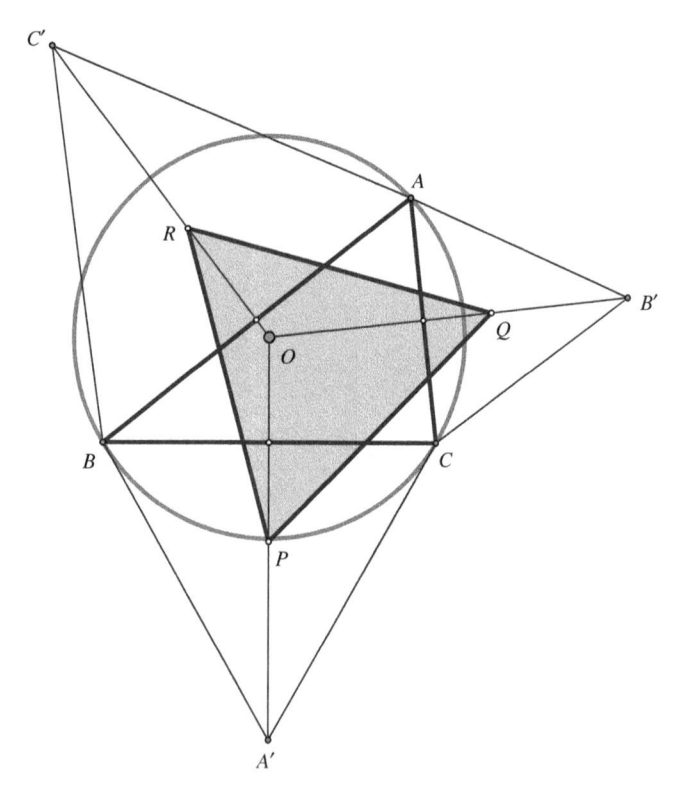

Figure 5-10

Similar Triangles on Triangle Sides **175**

is, the sum of the distances from that point to the three vertices of the triangle is less than the sum of the distances from any other point in the triangle to the vertices. In other words, this point has *two* important properties: it is the minimum-distance point (with respect to the vertices) and the equiangular point of the triangle.

Let's investigate how we can justify this last claim. We begin by considering △*ABC*, a triangle with all angles measuring less than 120°, as shown in Figure 5-11.

To show that the sum of the distances from point *F* to each of the three vertices of triangle *ABC* is less than that from any other point to the vertices, we need to take any other randomly selected point *D* and show that the sum of the distances from this point is greater than the sum of the distances from point *F* to the vertices of the triangle. The justification – or proof – of this relationship is quite interesting and a

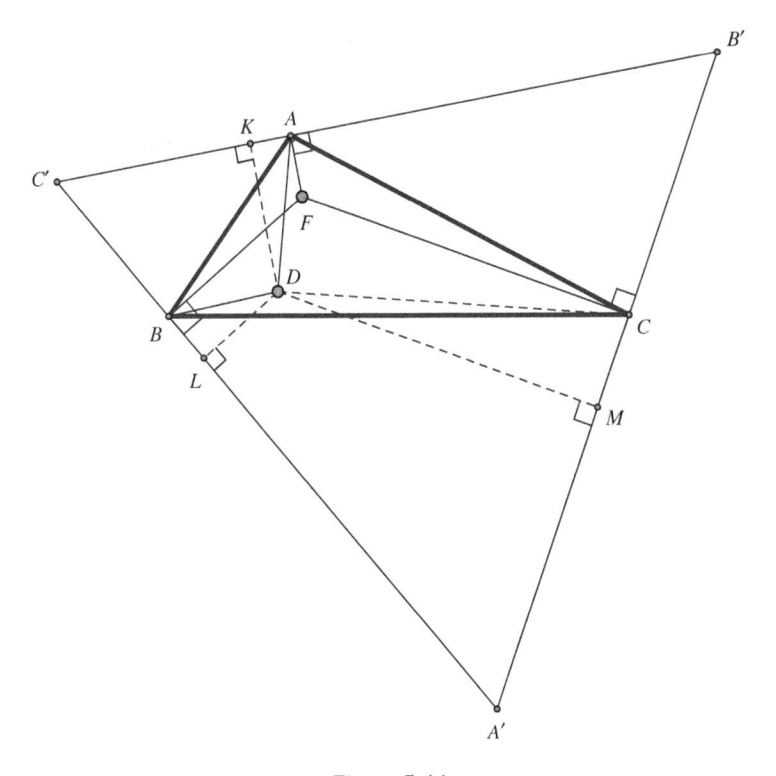

Figure 5-11

bit different from other geometric proofs. Follow along, and you will find it rewarding.

Let F be the equiangular point in the interior of $\triangle ABC$, that is, where $\angle AFB = \angle BFC = \angle AFC = 120°$. Draw lines through A, B, and C that are perpendicular to AF, BF, and CF, respectively. These lines meet to form yet another equilateral triangle, $A'B'C'$. (To prove $\triangle A'B'C'$ is equilateral, notice that each angle measures $60°$. This can be shown by considering, for example, quadrilateral $AFBC'$. Since $\angle C'AF = \angle C'BF = 90°$, and $\angle AFB = 120°$, it follows that $\angle AC'B = 60°$.) Let D be *any other* point in the interior of $\triangle ABC$. We must show that the sum of the distances from F to the vertices of triangle ABC is less than the sum of the distances from the randomly selected point D to the vertices of triangle ABC.

We can easily demonstrate that the sum of the distances from any point in the interior of an equilateral triangle to the sides is a constant, namely, the length of the altitude. Let's take a moment to review Viviani's theorem, named after the Italian mathematician Vincenzo Viviani (1622–1703).

In Figure 5-12, consider equilateral $\triangle ABC$, where $PQ \perp AB$, $PR \perp BC$, $PS \perp AC$, and $CD \perp AB$. Draw the line segments PA, PB, and PC. We have area$\triangle ABC$ = area$\triangle APB$ + area$\triangle BPC$ + area$\triangle CPA$ = $\frac{1}{2}(AB)(PQ) + \frac{1}{2}(BC)(PR) + \frac{1}{2}(AC)(PS)$. Since $AB = BC = AC$, we know

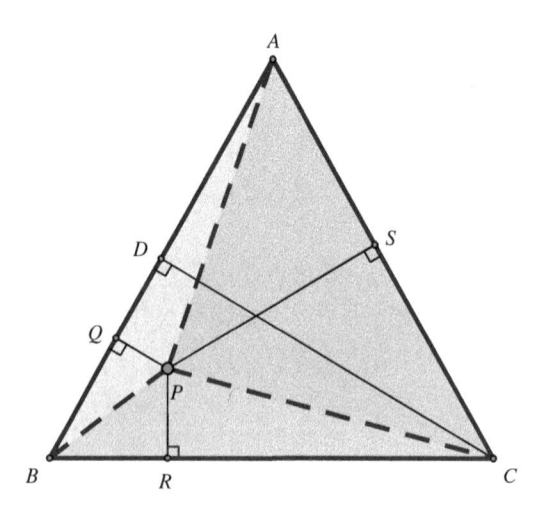

Figure 5-12

area of $\triangle ABC = \frac{1}{2}(AB) \cdot (PQ + PR + PS)$. However, the area of $\triangle ABC$ equals $\frac{1}{2}(AB)(CD)$. Therefore, $PQ + PR + PS = CD$ is a constant for the given triangle.

Using this constant relationship, we have in Figure 5-11 $FA + FB + FC = DK + DL + DM$, where DK, DL, and DM are the perpendiculars to $B'KC'$, $A'LC'$, $A'MB'$, respectively.

But $DK + DL + DM < DA + DB + DC$. (Recall: The shortest distance from an external point to a line is the length of the perpendicular segment from that point to the line.)

By substitution: $FA + FB + FC < DA + DB + DC$.

You may wonder why we restricted our discussion to triangles with angles of measure less than 120°. If you try to construct the Fermat point F in a triangle with one angle of measure of 150°, the reason for our restriction will become obvious. In Figure 5-13, we have $\angle BAC > 120°$, and we find that we find that the point F, the

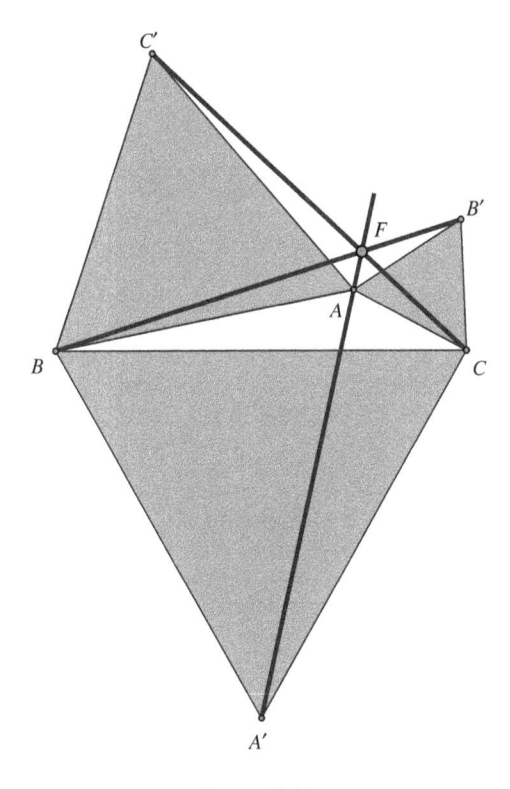

Figure 5-13

supposed minimum-distant point is outside of triangle ABC. When $\angle BAC = 120°$, as shown in Figure 5-14, the minimum-distance point is at vertex A. Therefore, the minimum-distance point *in* a triangle (with no angle of measure greater than 120°) is the equiangular point, which is the point at which the sides of the triangle subtend congruent angles.

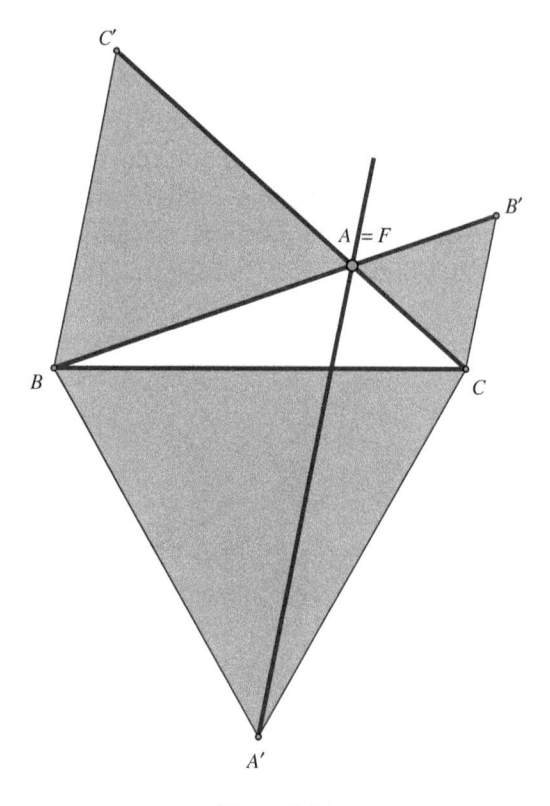

Figure 5-14

So far, we have dealt with *equilateral* triangles erected on arbitrary triangles. But this chapter is more general, and also discusses *similar* triangles placed on an arbitrary triangle. So, in the following, we will present some generalizations of Napoleon's theorem. There are many of them, and some are rather complicated. We will focus on four well-known generalizations that can be formulated and understood rather easily. For two of them, we supply a proof, and for the other two we give just a reference where interested readers can find a proof.

Generalization 1

If *similar* triangles *DBA*, *BEC*, and *ACF* are erected outwardly on the sides of any triangle *ABC*, and *P*, *Q*, and *R* are the respective *circumcenters*, as shown in Figure 5-15, then they form a triangle *RPQ* similar to the three erected triangles.

This means that, in Napoleon's theorem, the equilaterality of the erected triangles is not a necessary condition, and having similar triangles suffices. We will see that the proof also works analogously.

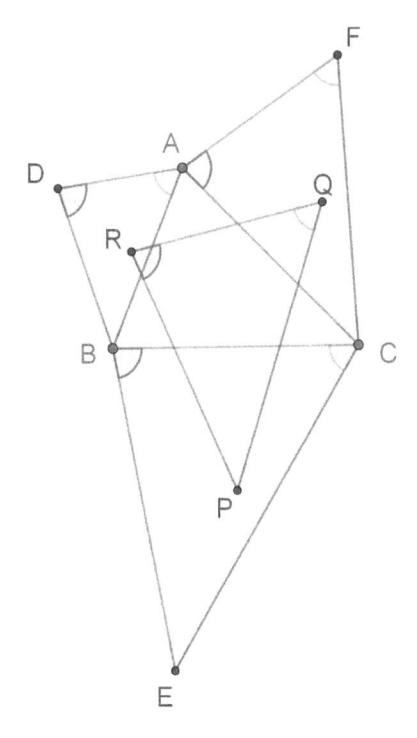

Figure 5-15

Proof

First, we note that $\angle D + \angle E + \angle F = 180°$ because these three angles appear as interior angles in all three erected similar triangles. Then, we construct the circumcircles of the outwardly erected similar triangles centered at P, Q, and R, respectively, and we notice that these circles are concurrent at point T, as shown in Figure 5-16. To prove this, let T be the intersection point of the circles Q and R. We have $\angle CTA = 180° - \angle F$ and $\angle BTA = 180° - \angle D$, which leaves $\angle BTC = 360° - \angle CTA - \angle BTA = \angle D + \angle F = 180° - \angle E$ (remember that we have $\angle D + \angle E + \angle F = 180°$). This, in turn, means that T lies also on circle P. Due to the right angles found at K, L, and M, we have

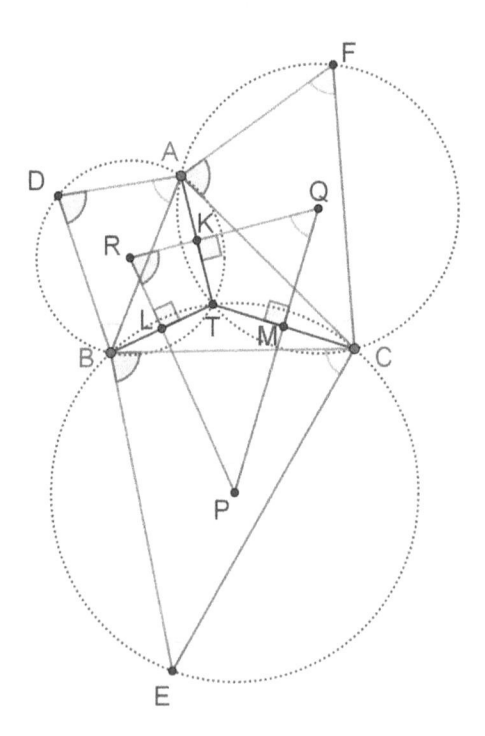

Figure 5-16

$\angle P = 180° - \underbrace{\angle LTM}_{=\angle BTC = 180° - \angle E} = \angle E$, and similarly, $\angle Q = \angle F$, and $\angle R = \angle D$.

These angles establish the claimed similarity.

We have seen in the above proof that the similarity of the erected triangles was not important to establish $\angle P = \angle E$, $\angle Q = \angle F$, and $\angle R = \angle D$. The only important issue was $\angle D + \angle E + \angle F = 180°$.

This brings us immediately to the next generalization of Napoleon's theorem, which is already shown by the above proof.

Generalization 2

In Figure 5-16, if triangles *DBA*, *BEC*, and *ACF* are erected on the sides of any triangle *ABC* so that $\angle D + \angle E + \angle F = 180°$, and *P*, *Q*, and *R* are the respective circumcenters, then $\angle P = \angle E$, $\angle Q = \angle F$, and $\angle R = \angle D$, and the three circumcircles of the three triangles are concurrent at point *T*.

For both above generalizations, interested readers can see and use a dynamic version by Michael de Villiers at http://dynamic mathematicslearning.com/napole-general.html. This also applies to Generalization 3.

Now, we come to a third noteworthy generalization of Napoleon's theorem, closely related to Generalization 1. There, *P*, *Q*, and *R* were the *circumcenters* of the erected similar triangles. We could have taken the *incenters* instead, but the triangle *PQR* would still be similar to the outwardly erected similar triangles. We also could have taken the *orthocenters* or the *centroids* instead, again yielding *PQR* similar to the outwardly erected similar triangles. This we will formulate as a further theorem in Generalization 3.

Generalization 3

If similar triangles *DBA*, *BEC*, and *ACF* are erected outwardly on sides of any triangle *ABC*, and any three points *P*, *Q*, and *R* are chosen so that they respectively lie in the *same relative positions* to these triangles, then *P*, *Q*, and *R* form a triangle similar to the three outwardly erected similar triangles.[2] For an example, see Figure 5-17, where *P*, *Q*, and *R* are the respective incenters.

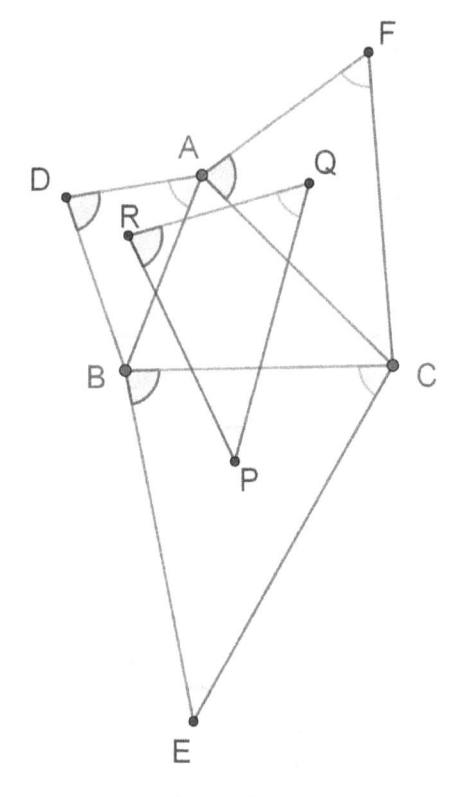

Figure 5-17

We have to explain what is meant by *same relative positions*. Assume two triangles *ABC* and *A'B'C'* are directly similar. Let *P* be any point in the plane. We can say that point *P'* is in the *same relative position* to *A'B'C'* as *P* is to *ABC* if the *similarity transformation* that maps triangle *ABC* onto triangle *A'B'C'* also maps point *P* onto *P'* (see Figure 5-18). Special cases of *same relative positions* are taken in each triangle: the *circumcenter*, the *incenter*, the *orthocenter*, or the *centroid*, respectively.

---

[2] M. De Villiers & J. H. Meyer: (1995). "A generalized dual of Napoleon's theorem and some further extensions." *Int. J. Math. Educ. Sci. Technol.*, 26, 2, 233–241 (1995).

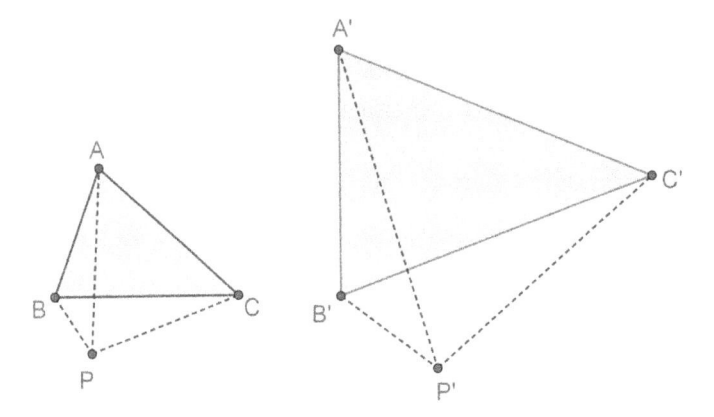

Figure 5-18   All angles in the left figure are equal to the corresponding angles in the right figure; all corresponding line segments have a constant ratio.

For a proof of Generalization 3, see the reference in footnote 2, on page 182.

Generalization 4

On the sides of an *affine-regular n*-gon, construct regular *n*-gons (all outwardly or all inwardly). The centers of these regular *n*-gons form a new regular *n*-gon. In Figure 5-21, we show the situation for an *n*-gon where $n = 5$, which are pentagons.

Let us briefly explain what an *affine-regular n*-gon is. This is an *n*-gon that is the image of a regular *n*-gon under an *affine transformation*. But what is an *affine transformation*? Roughly speaking, these are geometric transformations that preserve

1. lines (lines are mapped to lines),
2. parallelisms (parallel lines are mapped to parallel lines),
3. and ratios of distances of collinear points.

Euclidean distances and angles are not necessarily preserved. Of course, all congruence transformations (which even preserve Euclidean distances and angles) are affine transformations, as well as similarity transformations (for instance, *central dilations*, also called *homotheties*, preserve angles and are discussed in Chapter 11). But there are also other affine transformations like *shear mappings*, such as the mapping $\triangle ABC \mapsto \triangle A'B'C'$ shown in Figure 5-19, or *dilations from an axis* shown in Figure 5-20.

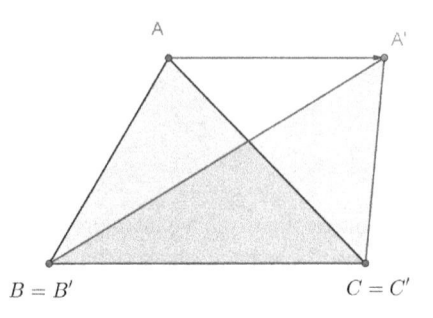

Figure 5-19   Shear mapping, A is moved on a parallel to BC

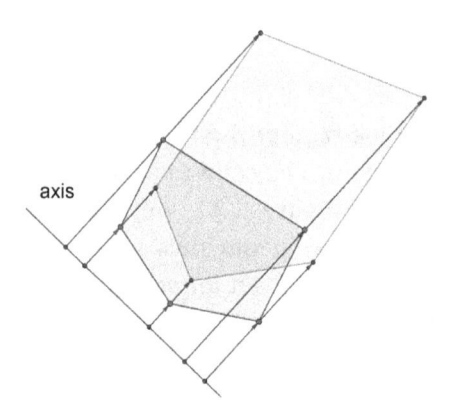

Figure 5-20   Dilation from an axis with factor 2

The statement of Generalization 4 is called the Napoleon–Barlotti theorem and extends both Napoleon's and Thébault's[3] theorem (see below) to an arbitrary *n*-gon. Note that every triangle is affine-regular (this means it is the image of an equilateral triangle under an

---

[3] Victor Thébault (1882–1960) was a French mathematician.

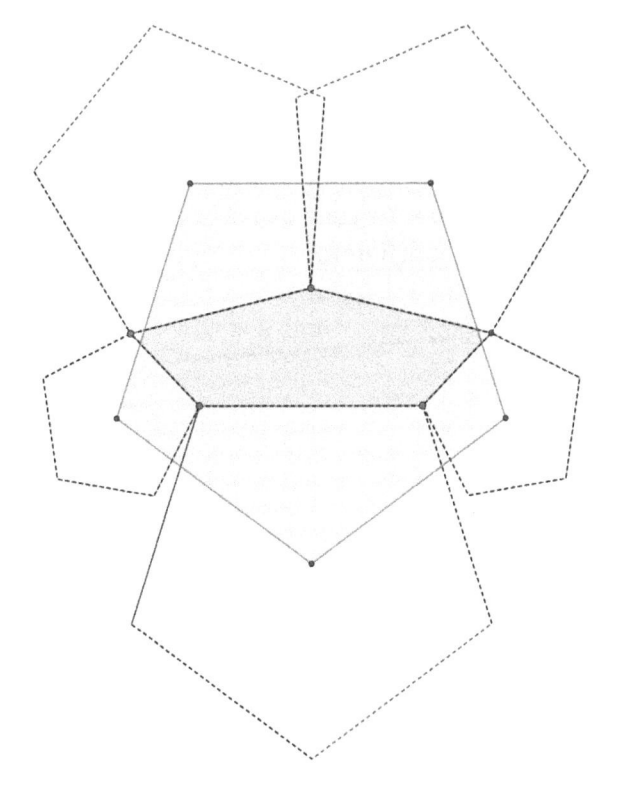

Figure 5-21 Napoleon–Barlotti theorem for pentagons ($n = 5$)

affine transformation), and the only affine-regular quadrilaterals (images of squares under affine transformations) are parallelograms. We refer the reader to one source for a proof of the Napoleon–Barlotti theorem.[4]

Although Generalizations 3 and 4 do not deal with similar triangles on triangle sides, we have presented them here in this chapter because they are well-known and amazing generalizations of Napoleon's theorem.

---

[4] L. Gerber. "Napoleon's Theorem and the Parallelogram Inequality for Affine-Regular Polygons." *American Mathematical Monthly*, 87, 8, 644–648 (1980).

Squares on the Sides of a Parallelogram – Thébault's theorem

Earlier, we considered squares placed on the sides of triangles. However, interesting results can be found by placing squares on the sides of special quadrilaterals such as parallelograms, which are the only affine-regular quadrilaterals. In Figure 5-22, we have a

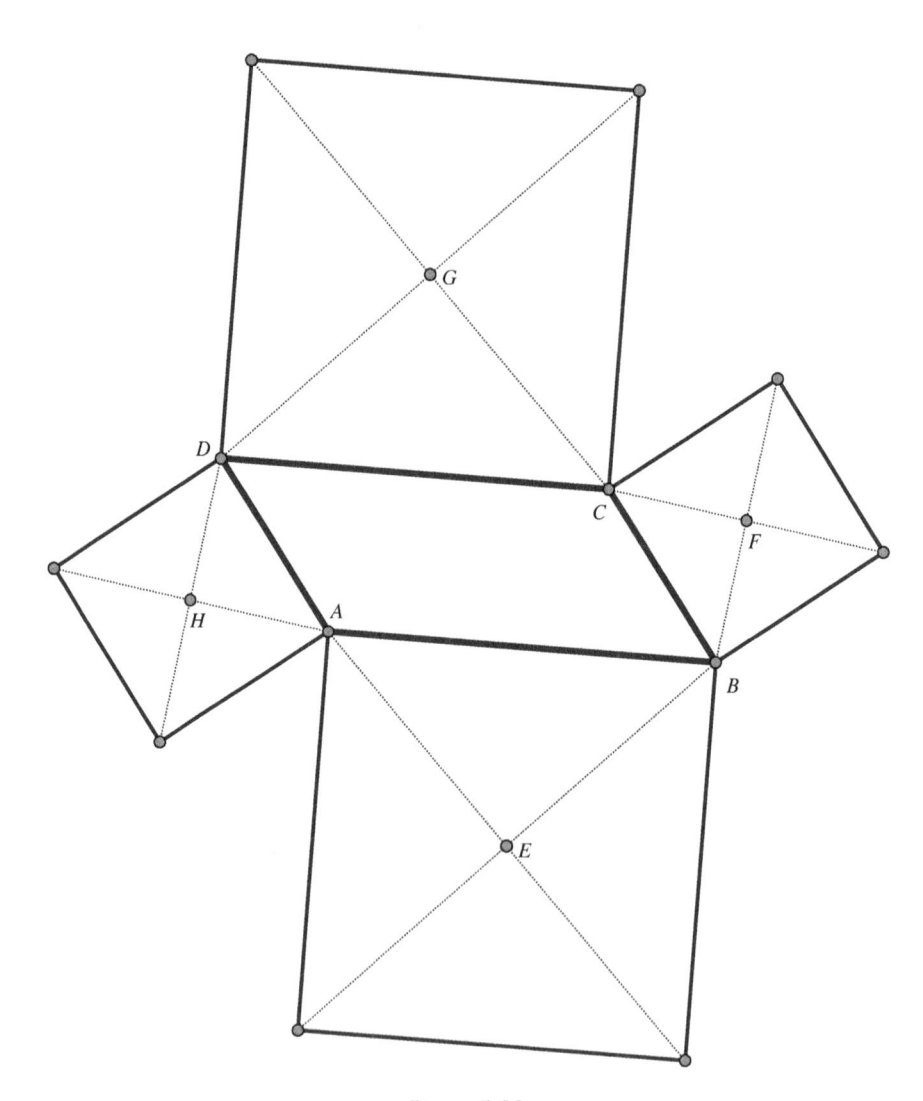

Figure 5-22

parallelogram *ABCD* with squares erected on each of the four sides. The centers of the four squares are *E*, *F*, *G*, and *H*. Curiously, regardless of the shape of the parallelogram, these four points *E*, *F*, *G*, and *H* turn out to be the vertices of a square. This is called *Thébault's theorem*, and we need to show how this can be proved.

Proof

Consider the four triangles △*HAE*, △*HDG*, △*EBF*, and △*FCG* shown in Figure 5-23. We can show that these four triangles are all congruent

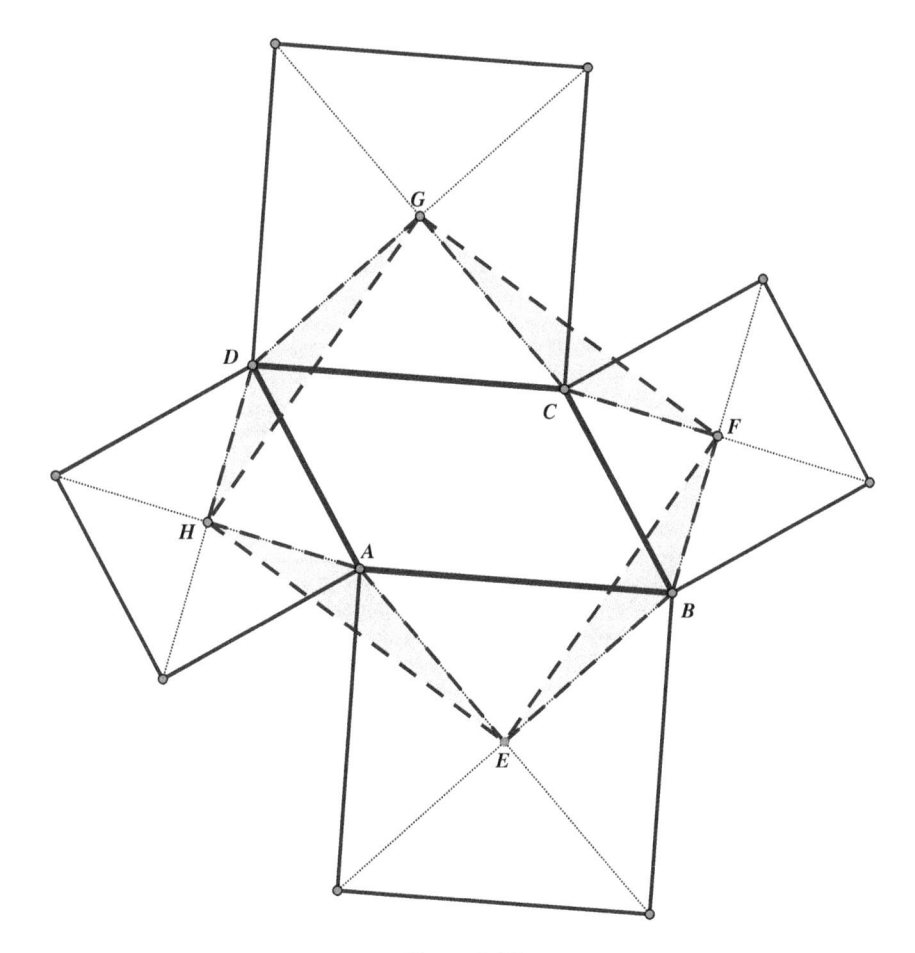

Figure 5-23

by simply choosing two of these triangles to prove their congruence and realizing that the procedure applies to the other two as well.

Consider the two triangles $\triangle FBE$ and $\triangle HDG$. We have many 45° angles in the diagram: $\angle GDC = \angle HDA = \angle FBC = \angle EBC = 45°$. We also know that $\angle ADC = \angle ABC$. Therefore, by addition, we find $\angle GDH = \angle EBF$. Furthermore, $DH = BF$ and $GD = BE$. Thus, $\triangle FBE \cong \triangle HDG$, and $GH = FE$. Using this procedure, we can show that the four triangles are all congruent to each other. This makes quadrilateral $GHEF$ a rhombus. However, we can also show that each of its angles is a right angle. Since $\angle DHG = \angle AHE$ and $\angle DHG + \angle AHG = 90°$, it follows that $\angle AHE + \angle AHG = 90°$, which is one of the angles of the rhombus, thus making it a square.

Chapter 6

Discovering Concyclic Points

It is common knowledge that two distinct points determine a unique line. We also know that three noncollinear points determine a unique plane, a unique triangle, and a unique circle. In other words, whenever you have three noncollinear points, exactly one circle will contain the three points. Might there be arrangements of four points that would determine a unique circle? Clearly, the four vertices of a square, the four vertices of a rectangle, and the four vertices of an isosceles trapezoid would determine a unique circle. In each case, these quadrilaterals have opposite angles that are supplementary, which is the criterion for a *cyclic quadrilateral*. Quadrilaterals that are cyclic enrich the field of geometry, as we will see with the first few examples provided here. However, concyclic points, that is, more than three points that lie on the same circle, also appear unexpectedly, as we will see throughout this chapter (though we have used this concept when needed earlier in this book).

Generating a Cyclic Quadrilateral

An interesting way to generate a cyclic quadrilateral is to draw angle bisectors for each vertex of any convex quadrilateral. The segments connecting the points of intersection of the adjacent angles form a cyclic quadrilateral. In Figure 6-1, angle bisectors *AG*, *BG*, *CE*, and *DE* intersect at points *E*, *F*, *G*, and *H*, respectively. The result is that quadrilateral *EFGH* is cyclic.

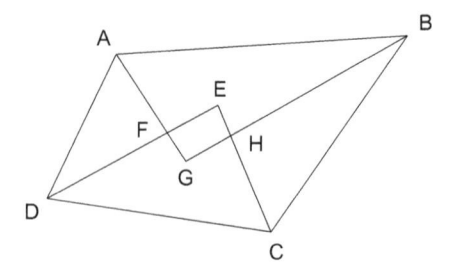

Figure 6-1

Proof

In Figure 6-1, the angle bisectors of quadrilateral *ABCD* meet to determine quadrilateral *EFGH*. We shall prove this latter quadrilateral to be cyclic. Since the sum of the angles of any convex quadrilateral is 360°, we have ∠*BAD* + ∠*ADC* + ∠*DCB* + ∠*CBA* = 360°. Therefore, $\frac{1}{2}$∠*BAD* + $\frac{1}{2}$∠*ADC* + $\frac{1}{2}$∠*DCB* + $\frac{1}{2}$∠*CBA* = $\frac{1}{2}$(360°) = 180°.

Substituting appropriately, we get

∠*EDC* + ∠*ECD* + ∠*GAB* + ∠*ABG* = 180°.   (I)

Consider △*ABG* and △*DEC*:

∠*EDC* + ∠*ECD* + ∠*GAB* + ∠*ABG* + ∠*AGB* + ∠*DEC* = 2(180°).   (II)

Now, subtracting (I) from (II), we find that ∠*AGB* + ∠*DEC* = 180°. Since one pair of opposite angles of quadrilateral *EFGH* are supplementary, the other pair must also be supplementary, and hence, quadrilateral *EFGH* is cyclic.

A Cyclic Quadrilateral Curiosity

When a cyclic quadrilateral with perpendicular diagonals has a line drawn through the point of intersection of the diagonals and perpendicular to one side of the quadrilateral, this line bisects the opposite side of the quadrilateral.

Proof

In Figure 6-2, the diagonals *AC* and *BD* of cyclic quadrilateral *ABCD* are perpendicular at point *G*, so that *GE* ⊥ *AED*. Our goal here is to justify that GE bisects BC at P. In right $\triangle AEG$, we find $\angle 5$ is complementary to $\angle 1$, and $\angle 2$ is complementary to $\angle 1$. Therefore, $\angle 5 = \angle 2$.

However, $\angle 2 = \angle 4$. Thus, $\angle 5 = \angle 4$. We also have $\angle 5 = \angle 6$ since both angles are measured by $\frac{1}{2}\overset{\frown}{DC}$. Therefore, $\angle 4 = \angle 6$, and $BP = GP$. Similarly, $\angle 7 = \angle 3$, and $\angle 7 = \angle 8$, so that $GP = PC$. Thus, $CP = PB$.

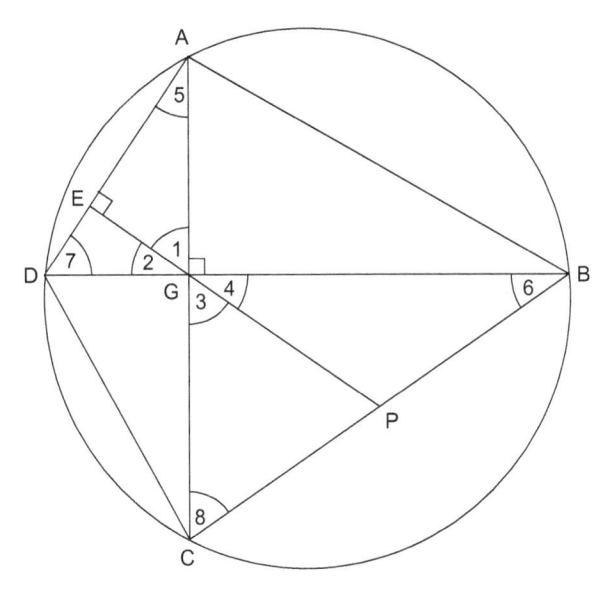

Figure 6-2

Area of Cyclic Quadrilaterals: Brahmagupta's Theorem

We present Heron's famous formula for finding the area of any triangle given only the lengths of its three sides. Heron's formula for the area of a triangle is $\sqrt{s(s-a)(s-b)(s-c)}$, where a, b, and c are the lengths of the sides and $s = \frac{a+b+c}{2}$ is the semi-perimeter.

It is natural to try to extend this formula to quadrilaterals. One common way is to consider the triangle as a quadrilateral with a zero-length side. Such an extension is credited to the Indian mathematician Brahmagupta (598–670 CE).[1] He developed the following formula to

---

[1] In 628, Brahmagupta wrote *Brahma-sphuta-siddhānta* ("The Revised System of Brahma") and devoted the twelfth and thirteenth chapters to mathematics.

find the area of a *cyclic quadrilateral* with sides of lengths a, b, c, and d, where s is the semi-perimeter $s = \frac{a+b+c+d}{2}$: Area = $\sqrt{(s-a)(s-b)(s-c)(s-d)}$. You will notice that Brahmagupta considered Heron's formula to be treating the triangle as a quadrilateral with d = 0.

Proof

Although the proof is somewhat complicated, it is still worth exploring. First, consider the case where quadrilateral ABCD is a rectangle with a = c and b = d. Accepting Brahmagupta's formula to be correct, we get for the area of rectangle ABCD the following expression $\sqrt{(s-a)(s-b)(s-c)(s-d)} = \sqrt{(a+b-a)(a+b-b)(a+b-a)(a+b-b)}$ $= \sqrt{a^2b^2} = ab$, which is the area of the rectangle as found by the usual method of taking the product, ab, of its length and width.

Now consider any non-rectangular cyclic quadrilateral ABCD, shown in Figure 6-3. Extend DA and CB to meet at P, forming △DCP. Let PC = x and PD = y. By Heron's formula, the area of △DCP

$$= \frac{1}{4}\sqrt{(x+y+c)(y-x+c)(x+y-c)(x-y+c)}. \qquad (I)$$

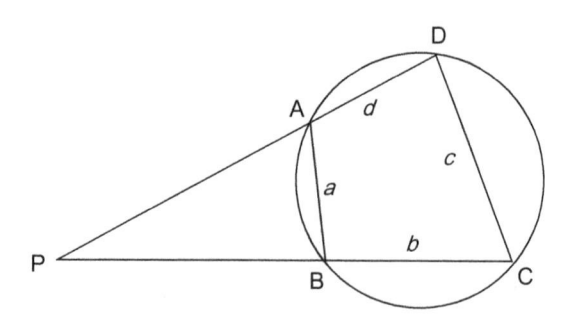

Figure 6-3

Since ∠CDA is supplementary to ∠CBA, and ∠ABP is also supplementary to ∠CBA, we have ∠CDA = ∠ABP so that △BAP ~ △DCP. (II)
From (II) we get

$$\frac{\text{area}\triangle BAP}{\text{area}\triangle DCP} = \frac{a^2}{c^2}, \frac{\text{area}\triangle DCP - \text{area}\triangle BAP}{\text{area}\triangle DCP} = \frac{c^2}{c^2} - \frac{a^2}{c^2},$$

$$\frac{\text{area}\triangle DCP - \text{area}\triangle BAP}{\text{area}\triangle DCP} = \frac{\text{area}\triangle ABCD}{\text{area}\triangle DCP} = \frac{c^2 - a^2}{c^2}. \qquad \text{(III)}$$

From (II) we also get

$$\frac{x}{c} = \frac{y-d}{a}, \qquad \text{(IV)}$$

and

$$\frac{y}{c} = \frac{x-b}{a}. \qquad \text{(V)}$$

By adding (IV) and (V), we have the following:

$$\frac{x+y}{c} = \frac{x+y-b-d}{a}, \quad x+y = \frac{c}{c-a}(b+d),$$

$$x+y+c = \frac{c}{c-a}(b+c+d-a). \qquad \text{(VI)}$$

The following relationships are found by using similar methods.

$$y-x+c = \frac{c}{c+a}(a+c+d-b). \qquad \text{(VII)}$$

$$x+y-c = \frac{c}{c-a}(a+b+d-c). \qquad \text{(VIII)}$$

$$x-y+c = \frac{c}{c+a}(a+b+c-d). \qquad \text{(IX)}$$

Substitute (VI), (VII), (VIII), and (IX) into (I).

Then the area of

$$\triangle DCP = \frac{c^2}{4(c^2-a^2)}\sqrt{(b+c+d-a)(a+c+d-b)(a+b+d-c)(a+b+c-d)}.$$

Since (III) may be read as area of $\triangle DCP = \frac{c^2}{c^2-a^2} \cdot (\text{area} ABCD)$, the area of cyclic quadrilateral $ABCD = \sqrt{(s-a)(s-b)(s-c)(s-d)}$.

Let's extend Brahmagupta's formula to the general quadrilateral (given without proof), which is called Bretschneider's formula, named after the German mathematician Carl Anton Bretschneider (1808–1878).

The formula states that area of any (convex) quadrilateral $= \sqrt{(s-a)(s-b)(s-c)(s-d) - abcd \cdot \cos^2\left(\frac{\alpha+\gamma}{2}\right)}$, where a, b, c, and d are the lengths of the sides, $s = \frac{a+b+c+d}{2}$, and α and γ are the measures of a pair of opposite angles of the quadrilateral.

This formula shows that of all quadrilaterals that can be formed from four given side lengths, the one with the maximum area is the cyclic quadrilateral. The maximum area is achieved when $abcd \cdot \cos^2\left(\frac{\alpha+\gamma}{2}\right) = 0$, which occurs when $\alpha + \gamma = 180°$, where the opposite angles are supplementary. This is true for cyclic quadrilaterals.

There are many interesting theorems about cyclic quadrilaterals, as we have seen at the beginning of this chapter. Brahmagupta also found that for a cyclic quadrilateral of consecutive side lengths a, b, c, and d, where m and n are the lengths of the diagonals, the following relationship holds true: $m^2 = \frac{(ab+cd)(ac+bd)}{ad+bc}$, $n^2 = \frac{(ac+bd)(ad+bc)}{ab+cd}$.

Ptolemy's Theorem

Perhaps the most famous theorem involving cyclic quadrilaterals is that attributed to Claudius Ptolemaeus of Alexandria, popularly known as Ptolemy (100–170 CE). In his major astronomical work, the *Almagest*[2] (ca. 150 CE), he states a most amazing theorem on cyclic quadrilaterals, which is that the product of the lengths of the diagonals of a cyclic quadrilateral equals the sum of the products of the lengths of the pairs of opposite sides.

We provide here two methods for proving Ptolemy's theorem, the second of which incorporates the proof of the converse.

Proof 1

In Figure 6-4, quadrilateral $ABCD$ is inscribed in circle O. A line is drawn through A to intersect CD at P, so that $\angle BAC = \angle DAP$. (I)

Since quadrilateral $ABCD$ is cyclic, $\angle ABC$ is supplementary to $\angle ADC$. However, $\angle ADP$ is also supplementary to $\angle ADC$. Therefore, $\angle ABC = \angle ADP$. (II)

Thus, $\triangle BAC \sim \triangle DAP$, and (III)

$$\frac{AB}{AD} = \frac{BC}{DP}, \text{ or } DP = \frac{(AD)(BC)}{AB}.$$ (IV)

---

[2] The Greek title, *Syntaxis Mathematica*, means "mathematical (or astronomical) compilation". The Arabic title, *Almagest*, is a renaming meaning "great collection (or compilation)". The book is a manual of all the mathematical astronomy that the ancients knew at the time of its composition. Book I (of the thirteen books that comprise this monumental work) contains the theorem (6.11) that now bears Ptolemy's name.

Discovering Concyclic Points 195

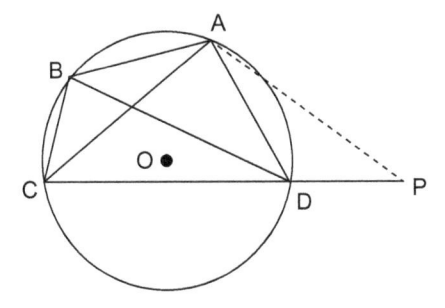

Figure 6-4

From (I), $\angle BAD = \angle CAP$, and from (III), $\frac{AB}{AD} = \frac{AC}{AP}$. Therefore, $\triangle ABD \sim \triangle ACP$ (SAS), and $\frac{BD}{CP} = \frac{AB}{AC}$, or $CP = \frac{(AC)(BD)}{AB}$. (V)

$$CP = CD + DP. \quad (VI)$$

Substituting (IV) and (V) into (VI), we get $\frac{(AC)(BD)}{AB} = CD + \frac{(AD)(BC)}{AB}$. Therefore, $(AC)(BD) = (AB)(CD) + (AD)(BC)$, which is Ptolemy's theorem.

Proof 2

In Figure 6-5, we draw $\triangle DAP$ on side AD of quadrilateral $ABCD$ similar to $\triangle CAB$ so that $\frac{AB}{AP} = \frac{AC}{AD} = \frac{BC}{PD}$ and (I)

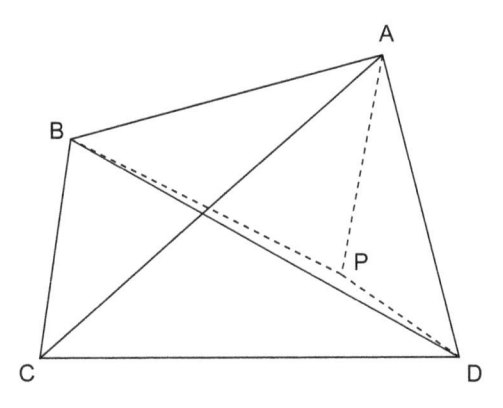

$$(AC)(PD) = (AD)(BC). \quad (II)$$

Figure 6-5

Since $\angle BAC = \angle PAD$, then $\angle BAP = \angle CAD$. Therefore, from (I), we have $\triangle BAP \sim \triangle CAD$, and

$$\frac{AB}{AC} = \frac{BP}{CD}, \text{ or } (AC)(BP) = (AB)(CD). \tag{III}$$

Adding (II) and (III), we have

$$(AC)(BP + PD) = (AD)(BC) + (AB)(CD). \tag{IV}$$

Now, $BP + PD > BD$ (triangle inequality) unless P is on BD. However, P will be on BD if and only if $\angle ADP = \angle ADB$. But we already know from the similar triangles that $\angle ADP = \angle ACB$. And if $ABCD$ were cyclic, then $\angle ADB$ would equal $\angle ACB$, and $\angle ADB$ would equal $\angle ADP$. Therefore, we can state that if, and only if, $ABCD$ is cyclic, then P lies on BD. This tells us that

$$BP + PD = BD. \tag{V}$$

Substituting (V) into (IV), $(AC)(BD) = (AD)(BC) + (AB)(CD)$. We have thus proved Ptolemy's theorem and its converse.

A further justification for the converse of Ptolemy's theorem is to assume quadrilateral $ABCD$ in Figure 6-5 is not cyclic. If C, D, and P are collinear, then $\angle ADP \neq \angle ACB$ because $\angle ADC$ would not be supplementary to $\angle ABC$. If C, D, and P are not collinear, then it is possible to have $\angle ADP = \angle ACB$. However, then $CP < CD + DP$, and from steps (IV) and (V) in Proof 1 (above), we have $(AC)(BD) < (AB)(CD) + (AD)(BC)$. But this contradicts the given information that $(AC)(BD) = (AB)(CD) + (AD)(BC)$. Therefore, quadrilateral $ABCD$ is cyclic.

Applications of Ptolemy's Theorem

Ptolemy's theorem allows us to create some rather unusual geometric relationships.

Ptolemy's Theorem Application 1

If any circle passing through vertex A of parallelogram $ABCD$ intersects sides AB and AD at points P and R, respectively, and intersects diagonal AC at point Q, the following result evolves: $(AQ)(AC) = (AP)(AB) + (AR)(AD)$.

Proof

We begin by drawing line segments RQ, QP, and RP, as shown in Figure 6-6. Since inscribed angles measured by the same intercepted arc are equal, we then have $\angle 4 = \angle 2$ and $\angle 1 = \angle 3$. From the parallel lines, $\angle 5 = \angle 3$, and therefore, $\angle 1 = \angle 5$. Furthermore, $\triangle RQP \sim \triangle ABC$, and since $\triangle ABC \cong \triangle CDA$, we have $\triangle RQP \sim \triangle ABC \sim \triangle CDA$.

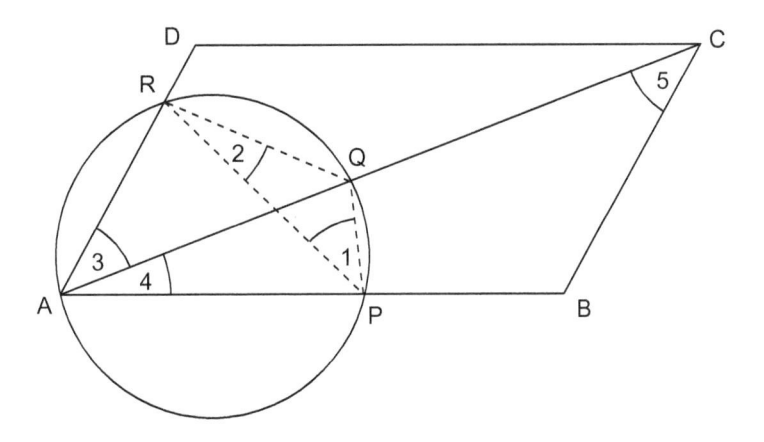

Then, $\dfrac{AC}{RP} = \dfrac{AB}{RQ} = \dfrac{AD}{PQ}$. (I)

Figure 6-6

By Ptolemy's theorem in quadrilateral $RQPA$:

$$(AQ)(RP) = (RQ)(AP) + (PQ)(AR). \quad \text{(II)}$$

Multiplying each of the three equal ratios in (I) by each member of (II) yields

$$(AQ)(RP)\left(\dfrac{AC}{RP}\right) = (RQ)(AP)\left(\dfrac{AB}{RQ}\right) + (PQ)(AR)\left(\dfrac{AD}{PQ}\right).$$

Thus, $(AQ)(AC) = (AP)(AB) + (AR)(AD)$.

Ptolemy's Theorem Application 2

The ratio of the lengths of the diagonals of a cyclic quadrilateral to the sides of the quadrilateral can be expressed as $\frac{AC}{BD} = \frac{(AB)(AD)+(BC)(DC)}{(DC)(AD)+(BC)(AB)}$, as shown in Figure 6-7.

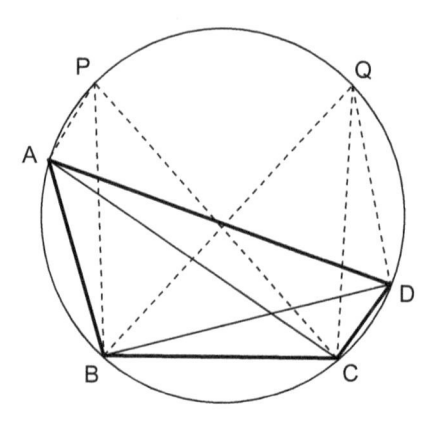

Figure 6-7

Proof

In Figure 6-7, on the circumcircle of quadrilateral $ABCD$, choose points P and Q so that $PA = DC$ and $QD = AB$. We apply Ptolemy's theorem to quadrilateral $ABCP$ to yield

$$(AC)(PB) = (AB)(PC) + (BC)(PA). \qquad (I)$$

Likewise, by applying Ptolemy's theorem to quadrilateral $BCDQ$, we have

$$(BD)(QC) = (DC)(QB) + (BC)(QD). \qquad (II)$$

Since $PA + AB = DC + QD$, $\widehat{PAB} = \widehat{QDC}$ and $PB = QC$.

Similarly, since $\widehat{PBC} = \widehat{DBA}$, $PC = AD$, and since $\widehat{QCB} = \widehat{ACD}$, $QB = AD$.

Finally, dividing (I) by (II) and substituting for all terms containing Q and P gives us $\frac{AC}{BD} = \frac{(AB)(AD)+(BC)(DC)}{(DC)(AD)+(BC)(AB)}$.

Ptolemy's Theorem Application 3

A point P is chosen inside parallelogram $ABCD$ such that $\angle APB$ is supplementary to $\angle CPD$. We find that $(AB)(AD) = (BP)(DP) + (AP)(CP)$, as shown in Figure 6-8.

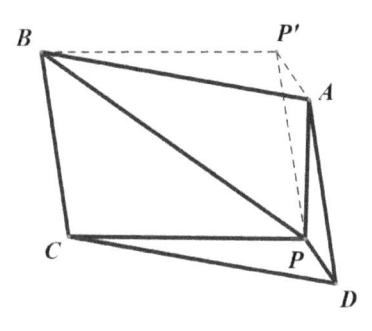

Figure 6-8

Proof

In Figure 6-8, on side AB of parallelogram $ABCD$, we draw $\triangle AP'B \cong \triangle DPC$ so that

$$DP = AP' \text{ and } CP = BP'. \qquad (I)$$

We know that $\angle APB$ is supplementary to $\angle CPD$, and $\angle BP'A = \angle CPD$, and it follows that $\angle APB$ is supplementary to $\angle BP'A$. Therefore, quadrilateral $BP'AP$ is cyclic. Now applying Ptolemy's theorem to cyclic quadrilateral $BP'AP$, we have

$$(AB)(P'P) = (BP)(AP') + (AP)(BP').$$

From (I), where $CP = BP'$, we then have

$$(AB)(P'P) = (BP)(DP) + (AP)(CP). \qquad (II)$$

Since $\angle BAP' = \angle CDP$, $CD \| AB$, and $PD \| P'A$, therefore, $PDAP'$ is a parallelogram, and $P'P = AD$.

Thus, from (II), $(AB)(AD) = (BP)(DP) + (AP)(CP)$.

Ptolemy's theorem simplifies a rather nice pattern about regular polygons, which we will provide in the following applications.

Ptolemy's Theorem Application 4

We begin with isosceles $\triangle ABC$ ($AB = AC$), which is inscribed in a circle as shown in Figure 6-9, and point P on $\overset{\frown}{BC}$.

Curiously, we find $\frac{PA}{PB+PC} = \frac{AC}{BC}$ is a constant for the given triangle.

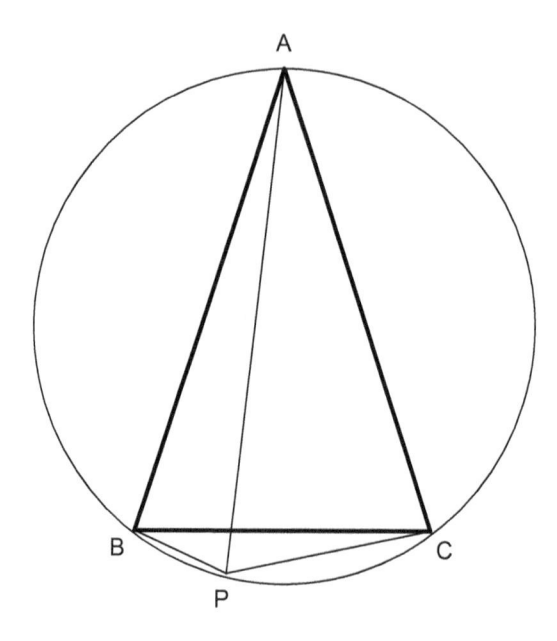

Figure 6-9

Proof

We apply Ptolemy's theorem to cyclic quadrilateral $ABPC$, shown in Figure 6-9, to get $(PA)(BC) = (PB)(AC) + (PC)(AB)$.

Since $AB = AC$, we have $(PA)(BC) = (AC)(PB + PC)$, and it follows that $\frac{PA}{PB+PC} = \frac{AC}{BC}$.

Ptolemy's Theorem Application 5

In Figure 6-10, equilateral $\triangle ABC$ is inscribed in a circle, and a point P is on $\overparen{BC}$. We find that $PA = PB + PC$.

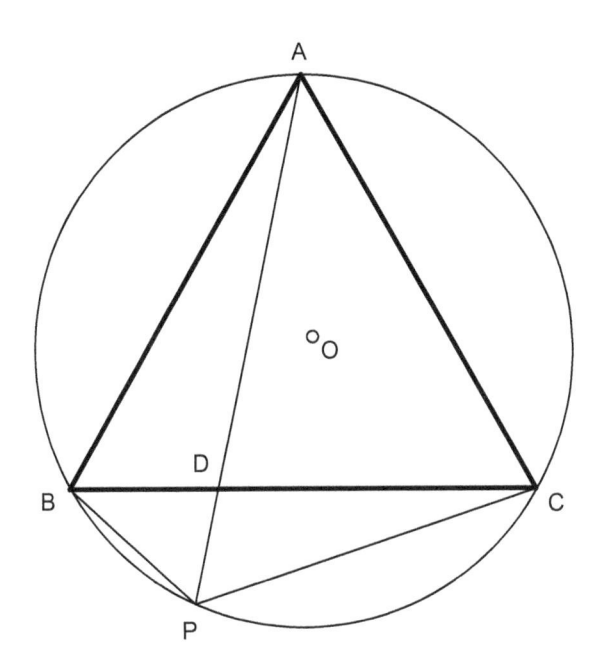

Figure 6-10

Proof

Since quadrilateral $ABPC$ is cyclic, as shown in Figure 6-10, we apply Ptolemy's theorem to the cyclic quadrilateral to get $(PA)(BC) = (PB)(AC) + (PC)(AB)$. However, because $\triangle ABC$ is equilateral, $BC = AC = AB$.

Therefore, from the above equation, we find that $PA = PB + PC$.

Ptolemy's Theorem Application 6

If square *ABCD,* shown in Figure 6-11, is inscribed in a circle, and a point *P* is on $\widehat{BC}$, we find that $\frac{PA+PC}{PB+PD} = \frac{PD}{PA}$.

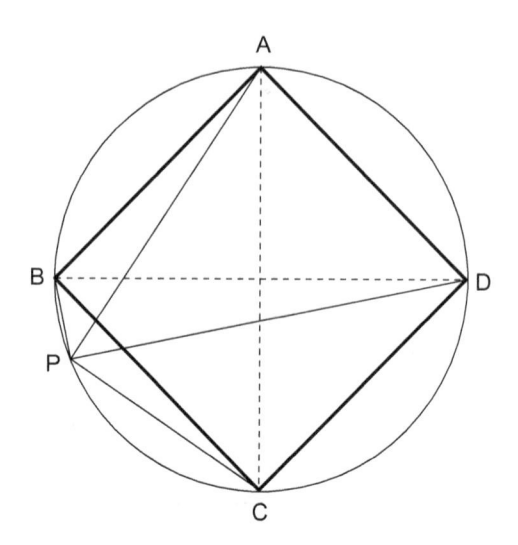

Figure 6-11

Proof

In Figure 6-11, consider isosceles $\triangle ABD$, where $AB = AD$.
Using the results of Application 4, we have

$$\frac{PA}{PB+PD} = \frac{AD}{DB}. \quad \text{(I)}$$

Similarly, for isosceles $\triangle ADC$, we find that

$$\frac{PD}{PA+PC} = \frac{DC}{AC}. \quad \text{(II)}$$

Since $AD = DC$ and $DB = AC$, we get

$$\frac{AD}{DB} = \frac{DC}{AC}. \quad \text{(III)}$$

Therefore, from (I), (II), and (III), we have

$$\frac{PA}{PB+PD} = \frac{PD}{PA+PC}, \quad \text{or} \quad \frac{PA+PC}{PB+PD} = \frac{PD}{PA}.$$

Ptolemy's Theorem Application 7

If regular pentagon ABCDE, shown in Figure 6-12, is inscribed in a circle, and point P is on $\widehat{BC}$, we find a surprising result, namely, that $PA + PD = PB + PC + PE$.

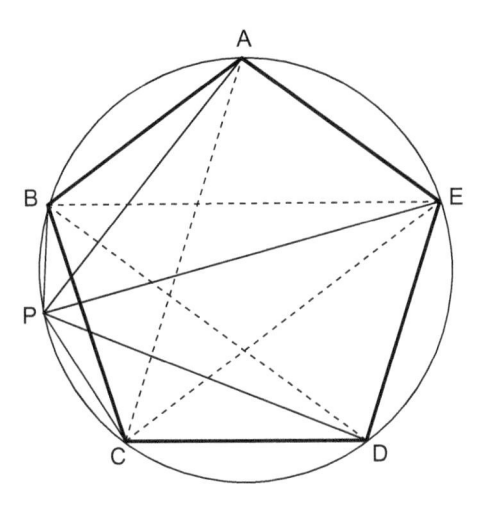

Figure 6-12

Proof

In Figure 6-12, applying Ptolemy's theorem to quadrilateral ABPC, we get

$$(PA)(BC) = (BA)(PC) + (PB)(AC), \qquad (I)$$

When Ptolemy's theorem is applied to quadrilateral BPCD, we get

$$(PD)(BC) = (CD)(PB) + (PC)(BD). \qquad (II)$$

Because $BA = CD$ and $AC = BD$, by adding equations (I) and (II), we obtain

$$(BC)(PA + PD) = (BA)(PB + PC) + (AC)(PB + PC). \qquad (III)$$

However, since $\triangle BEC$ is isosceles, and using results of Application 4, we get

$$\frac{CE}{BC} = \frac{PE}{PB+PC}, \text{ or } \frac{(PE)(BC)}{PB+PC} = CE = AC. \tag{IV}$$

Substituting (IV) into (III),

$$BC(PA+PD) = (BA)(PB+PC) + \frac{(PE)(BC)}{PB+PC}(PB+PC).$$

But since $BC = BA$, we can establish our claimed result

$$PA + PD = PB + PC + PE.$$

Ptolemy's Theorem Application 8

If a regular hexagon *ABCDEF*, shown in Figure 6-13, is inscribed in a circle, and point *P* is on $\overarc{BC}$, the surprising result is that

$$PE + PF = PA + PB + PC + PD.$$

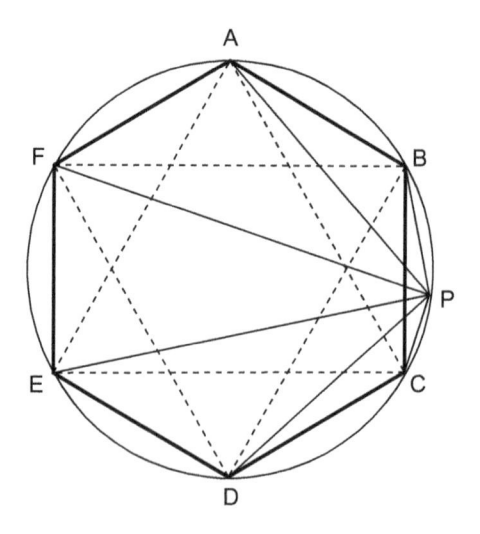

Figure 6-13

Proof

We begin by drawing lines between points *A*, *E*, and *C* to make equilateral $\triangle AEC$, which is shown in Figure 6-13.

Using the results from Application 5, we have

$$PE = PA + PC. \quad \text{(I)}$$

In the same way, in equilateral $\triangle BFD$, we have

$$PF = PB + PD. \quad \text{(II)}$$

By adding equations (I) and (II), we get $PE + PF = PA + PB + PC + PD$.

Ptolemy's Theorem Application 9

In Figure 6-14, triangle $ABC(C')$ is inscribed in a circle with radius $AO = 5$ and has sides $AB = 5$ and $AC = AC' = 6$. We seek the measure of the third side of triangle ABC.

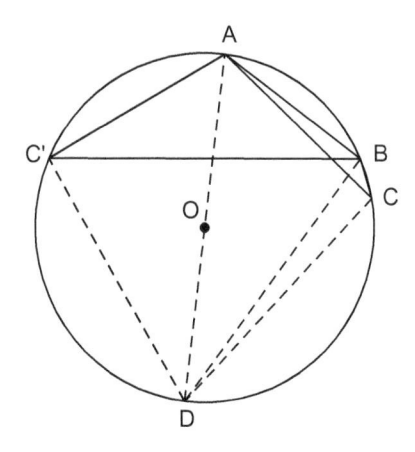

Figure 6-14

Solution: In Figure 6-14, we notice that there are two cases to consider regarding this problem. Both $\triangle ABC$ and $\triangle ABC'$ are inscribed in circle O, with $AB = 5$ and $AC = AC' = 6$. We want to find BC and BC'. We begin by drawing diameter $AOD = 10$, and then draw DC, DB, and DC'. We then have angles inscribed in a semicircle so that $\angle AC'D = \angle ACD = \angle ABD = 90°$. We now consider the case where $\angle A$ of $\triangle ABC$ is acute. We apply the Pythagorean theorem to right $\triangle ACD$ to get $DC = 8$, and to right $\triangle ABD$ to get $BD = 5\sqrt{3}$. By Ptolemy's theorem applied to quadrilateral $ABCD$, we get $(AC)(BD) = (AB)(DC) + (AD)(BC)$, or $(6)(5\sqrt{3}) = (5)(8) + (10)(BC)$, and $BC = 3\sqrt{3} - 4$.

Now consider the case where $\angle A$ is obtuse, as in $\triangle ABC'$. In right $\triangle AC'D$, we have $DC' = 8$. We apply Ptolemy's theorem to quadrilateral $ABDC'$ and get $(AC')(BD) + (AB)(DC') = (AD)(BC')$, which is $(6)(5\sqrt{3}) = (5)(8) = (10)(BC')$, and $BC' = 3\sqrt{3} + 4$.

Four Concyclic Points

Finding four or more points that all lie on the same circle is always noteworthy in geometry. The extension of an altitude of a triangle can have a surprising result. In Figure 6-15, where H is the orthocenter (the point of intersection of the altitudes) of triangle ABC, we extend AE to point K so that $HE = KE$. We will show that point K is a fourth point on the circumscribed circle of triangle ABC, thus determining, in a rather unusual way, four points that lie on the same circle, namely, A, B, K, and C. Two more points can be added to the circle by extending the other altitudes similarly.

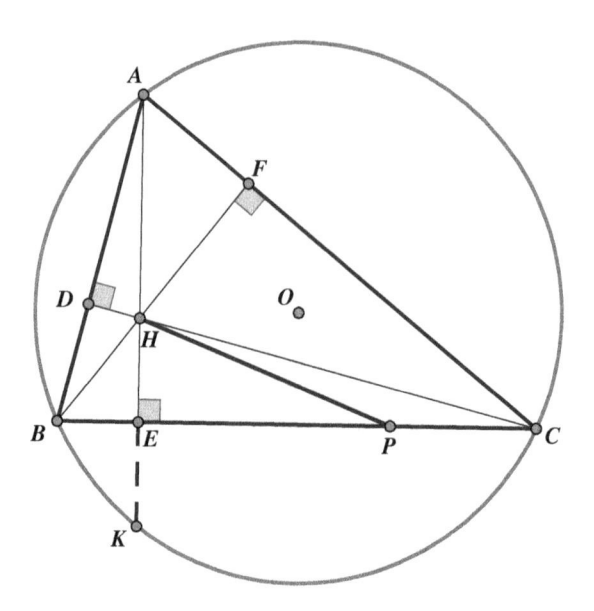

Figure 6-15

Proof

Our task is to prove that these four points actually lie on the same circle with center O. We begin by noting point K is the second intersection of line AHE and the circle with center P on side BC, shown in Figure 6-16. But how far must AHE be extended so that K is on circle O? The equal vertical angles AHF and BHE generate similar right triangles AFH and BEH. Therefore, $\angle HAF = \angle HBE$. We now select point K' on circle O so that $\angle K'BC = \angle HBC$. Therefore, $\angle K'BC = \angle CAK$, and since each is measured by one-half arc KC or $K'C$, we have $K' = K$, and $\angle KBC = \angle CAK$. We then have $\triangle HBE \cong \triangle BEK$. This gives us $HE = KE$, which tells us that extending HE its own length to point K places a point on circle O.

Point K can also be located by selecting any point P on side BC of triangle ABC, as we can see in Figure 6-16, where we then construct a

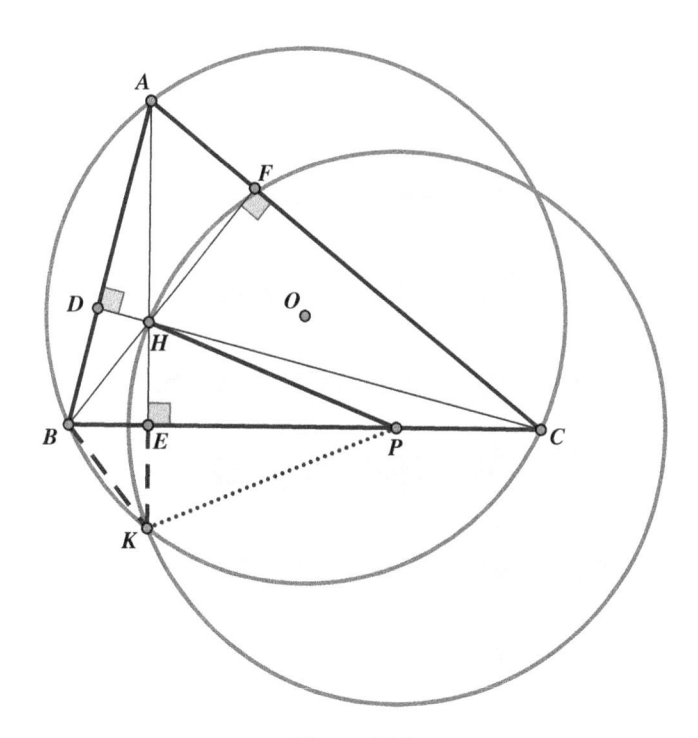

Figure 6-16

circle with center *P* and radius *HP*. This circle *P* intersects circle *O* at a point that we can see is point *K*, since *HE* = *KE*. We draw the line *PK* to create congruent right triangles △*HPE* ≅ △*EPK*, so that *PH* = *PK*. Since *PH* and *PK* are the equal radii of circle with center *P*, the intersection of the two circles is at point *K*. Therefore, we can locate a fourth point on circle *O* generated by random point *P* on side *BC* of triangle *ABC*.

Another Four Concyclic Points

Concyclic points tend to appear when least expected. Consider triangle *ABC* in Figure 6-17, with its circumscribed circle's center at point *O*. The points *E* and *F* are placed on sides *AB* and *AC* so that

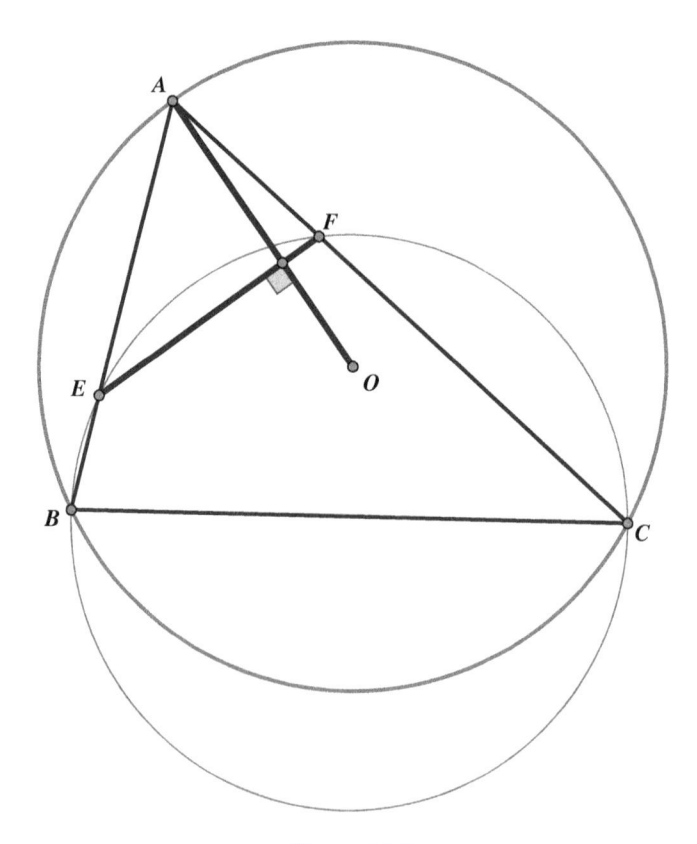

Figure 6-17

EF will be perpendicular to *AO*. Once that is set, we find that the points *B*, *E*, *F*, and *C* are concyclic, thus providing us with another four-point circle.

Proof

We begin by drawing *BO* and altitude *AG* of triangle *ABC*, as shown in Figure 6-18. We then notice that $\angle AOB = 2\angle C$, since both the inscribed angle in the central angle are measured by the same intercepted arc *AB*. In isosceles triangle *ABO*, we have $\angle BAO = \frac{1}{2}(180° - \angle AOB) = 90° - \angle C$. In right triangle *AGC*, we have $\angle GAC = 90° - \angle C$. Therefore, $\angle BAO = \angle GAC$. Thus, since $\angle AEH = 90° - \angle BOA$ and $\angle C = 90° - \angle GAC$, we have $\angle AEH = \angle C$. However, $\angle AEH$ is supplementary to $\angle BEF$.

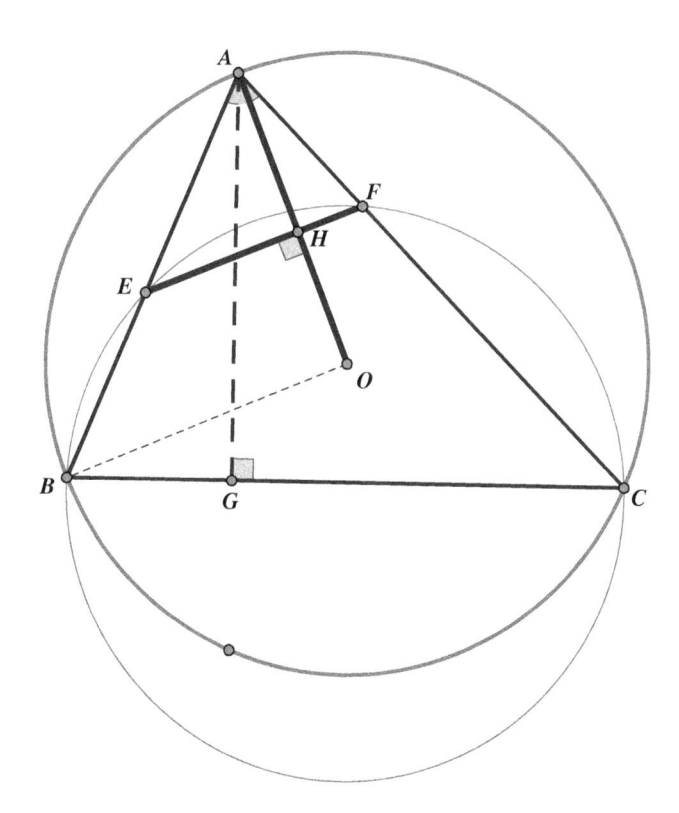

Figure 6-18

Therefore, ∠BEF is also supplementary to ∠C, and since the opposite angles of quadrilateral DEFC are supplementary, then DEFC must be a cyclic quadrilateral. Therefore, points B, E, F, and C lie on the same circle.

Four Other Unexpected Concyclic Points

To reitierate, finding four or more points that lie on the same circle is always a challenge in geometry. One such example is shown in Figure 6-19, where H is the orthocenter of triangle ABC and points M and N are the midpoints of sides AB and BC, respectively. We extend

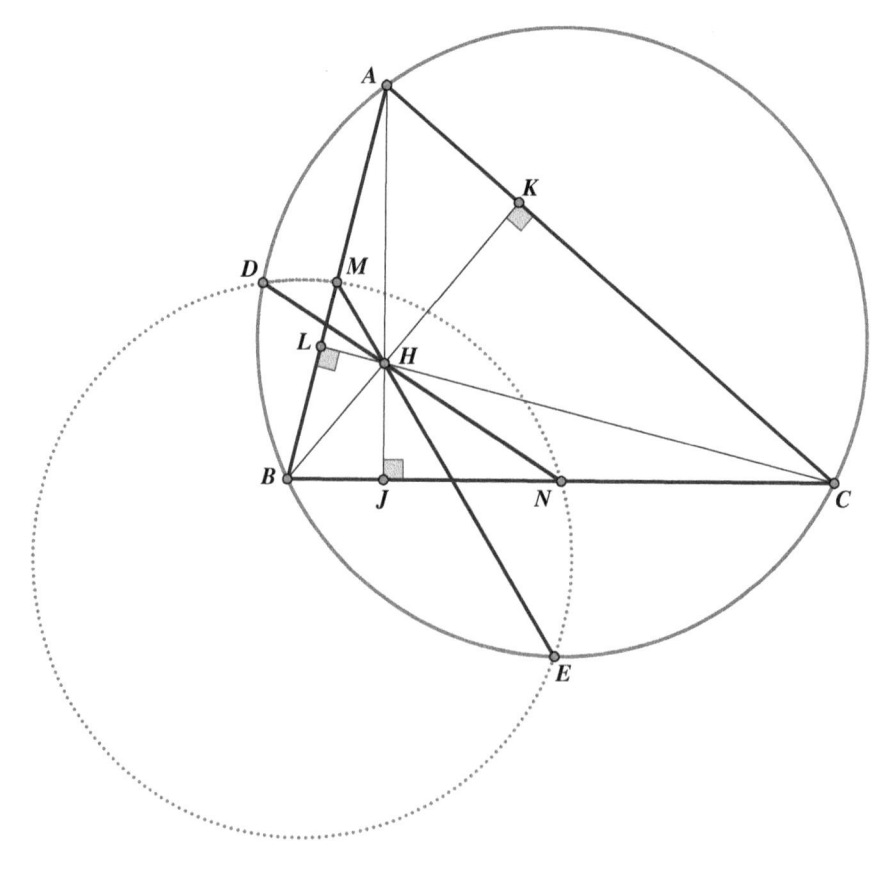

Figure 6-19

NH to intersect the circumscribed circle of triangle *ABC* at point *D*. We extend *MH* to intersect the circumscribed circle at point *E*. Astonishingly, the points *D*, *M*, *N*, and *E* lie on the same circle, making them concyclic.

Proof

When we construct the diameter of the circumcircle of triangle *ABC* from point *A*, it meets the circle at point *S*, as shown in Figure 6-20. Therefore, ∠*ACS* and ∠*ABS* are both right angles, since they are inscribed in semicircles. Thus, *BH* is parallel to *CS* and *BS* is parallel to

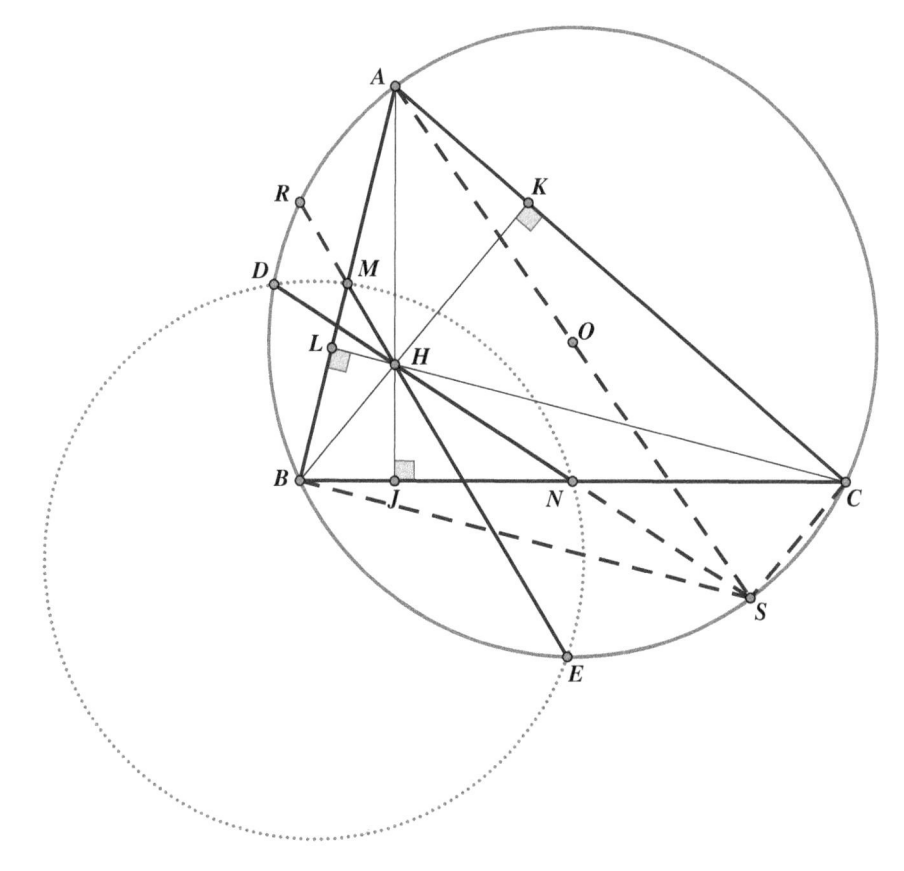

Figure 6-20

CH, which establishes *BHCS* as a parallelogram. We then state that *HN* and *NS* are portions of the diagonal of *HS*, which determines that points *D*, *H*, *N*, and *S* are collinear. Since the diagonals of a parallelogram bisect each other, *HN* = *SN*. As we have now established that the line segment from the orthocenter through the midpoint of a side to the circumscribed circle is bisected by the midpoint, we know that *MH* = *RM*. Using the well-known relationship of the equal products of the segments of intersecting chords of the circle, we have *EH*·*HR* = *SH*·*HD*. Since 2*MH* = *HR* and 2*NH* = *SH*, by substitution we have *EH* · 2*MH* = 2*NH* · *HD*, or *EH* · *MH* = *NH* · *HD*. This implies that the products of the two intersected segments are equal, which in turn would indicate that their endpoints *D*, *M*, *N*, and *E* lie on the same circle.

A Variety of Perpendiculars can Generate Concyclic Points

Yet another arrangement of perpendicular lines in a triangle will generate concyclic points. In Figure 6-21, right triangle *ABC*, with right angle at point *A*, has two randomly selected points *D* and *F* on sides *AB* and *AC*, respectively. We now construct perpendicular lines from

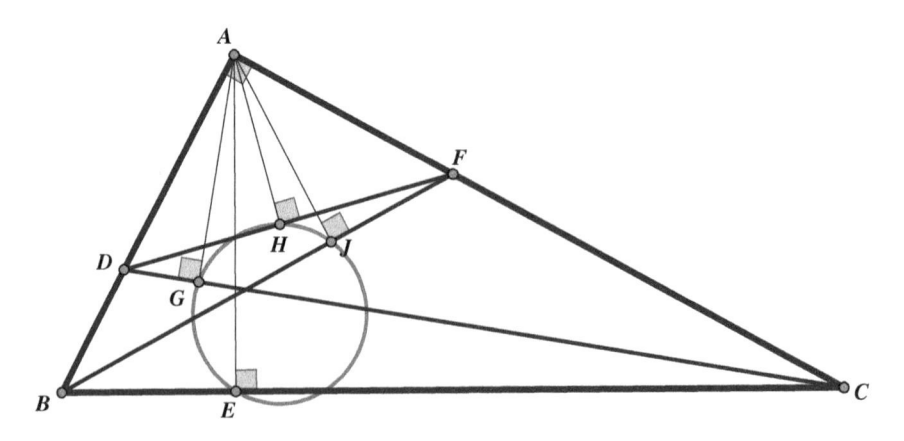

Figure 6-21

Discovering Concyclic Points **213**

point A to lines BC, DC, BF, and DF, intersecting these lines at points E, G, J, and H, respectively. Surprisingly, we find that these points E, G, H, and J all lie on the same circle.

Proof

In order to show that quadrilateral $EGHJ$ in Figure 6-22 is cyclic, and thereby establish the four concyclic points, we will show that $\angle EGH$ and $\angle EJH$ are supplementary. Triangle ADF is a right triangle, therefore, $\angle ADH + \angle AFD = 90°$. Since $\angle AHD$ and $\angle AGD$ are right angles, we have cyclic quadrilateral $ADGH$, with diameter AD. It follows that $\angle ADH = \angle AGH$, as they are both measured by one-half the measure of arc AH. Analogously, quadrilateral $AFJH$ is also cyclic, with diameter AF, and therefore, $\angle AFH = \angle AJH$, since they are both measured by arc AH. Thus, $\angle AGH + \angle AJH = 90°$. Another cyclic quadrilateral is $AGEC$, which analogously determines that $\angle EGC = \angle EAC$. Thus, since the right angle $\angle BAC$ can be expressed as $\angle BAE + \angle CAE = 90°$, we have $\angle EGC + \angle EJB = 90°$. Considering right angles AGB and AEB, we have yet another cyclic quadrilateral, $AJEB$, which gives us $\angle BJE = \angle BAE$. The sum of the two right angles $\angle AGC + \angle AJB = 180°$, from which we can subtract $\angle AGH + \angle AJH = 90°$ so that we now have $\angle HGC + \angle HJB = 90°$. By adding this last equation to the previously established

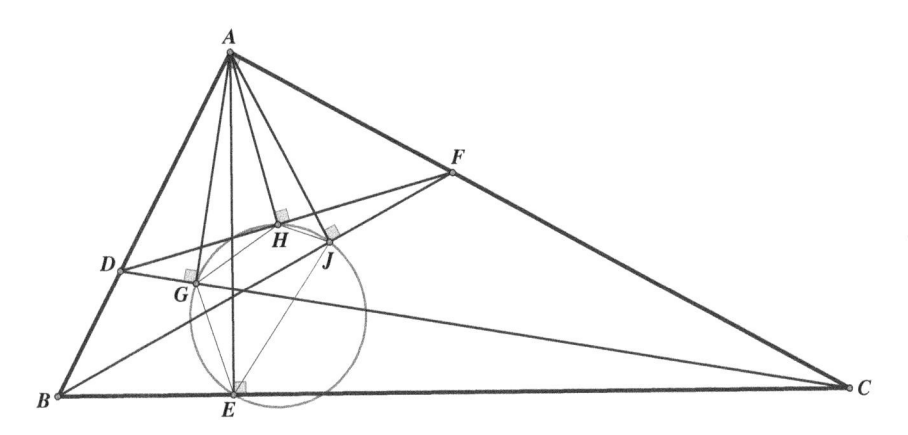

Figure 6-22

∠EGC + ∠EJB = 90°, we get ∠EGC + ∠HGC + ∠EJB + ∠HJB = 180°. Quadrilateral EGHJ is therefore cyclic because the opposite angles ∠EGH + ∠EJH = 180°, and so we have shown that the points E, G, H, and J all lie on the same circle.

Altitudes and Circles Generate Another Circle

Intersecting circles can also produce concyclic points, as shown in Figure 6-23, where sides AC and BC of triangle ABC are the diameters for circles with center at points M and N, respectively. Altitude AD extended intersects the circle with center N at points R and S. Altitude BE intersects the circle with center M at points H and K. These four intersection points S, H, R, and K all lie on the same circle and

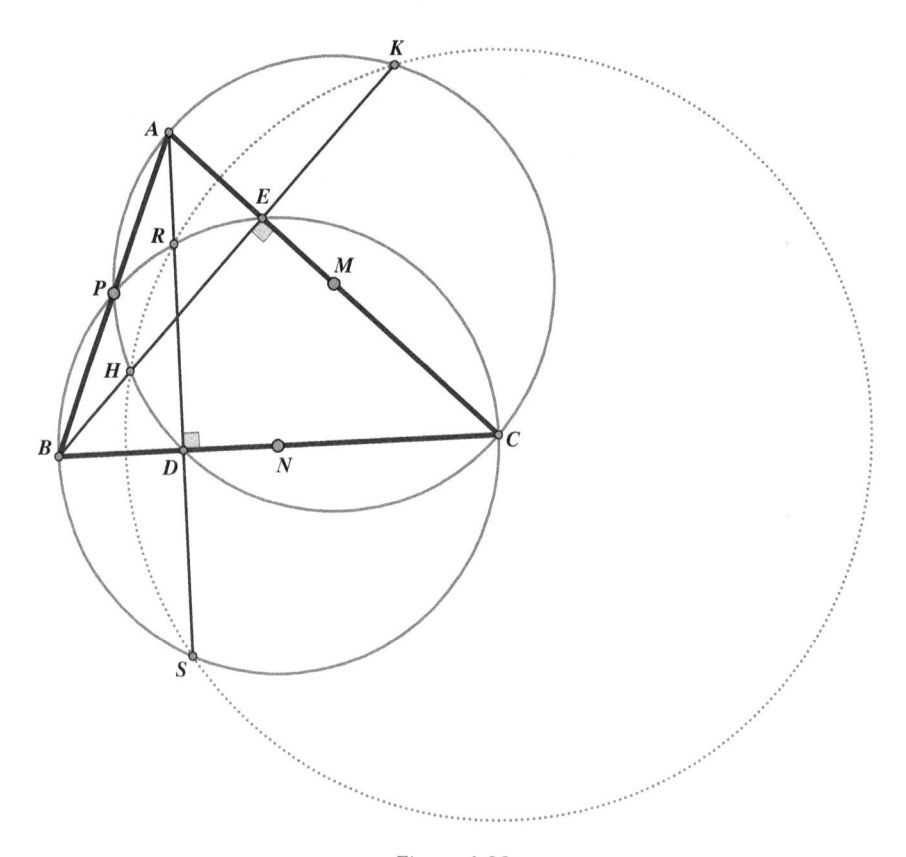

Figure 6-23

therefore are concyclic. It should also be noted that the two circles with centers at *M* and *N* intersect on side *AB* at point *P*.

Proof

The basis for proving this relationship is the fact that the product of the segments of intersecting chords are equal. In Figure 6-24, we have circle *M*, where $HL \cdot KL = AL \cdot LD$, and in circle *N*, we have $BL \cdot LE = AL \cdot LD$. Therefore, $HL \cdot KL = BL \cdot LE$. Yet, in circle *N*, we have $RL \cdot SL = BL \cdot LE$. Thus, $HL \cdot KL = RL \cdot SL$, which indicates that the points *S*, *H*, *R*, and *K* all lie on the same circle.

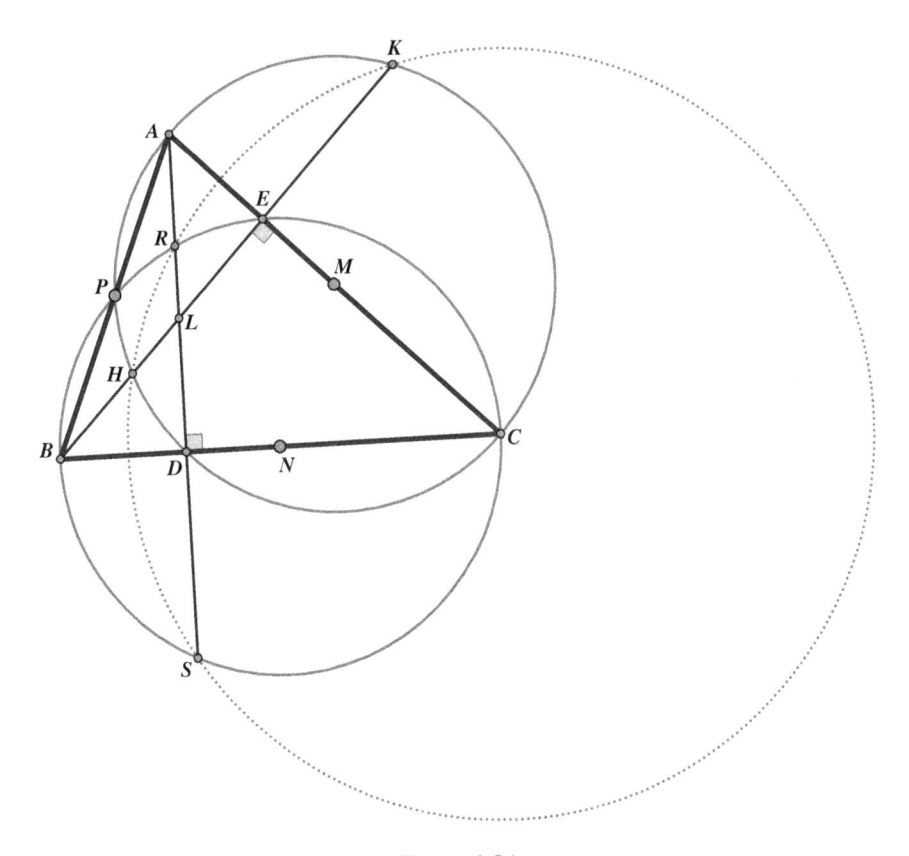

Figure 6-24

Unexpected Concyclic Points

In Figure 6-25, a circle is tangent to side AC of triangle ABC at point Q while intersecting side AB at points P and R so that line PQ is parallel to side BC. Quite unexpectedly, we find that points B, C, Q, and R are concyclic.

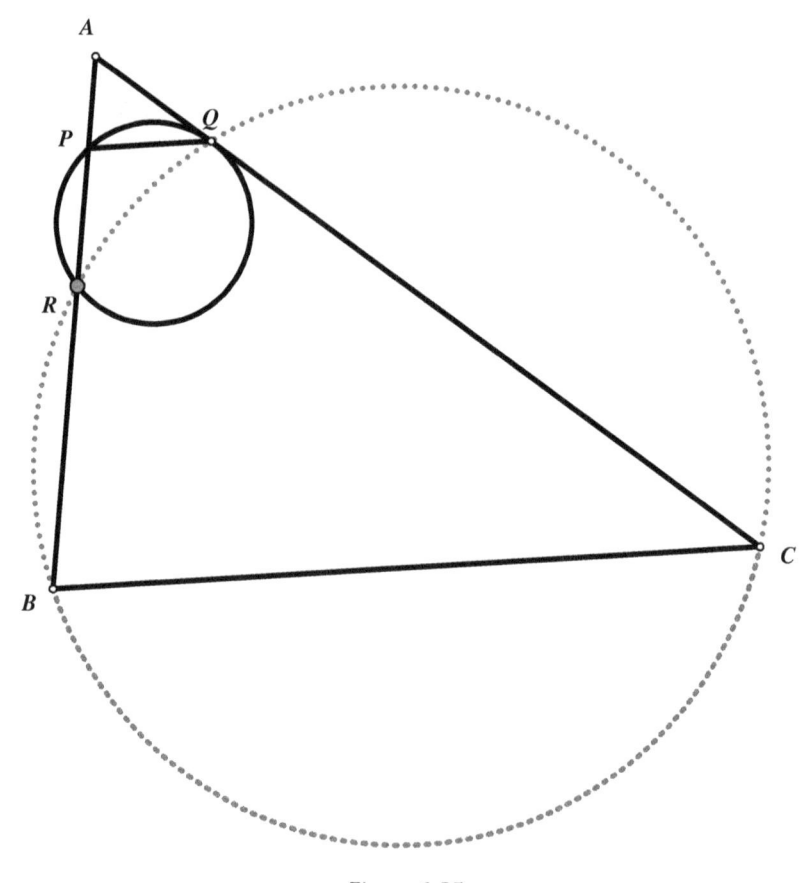

Figure 6-25

Proof

With PQ parallel to BC, in Figure 6-26, we have $\angle AQP = \angle C$. However, $\angle PQA = \frac{1}{2}\overset{\frown}{PQ} = \angle QRA$. Furthermore, $\angle QRA = 180° - \angle QRB$.

That implies that $\angle C = 180° - \angle QRB$. Therefore, the opposite angles of quadrilateral BCQR are supplementary so that the quadrilateral is cyclic.

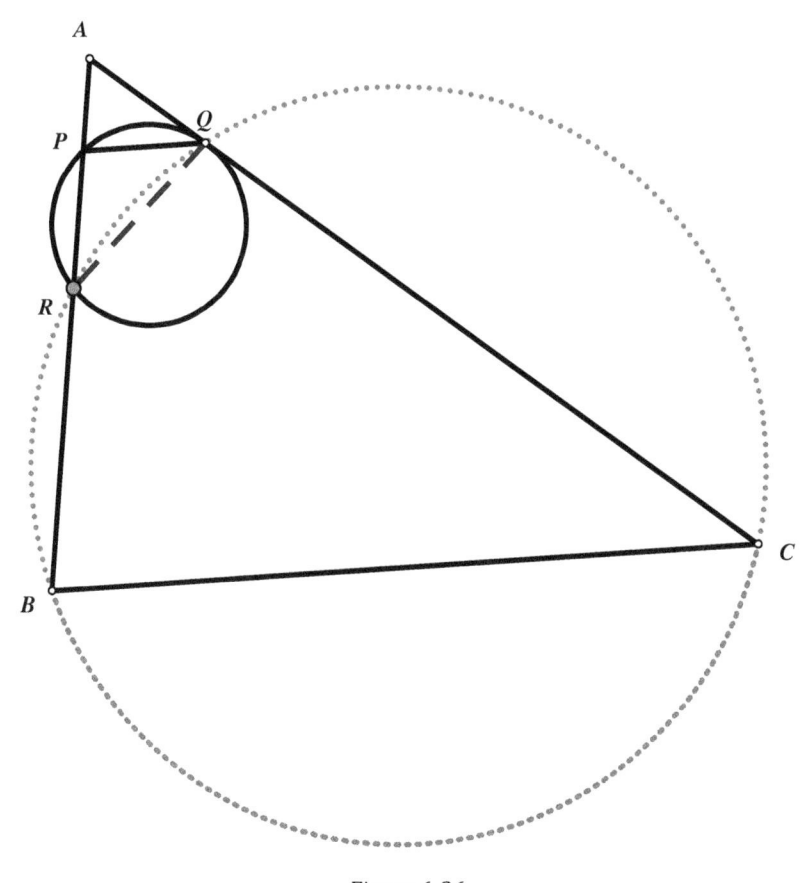

Figure 6-26

Unexpected Curious Concyclic Points

When chord *AB* of circle *O* is drawn as shown in Figure 6-27, the midpoint, *P*, of arc *AB* can produce four concyclic points. These are determined by two randomly drawn chords emanating from point *P*, which intersect *AB* at points *G* and *H*, and intersect circle *O* at points *C* and *D*, respectively. Regardless of where these chords emanating from point *P* are drawn to intersect chord *AB* and circle *O*, points *C*, *D*, *H*, and *G* will always be concyclic.

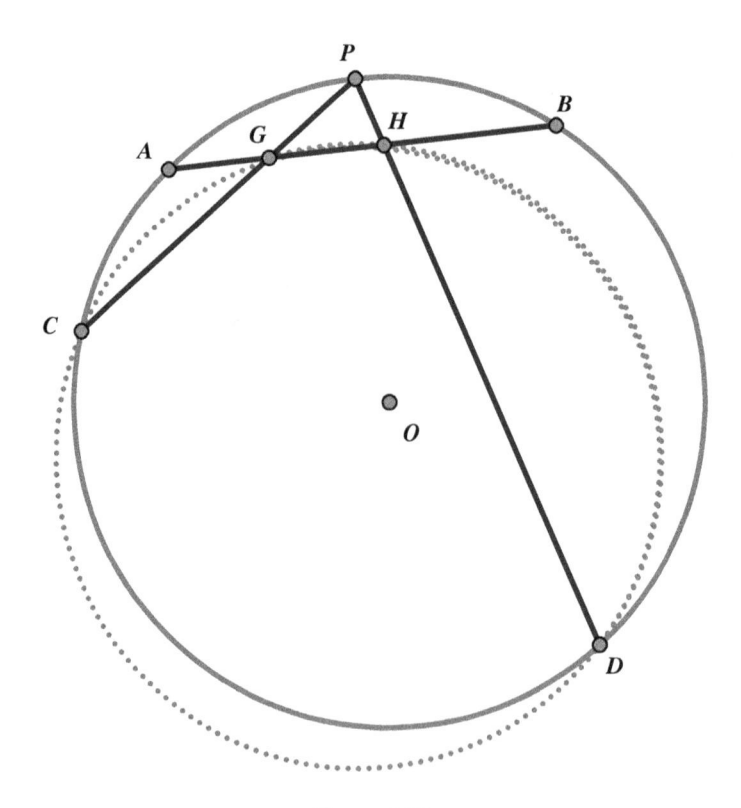

Figure 6-27

Proof

Begin by drawing chord CD, as shown in Figure 6-28. We know that $\overset{\frown}{AP}=\overset{\frown}{PB}$. An angle formed by two intersecting chords in a circle is equal to one-half the sum of the intercepted arcs. Therefore, $\angle CGB=\tfrac{1}{2}\left(\overset{\frown}{BDC}+\overset{\frown}{AP}\right)$ and $\angle CDP=\tfrac{1}{2}\overset{\frown}{PAC}$, and by addition $\angle CGB+\angle CDP=\tfrac{1}{2}\left(\overset{\frown}{BDC}+\overset{\frown}{AP}+\overset{\frown}{PAC}\right)=\tfrac{1}{2}\left(\overset{\frown}{BD}+\overset{\frown}{DC}+\overset{\frown}{AP}+\overset{\frown}{AP}+\overset{\frown}{AC}\right)=\tfrac{1}{2}\left(\overset{\frown}{BD}+\overset{\frown}{DC}+\overset{\frown}{PB}+\overset{\frown}{AP}+\overset{\frown}{AC}\right)$, since $\overset{\frown}{PA}=\overset{\frown}{PB}$. Thus, $\angle CGB+\angle CDP=\tfrac{1}{2}\left(\overset{\frown}{BD}+\overset{\frown}{DC}+\overset{\frown}{PB}+\overset{\frown}{AP}+\overset{\frown}{AC}\right)=\tfrac{1}{2}360°=180°$. When the opposite angles $\angle CGB+\angle CDP$ of a quadrilateral $DCGH$ are supplementary, the quadrilateral is cyclic. Thus, points C, D, H, and G are concyclic.

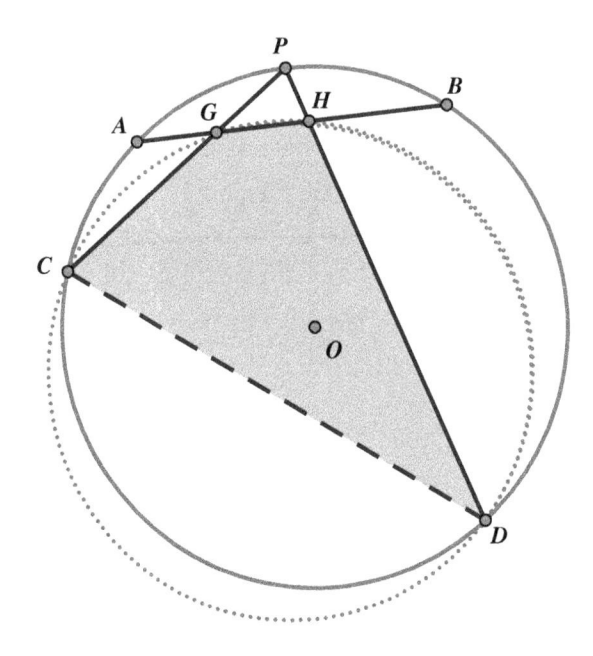

Figure 6-28

Unexpected Concyclic Points Produce Unexpected Segment Products

The acute triangle ABC, shown in Figure 6-29, has point H as the orthocenter, and M as the midpoint of side BC. The line HF is perpendicular to AM at point F. Unexpectedly, we get the following segment product: $AM \cdot FM = DM^2$.

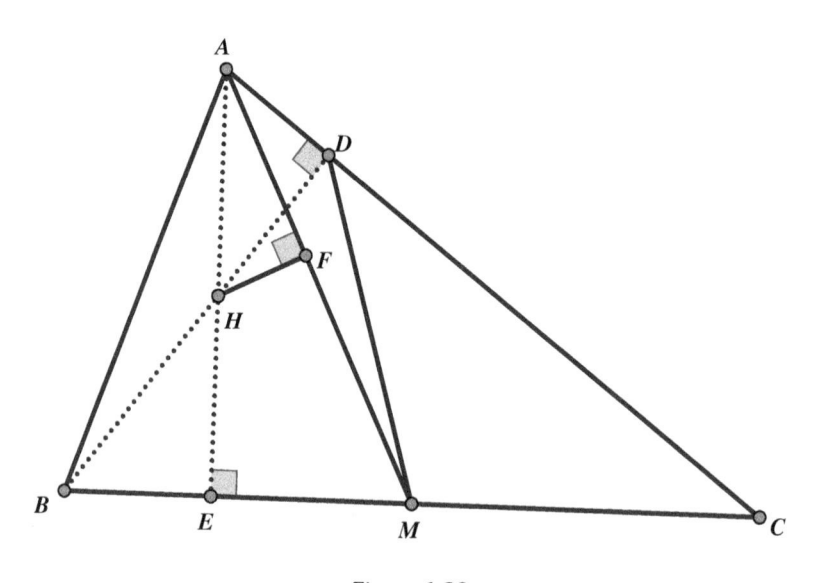

Figure 6-29

Proof

In Figure 6-30, the altitudes BD and AE of triangle ABC intersect at the orthocenter H. In triangle BDC, DM is the median to its hypotenuse BC, whereupon $DM = \frac{1}{2}BC = BM$. Because $\angle ADH = 90° = \angle AFH$, we can

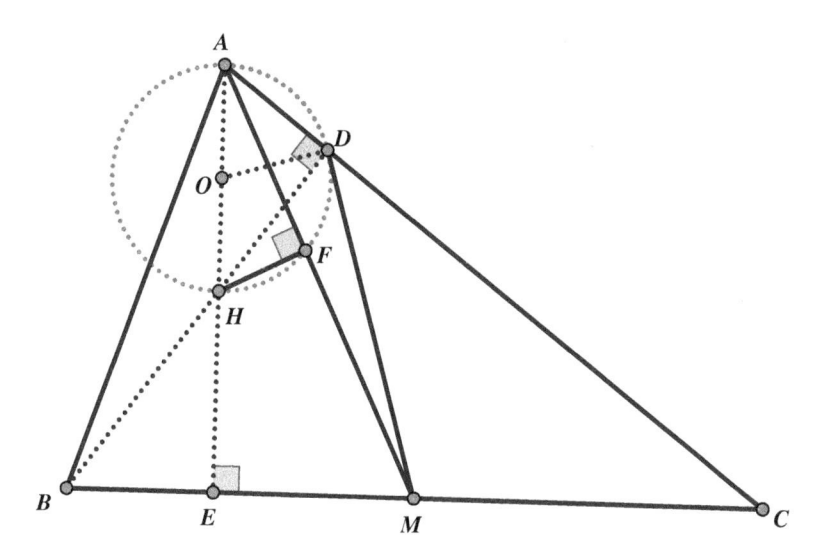

Figure 6-30

consider quadrilateral *ADFH* to be cyclic and the points *A*, *D*, *F*, and *H* thus lie on the same circle. Now, in triangle *AEC*, we find ∠*CAE* = 90° − ∠*C*. Furthermore, since right triangles *ADH* and *BEH* are similar, we have ∠*EAD* = ∠*DBC*, which is the same as ∠*CAE* = ∠*DBM* . We already know that *DM* = *BM*, and in isosceles triangle *DBM* we have ∠*BDM* = ∠*DBC*. Therefore, ∠*CAE* = ∠*BDM*. We also know that ∠*CAE* + ∠*DHA* = 90°, and in isosceles triangle *DOH*, we have ∠*HDO* = ∠*DHA*. Therefore, ∠*HDO* + ∠*BDM* = 90° so that *OD* ⊥ *DM*. Since in right triangle *ADH*, the hypotenuse *AH* is a diameter of the circle with center *O*, and since the radius *OD* is perpendicular to *DM* on circle *O*, we can conclude that *DM* is tangent to circle *O*, which, by the secant tangent relationship, gives us $AM \cdot FM = DM^2$.

Surprising Concyclic Points Generated by a Triangle and its Circumscribed Circle

As shown in Figure 6-31, point P is a randomly selected point in triangle ABC, where AP, BP, and CP intersect the circumscribed circle at points S, Q, and R, respectively. D is a point on AS and F is a point on BQ so that DF∥AB. Point E is on CR so that DE∥AC.

Quite unpredictably, we find that the points R, E, F, and Q are on the same circle, that is, they are concyclic points.

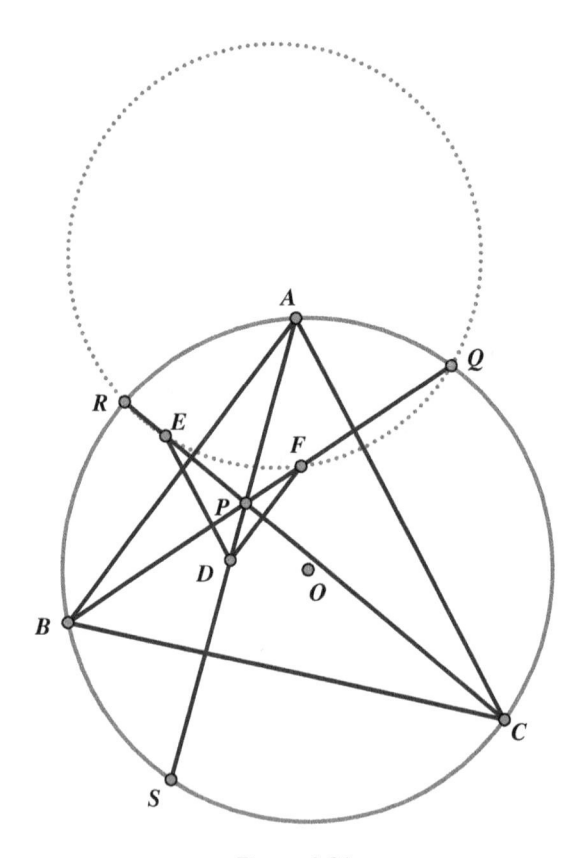

Figure 6-31

Proof

To begin, draw line *SQ*, as shown in Figure 6-32. Since *DF*∥*AB*, we have ∠*DFP* = ∠*ABQ* and ∠*ABQ* = ∠*ASQ* since both angles are one-half the measure of arc *AQ*. We then know that points *D*, *F*, *Q*, and *S* are concyclic points, since ∠*DFQ* is supplementary to ∠*DFP*, thus making ∠*DFQ* supplementary to ∠*ASQ*. As we have secants to this cyclic quadrilateral (note that *P* must lie outside the circumcircle of *DFQS*), we get *PD·PS* = *PF·PQ*. Analogously, we can show that the points *D*, *E*, *R*, and *S* are also concyclic, and *PD·PS* = *PE·PR*. Therefore, *PF·PQ* = *PE·PR*, which allows us to conclude that the points *F*, *E*, *R*, and *Q* all lie on the same circle.

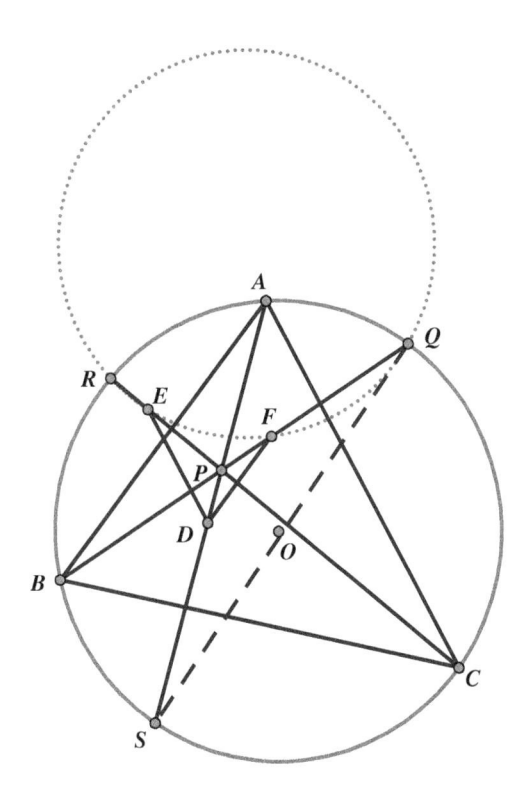

Figure 6-32

Three Related Circle Centers and Their Common Intersection Point are Concyclic

In Figure 6-33, point P lies above three collinear points A, B, and C and determines three triangles PAB, PBC, and PAC, whose circumscribed circle centers are points R, S, and T. From this relatively simple configuration, we find that these three centers are concyclic with the point P, thereby providing us with four concyclic points.

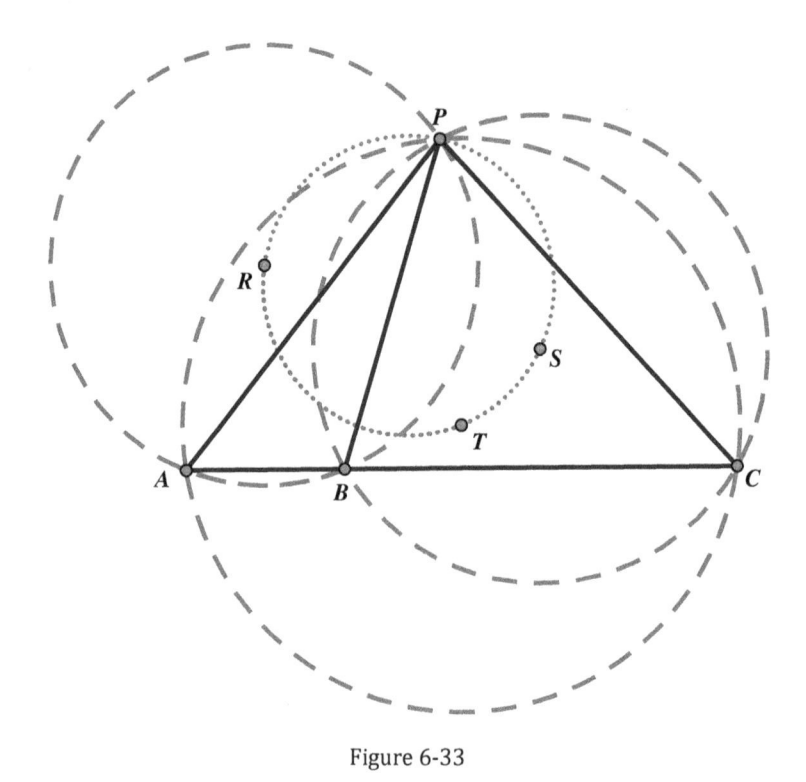

Figure 6-33

Proof

The goal here is to show that quadrilateral PRTS in Figure 6-34 is a cyclic quadrilateral. We need to show that one pair of opposite angles are supplementary. In the circle with center T, radius $TR \perp PA$ and bisects PA at point J. Therefore, $\triangle RAJ \cong \triangle RPJ$ so that $\angle PRT = \frac{1}{2}\angle PRA$. Similarly, $TSQ \perp PC$, and we have $\angle PSQ = \frac{1}{2}\angle PSC$. Furthermore, in circle with center S, we have $\angle PSC = 2\angle PBC$, as they are both measured by arc PC. Considering cyclic quadrilateral $PUAB$, where U is on the circumcircle of triangle ABP, we find that $\angle PBC = \angle PUA$ since both angles are supplementary to $\angle ABP$. In the circle with center R, $\angle PUA = \frac{1}{2}\angle PRA$, as they are both measured by arc PBA. Therefore, $\angle PRT = \angle PSQ$, and since $\angle PST$ is supplementary to $\angle PSQ$, we can conclude that the angles $\angle PRT$ and $\angle PST$ are supplementary. Therefore, quadrilateral $PRTS$ is a cyclic quadrilateral, and the points $P, R, T,$ and S are concyclic.

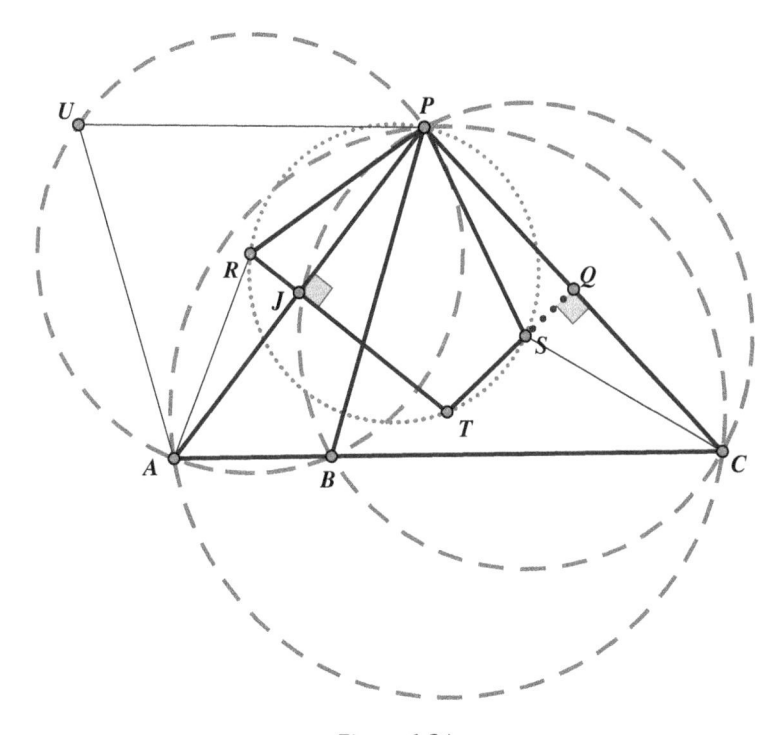

Figure 6-34

A Quadrilateral with Two Intersecting Circles Generates Another Circle with Four Points

In Figure 6-35, quadrilateral *ABCD* is partitioned into two triangles *ABC* and *ADC*, each of which is circumscribed by a circle. The quadrilateral is such that the circumscribed circle about triangle *ABC* intersects *AD* and *DC* in points *F* and *E*, respectively. This quadrilateral also enables the circumscribed circle about triangle *ADC* to intersect *AB* and *BC* in points *P* and *Q*, respectively. Lines *BF* and *BE* intersect *PQ* at points *R* and *S*, respectively. Quite astonishingly, the points *E*, *F*, *R*, and *S* turn out to be concyclic.

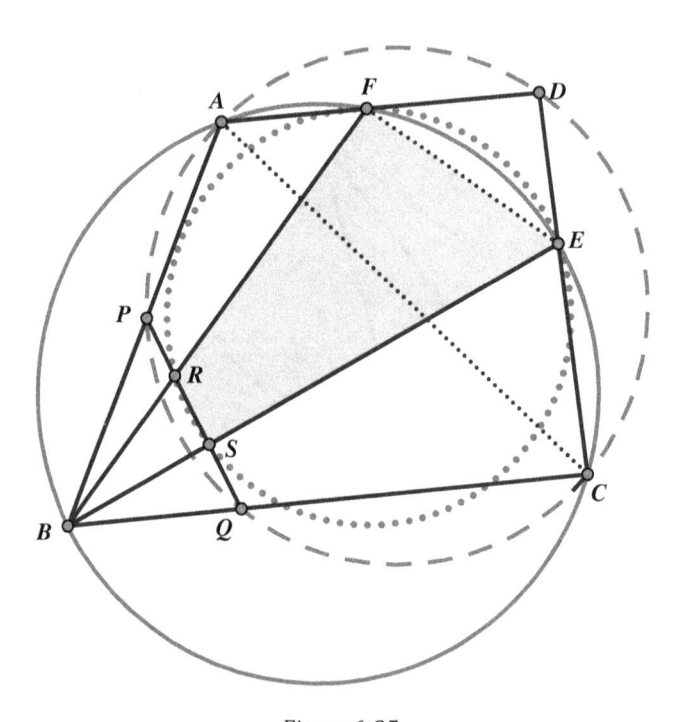

Figure 6-35

Proof

Be cautioned that not all quadrilaterals allow circle intersections as we show for quadrilateral *ABCD* in Figure 6-36. We notice that points *A*, *B*, *E*, and *F* all lie on the circumscribed circle about triangle *ABC*. This enables us to establish the following. Since the opposite angles of quadrilateral *AFEB* are supplementary, we have $\angle FEB = 180° - \angle FAB$. In triangle *ABF*, we have $\angle FBA + \angle AFB = 180° - \angle FAB$. Since points *A*, *P*, *Q*, and *C* are concyclic, $\angle APQ + \angle ACQ = 180°$ as we have $\angle APQ + \angle BPQ = 180°$. Therefore, $\angle ACQ = \angle BPQ$. In the circumscribed circle of triangle *ADC*, we have $\angle ACQ = \angle AFB$ since they are both one-half the measure of arc *APQ*. Thus, we now have $\angle FEB = \angle FBA + \angle BPQ = \angle PRF$, which is the exterior angle of triangle *BPR*. In other words, $\angle SEF = \angle PRF$, which then enables us to show $\angle FRS + \angle SEF = 180°$. We can conclude that the points *F*, *E*, *S*, and *R* are concyclic, since the opposite angles of quadrilateral *FESR* are supplementary.

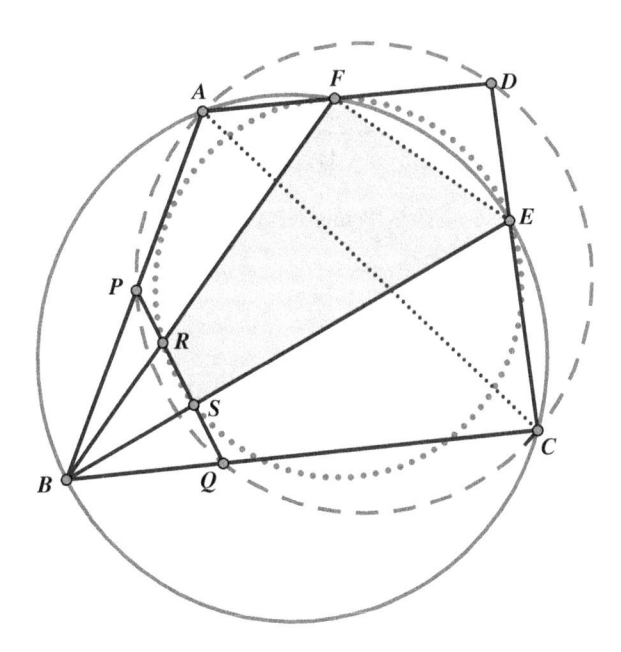

Figure 6-36

A Surprising Five-Point Circle

Having five points on the same circle is quite astonishing. We have such a curiosity in Figure 6-37, where the altitudes *AE*, *BF*, and *CD* of $\triangle ABC$ intersect at the orthocenter *H*. The points *P*, *Q*, and *M* are the midpoints of *BF*, *CD*, and *BC*, respectively. Surprisingly, we find that the points *M*, *E*, *Q*, *H*, and *P* all lie on the same circle, thus providing a five-point circle. As an extra marvel in this configuration, we find that $\triangle EPQ$ is similar to $\triangle ABC$.

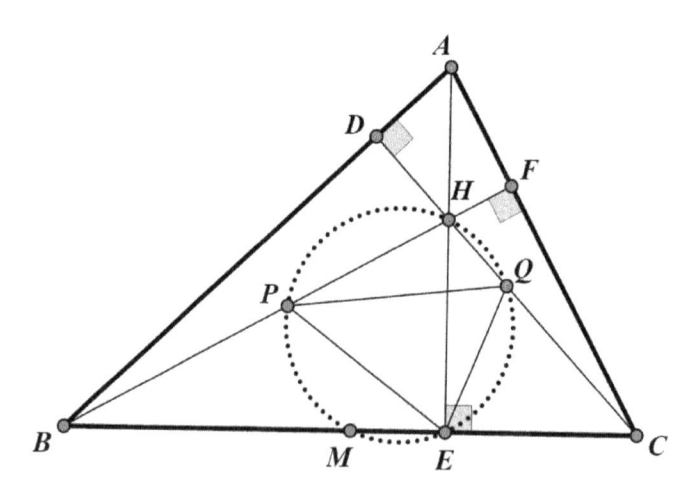

Figure 6-37

Proof

We begin by drawing lines *PM*, *QM*, and *HM*, as we show in Figure 6-38. The circumscribed circle around triangle *HPM* has its diameter *HM*, since $\angle HPM = 90°$. Since *M* is the midpoint of *BC* and *P* is the midpoint of *BF* in triangle *BFC*, the line segment joining the midpoints of these

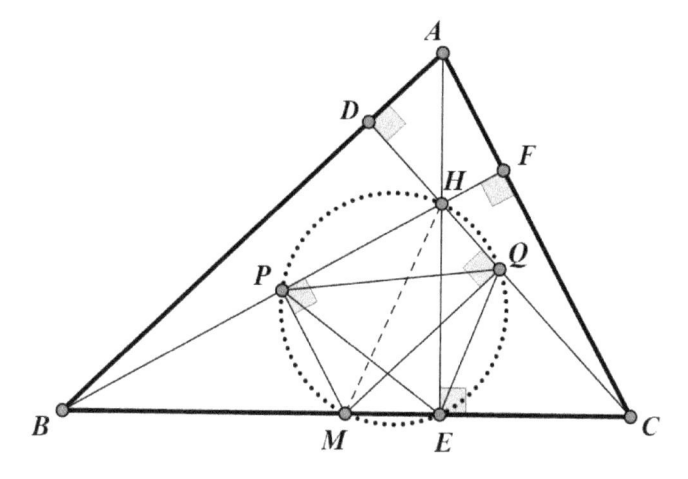

Figure 6-38

two sides of a triangle is parallel to the third side so that *MP* is parallel to *CF*. Since ∠*BFC* = 90°, then ∠*FPM* = 90°. Since ∠*HEM* = 90°, we now have points *E*, *M*, *P*, and *H* on the circle with diameter *MH*. An analogous argument can be made for point *Q*, since *QM* is parallel to *AB*. Consequently, ∠*HQM* = 90° so that this right triangle is also inscribed in the circle with diameter *MH*. We therefore have the five points *E*, *M*, *P*, *H*, and *Q* all on the same circle.

To address the second wonder in this configuration, we will now prove that triangles *EPQ* and *ABC* are similar. We know that the triangles *ECH* and *BCD* are right triangles sharing the angle at *C*. Thus,

$$\angle ABC = \angle DBC = \angle EHC = \angle EHQ,$$

and by the inscribed angle theorem, this equals ∠*EPQ*, so altogether ∠*ABC* = ∠*EPQ*. In an analogous fashion, we obtain ∠*EQP* = ∠*ACB*, whereupon triangles *EPQ* and *ABC* are similar.

Intersecting Circles Create Five Concyclic Points

Getting more than four points on one circle is always a delightful challenge. Consider circles with centers O and Q intersecting at points P and R, as shown in Figure 6-39. Extend QP to intersect circle O at point A, and then extend OP to intersect circle Q at point B. Quite astonishingly, we find that the points A, O, R, Q, and B are all on the same circle.

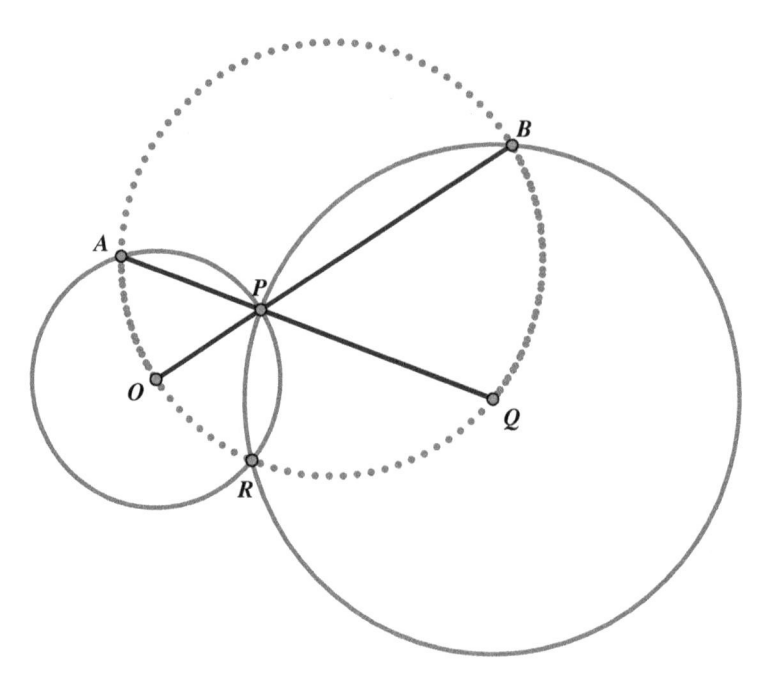

Figure 6-39

Proof

As shown in Figure 6-40, the four radii determine two isosceles triangles, $\triangle APO$ and $\triangle BPQ$, which provide equal angles $\angle OAP = \angle OPA$ and $\angle QBP = \angle QPB$. With the vertical angles $\angle OPA = \angle QPB$, we get $\angle OAP = \angle QBP$, which is also $\angle OAQ = \angle QBO$. Since these two angles are measured by one-half $\overarc{ORQ}$, we know that quadrilateral $OABQ$ must be cyclic, and the points O, A, B, and Q therefore all lie on the same circle. We now need to show that point R also lies on that circle. Drawing QO creates two congruent triangles, $\triangle OPQ \cong \triangle ORQ$, and therefore, $\angle ORQ = \angle OPQ$. Considering isosceles triangle OAP with $\angle OAP = \angle OPA$, then $\angle OPQ = 180° - \angle OPA = 180° - \angle OAP$. Thus the supplementary angles ($\angle OAQ$ and $\angle ORQ$) in quadrilateral $AORQ$ make it a cyclic quadrilateral, so that points O, A, Q, and R are concyclic. Hence, we have the points O, A, B, Q, and R all on the same circle.

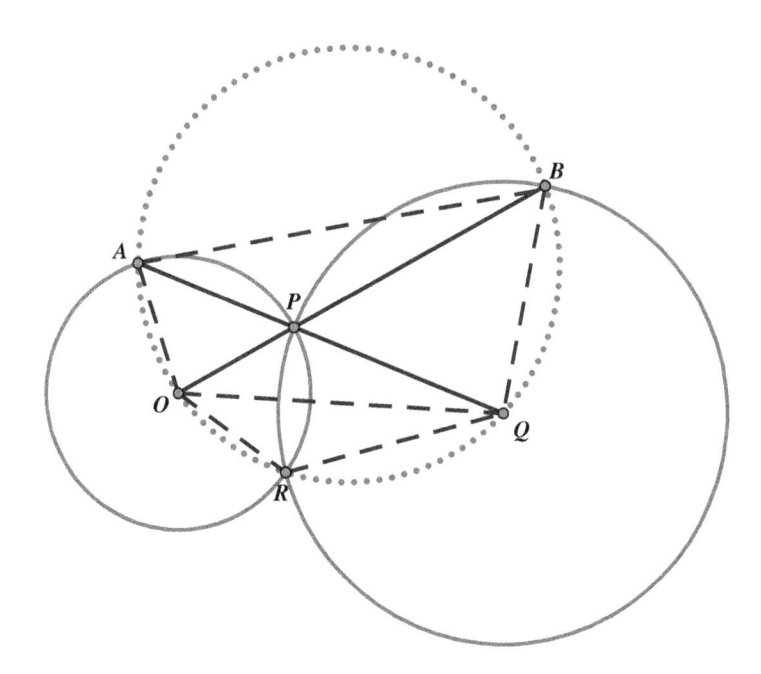

Figure 6-40

The Euler Six-Point Circle

One of the most famous theorems in plane geometry is known as the nine-point circle, which, as the name indicates, establishes a circle that has nine specific triangle-related points on it. In 1765, before the discovery of the nine-point circle, the famous Swiss mathematician Leonhard Euler (1707–1783) discovered that there are six triangle-related points that lie on a circle. These six points are the midpoints of the sides of triangle ABC, which in Figure 6-41 are the points M, N, and K, as well as the feet of the three altitudes, points D, E, and F. This is already quite an amazing discovery – but there is more to come!

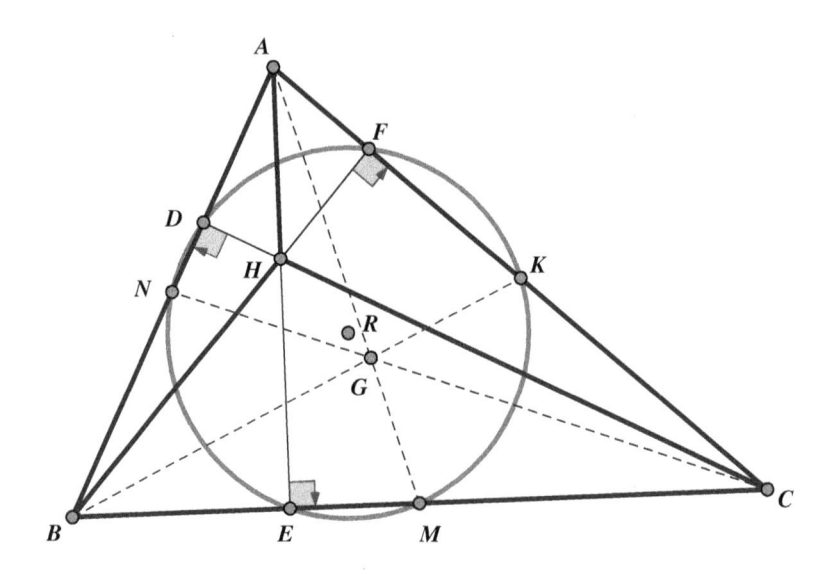

Figure 6-41

The Famous Nine-Point Circle

In 1822, the German mathematician Karl Wilhelm Feuerbach (1800–1834) discovered that another three points also lie on the previously presented six-point Euler circle, thus making a nine-point circle. In Figure 6-42, we see that these additional three points, X, Y, and Z, are the midpoints of the segments AH, BH, and CH, each of

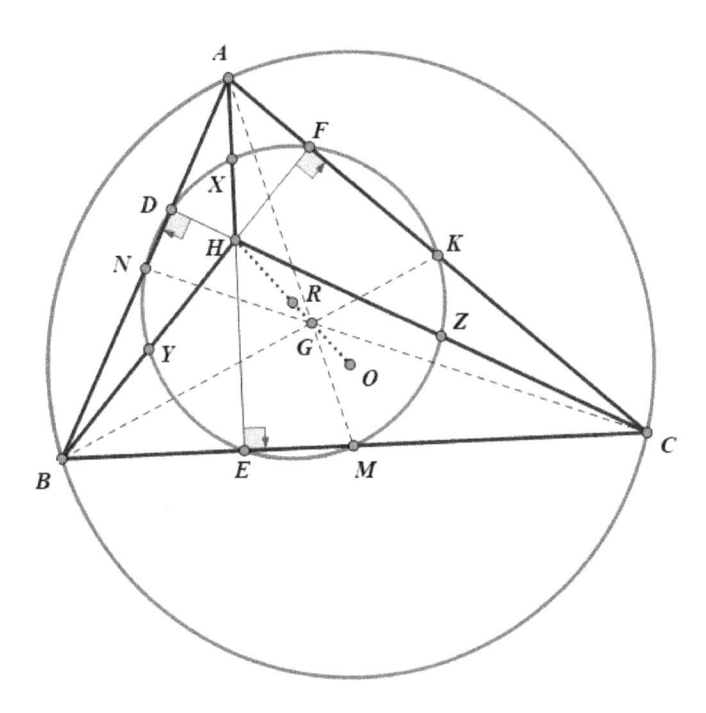

Figure 6-42

234 *A Journey Through the Wonders of Plane Geometry*

which joins a vertex with the orthocenter of the triangle *ABC*. These nine concyclic points are:

- The feet of the altitudes: *D, E,* and *F*
- The midpoints of the sides: *M, N,* and *K*
- The midpoints of the segments joining the vertex with the orthocenter: *X, Y,* and *Z*

If this weren't enough, another spectacular result, shown in Figure 6-44, is that the following four points are collinear:

- The center of the circumscribed circle, *O*
- The center of the nine-point circle, *R*
- The orthocenter, *H*
- The centroid, *G*

Proof

To simplify this proof, we shall consider each part with a separate diagram. Bear in mind that each of the Figures from 6-43 to 6-46 is extracted from Figure 6-42, which is the complete diagram of the nine-point circle.

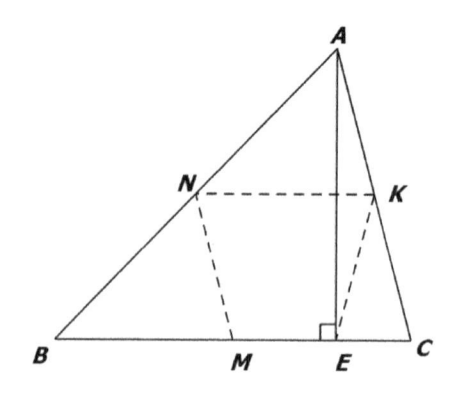

Figure 6-43

Discovering Concyclic Points **235**

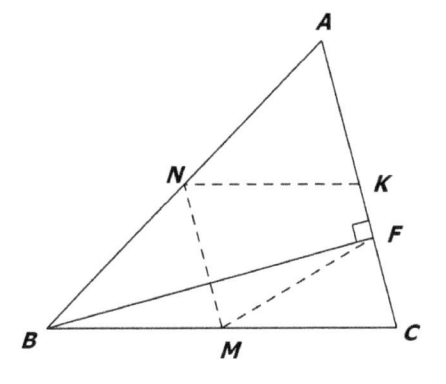

Figure 6-44

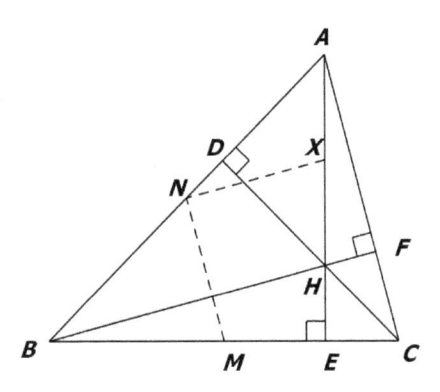

Figure 6-45

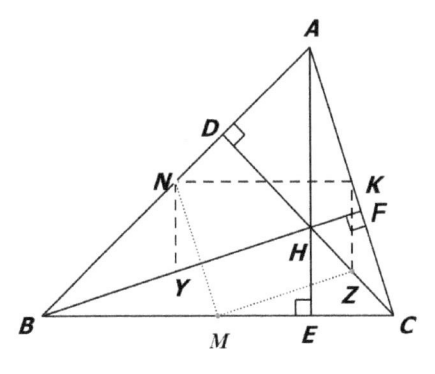

Figure 6-46

In Figure 6-43, points *K*, *N*, and *M* are the midpoints of the three sides of △*ABC* opposite the respective vertex. We have *AE* as an altitude of △*ABC*. Since *NK* is a midline of △*ABC*, *BC*||*NK*. Therefore, quadrilateral *KNME* is a trapezoid. We also have *MN* as a midline of △*ABC* so that $MN = \frac{1}{2}AC$. Since *KE* is the median to the hypotenuse of right △*ACE*, $KE = \frac{1}{2}AC$. Therefore, *MN* = *KE*, and trapezoid *KNME* is isosceles.

When the opposite angles of a quadrilateral are supplementary, as in the case of an isosceles trapezoid, the quadrilateral is cyclic. Therefore, quadrilateral *KNME* is cyclic.

So far, we have four of the nine points we seek on one circle.

To avoid any confusion, we redraw △*ABC* as shown in Figure 6-44 and include altitude *BF*. Using the same argument as before, we find that quadrilateral *KNMF* is an isosceles trapezoid and therefore is a cyclic quadrilateral. We now have five of the nine points on one circle (they are points *K*, *N*, *M*, *F*, and *E*). By repeating the same argument for altitude *CD*, we can then state that points *D*, *E*, and *F* lie on the same circle as points *K*, *N*, and *M*. These six points are as far as Euler got with this configuration.

With *H* as the orthocenter (the point of intersection of the altitudes), *X* is the midpoint of *AH*, as shown in Figure 6-45. Therefore, *NX* is a midline of △*ABH* and parallel to *BH* and altitude *BF*. Since *NM* is a midline of △*ABC*, *NM*||*AC*. Therefore, since ∠*AFB* is a right angle, ∠*MNX* is also a right angle. Thus, quadrilateral *XNME* is cyclic (recall that the opposite angles are supplementary). This then places point *X* on the circle determined by points *N*, *M*, and *E*. We now have a seven-point circle.

We repeat this procedure with point *Z*, the midpoint of *CH*, as shown in Figure 6-46. As before, ∠*NMZ* is a right angle. Therefore, points *N*, *M*, *Z*, and *D* are concyclic (opposite angles supplementary). We now have *Z* as an additional point on our circle, making it an eight-point circle.

To locate our final point on the circle, consider point *Y*, the midpoint of *BH*. As we did earlier, we find ∠*KNY* to be a right angle. Therefore, quadrilateral *KNYF* is cyclic, and point *Y* is on the same circle as points *N*, *K*, and *F*. We have therefore proved that nine specific points lie on this circle. We can see the nine-point circle in Figure 6-47, where the nine points *M*, *E*, *Y*, *N*, *D*, *X*, *F*, *K*, and *Z* are concyclic.

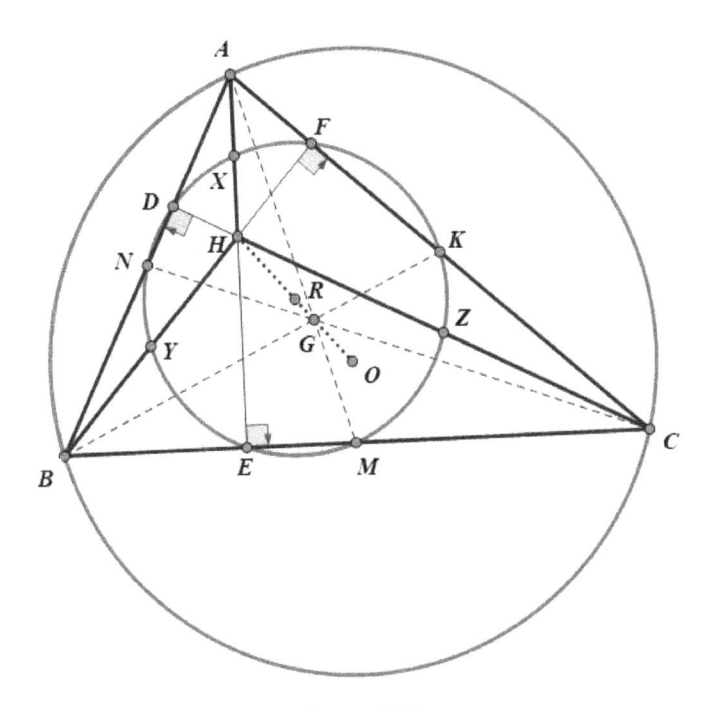

Figure 6-47

A Property of the Center *R* of the Nine-Point Circle

Not only does the nine-point circle have astonishing properties, its center *R* has a very special position, namely, it is the midpoint of the line segment *HO*, joining the orthocenter *H* and the circumcenter *O*, shown in Figure 6-48. It also follows that *R* lies on the *Euler line* (see Chapter 3).

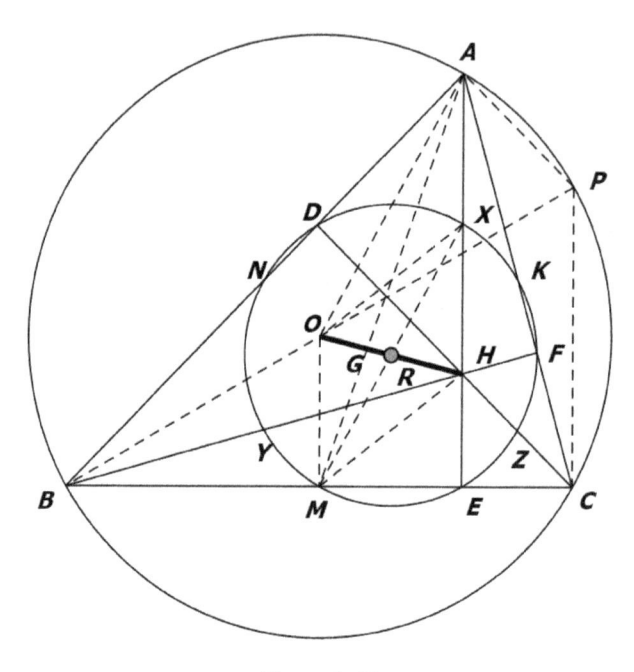

Figure 6-48

Proof

Since *XM* subtends a right angle at point *E*, it must be a diameter of the nine-point circle. Therefore, the midpoint, *R*, of *XM* is the center of the nine-point circle, as shown in Figure 6-48.

We now draw *BO* to intersect circumcircle *O* at point *P*. Then we draw *PC* and *PA* so that *OM* is a midline of △*BPC* and *OM*∥*PC*. Since ∠*BCP* is inscribed in a semicircle, it is a right angle. Now both *PC* and *AE* are perpendicular to *BC*, so that *PC*∥*AE*. Similarly, *AP*∥*CD*.

We therefore have parallelogram $APCH$ so that $PC = AH$. Furthermore, since OM is the midline of $\triangle BPC$, we have $OM = \frac{1}{2}PC$. Therefore, $OM = \frac{1}{2}AH = XH$, and, thus, $OMHX$ is a parallelogram with one pair of sides both congruent and parallel. Finally, since the diagonals of a parallelogram bisect each other, the midpoint R of MX is also the midpoint of OH.

Another Curious Collinearity with the Center of the Nine-Point Circle

What makes this nine-point circle even more special is that it generates collinearities such those shown in Figure 6-49, where for triangle ABC, the following three points S, R, and M are collinear:

- The point S is the intersection point of the perpendicular from the orthocenter H to the angle bisector AT of angle $\angle BAC$.
- Point R is the center of the nine-point circle.
- Point K is the midpoint of the opposite side, BC.

Thus, this amazing concyclical arrangement of nine points also exposes a collinearity of relatively unrelated points, namely, points S, R, and M.

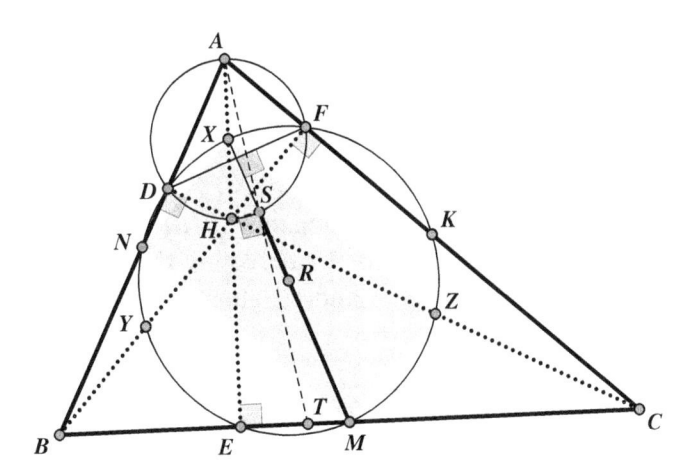

Figure 6-49

Proof

In Figure 6-49, we begin by recalling that XM is a diameter of the nine-point circle, and point R is therefore the midpoint of XM. This we have proved above already but here we will present another possible proof. It results from the fact that $XNMZ$ is a rectangle, and the diagonal XM is a diameter of its circumcircle, which also happens to be the nine-point circle. In order to show this, we note that N and M are the midpoints of BA and BC, respectively, which means that NM is parallel to AC, and NM is half as long as AC. Similarly, since Z and X are the midpoints of HC and HA, respectively, ZX is also parallel to AC and half as long as AC. This results in recognizing $XNMZ$ as a parallelogram. In addition to this, since N and X are the midpoints of AB and AH, respectively, NX is parallel to BH. Since $BH \perp AC$, we also have $NX \perp ZX$, and parallelogram $XNMZ$ is therefore a rectangle.

Now that we have (again) established that XM is a diameter of the nine-point circle, and X, R and M are thus collinear, we wish to show that S lies on this diameter.

From the definition of the orthocenter H in ABC, and the fact that point S is the foot of the perpendicular from the orthocenter H to the bisector AT of $\angle BAC$, we have $DH \perp DA$, $FH \perp FA$, and $SH \perp SA$. This means that points D, F, and S all lie on the circle with diameter AH. Since X is the midpoint of AH, it is the center of this circle with radius XH. Since AT is the angle bisector of $\angle BAC$, we have $\angle DAS = \angle BAT = \angle TAC = \angle SAF$, and thus, $SD = SF$. This means that $XDSF$ is a kite, since we have both $XD = XF$ and $SD = SF$, and its diagonals are therefore perpendicular, giving us $XS \perp DF$. Since F and D are both points of the nine-point circle, we also have $RD = RF$, and $XDRF$ is therefore also a kite, since we have both $XD = XF$ and $RD = RF$. Its diagonals are also perpendicular, giving us $XR \perp DF$, which means that X, S, and R are collinear. Since we already know that X, R, and M are collinear, this is also true for S, R, and M, which is what we had originaslly set out to prove.

A Further Unexpected Collinearity

The feet of the altitudes of a triangle can sometimes lead to a most unexpected result, as we will see in Figure 6-50, where a circle is drawn containing vertex A and the feet of altitudes CD and BE of triangle ABC. A circle that goes through points D, A, and E must also contain point H, since quadrilateral $ADHE$ must be cyclic as it has one pair of supplementary opposite angles ($\angle ADH = \angle AEH = 90°$). However, the amazing surprise is that the tangents to the circle at points E and D will intersect at point F, which is on side BC, regardless of the shape of the original triangle. We can then say that the points B, F, and C are collinear. Furthermore, we also unexpectedly find that point F is always the midpoint of BC.

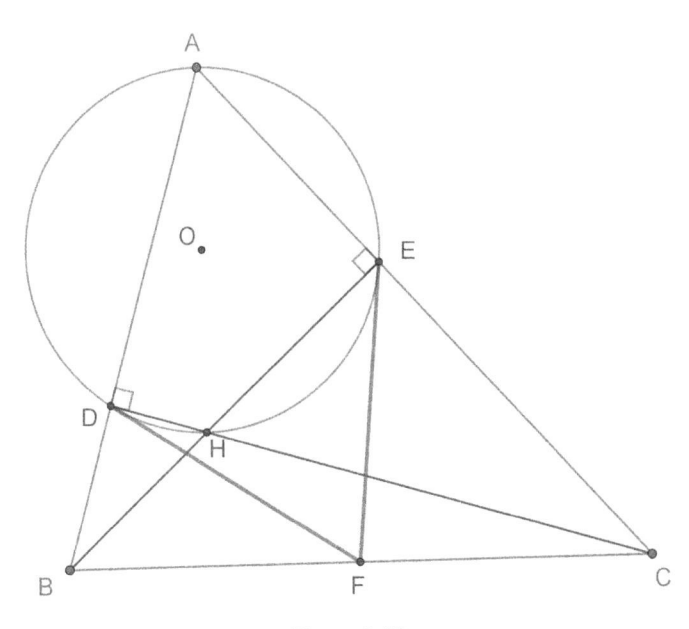

Figure 6-50

Proof

In Figure 6-51, point H must be the orthocenter of $\triangle ABC$ and point O is the midpoint of AH. Thus, the unique circle passing through D, O, and E must be the nine-point circle of $\triangle ABC$. Let F be the midpoint of BC. We know that the nine-point circle passes also through the midpoints of the triangle sides and that OF is a diameter of the nine-point circle. Hence, since an angle inscribed in a semicircle is a right angle, we have $\angle OEF = 90° = \angle ODF$, which determines that DF and EF are tangents to the circle through A, D, and E. This was our claim and completes the proof.

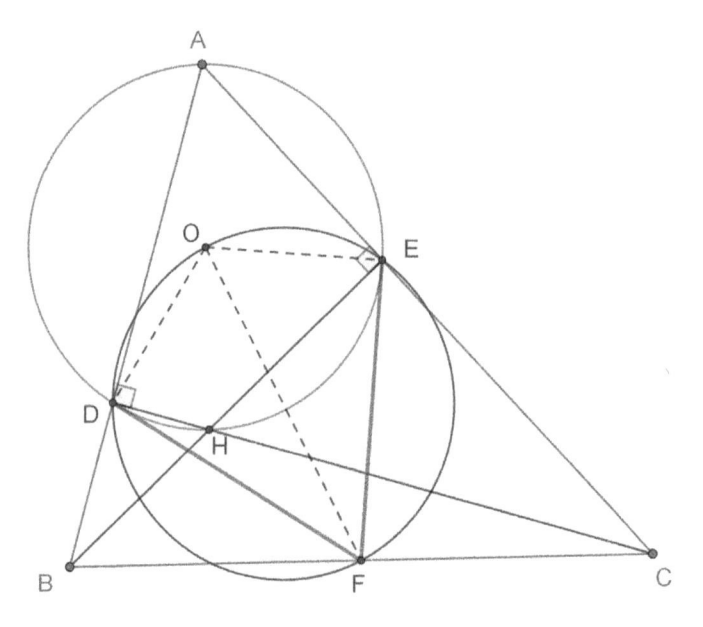

Figure 6-51

Final Interesting Properties of the Nine-Point Circle

Finally, we should consider also other spectacular features that the nine-point circle provides for us. Any line segment from the orthocenter to the circumscribed circle will be bisected by the nine-point circle. Suppose we take a random line segment *HJ*, in Figure 6-52, which joins the orthocenter of triangle *ABC* with a point on the circumscribed circle. The point *L* is where it intersects the nine-point circle. We find that *HJ* is bisected at point *L*, so that *HL* = *LJ*. Furthermore, the radius of the circumcircle is twice the radius of the nine-point circle.

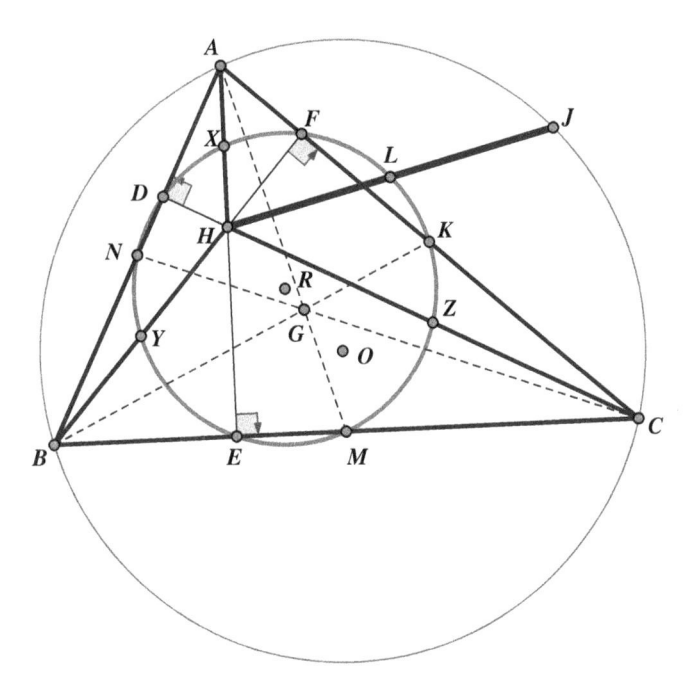

Figure 6-52

Proof

From above, we know that R is the midpoint of HO. Now we consider the concept of *homothety*, which will be more precisely dealt with in Chapter 11. The homothety centered at H with factor 2 maps R to O and (more importantly) every circle centered at R to a circle centered at O with the radius doubled. The homothety also maps $X \mapsto A$, $Y \mapsto B$, $Z \mapsto C$; thus, the image of the nine-point circle under this homothety must be the circumcircle of $\triangle ABC$. As a bonus of these considerations, we verify our second claim: The radius of the circumcircle is twice the radius of the nine-point circle, another beautiful result that shows the rich relationships of the nine-point circle.

Chapter 7

Circle Wonders

We begin our journey into the realm of circles by using our skill with linear figures to introduce the wonders that circles provide. Two intersecting equal circles can generate an equilateral triangle when each contains the other circle's center. In Figure 7-1, the centers O and Q of the two circles are on the other circle, so that triangle ODQ is equilateral. Yet, the amazing additional aspect that these two equal intersection circles offer is as follows.

Two Circles Generating an Unexpected Equilateral Triangle

To construct another equilateral triangle here, we draw Figure 7-1. Let point C be an arbitrary point on the smaller $\overset{\frown}{AD}$ of circle Q, and point B be the intersection point of DC with circle O. By drawing AB and AC, we have created equilateral triangle ABC.

Proof

We begin the proof by drawing radii OD, OA, QA, and QD in Figure 7-2. These radii generate two equilateral triangles: $\triangle DOQ$ and $\triangle AOQ$. Therefore, $\overset{\frown}{AQD} = \angle AOD = 120°$. In circle O, $\angle ABD = \frac{1}{2}\overset{\frown}{ADO} = 60°$.

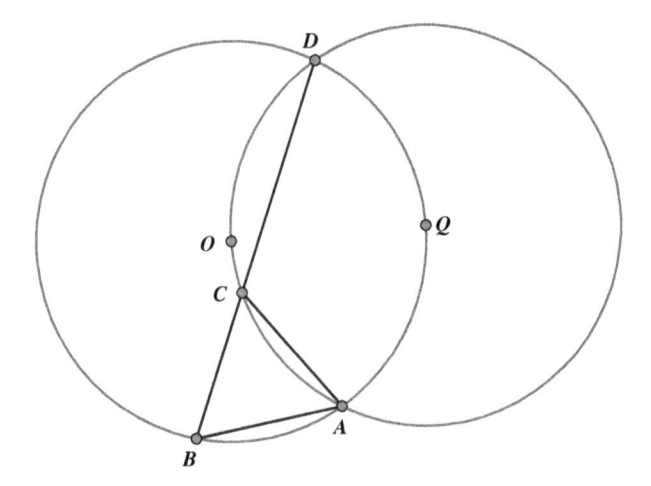

Figure 7-1

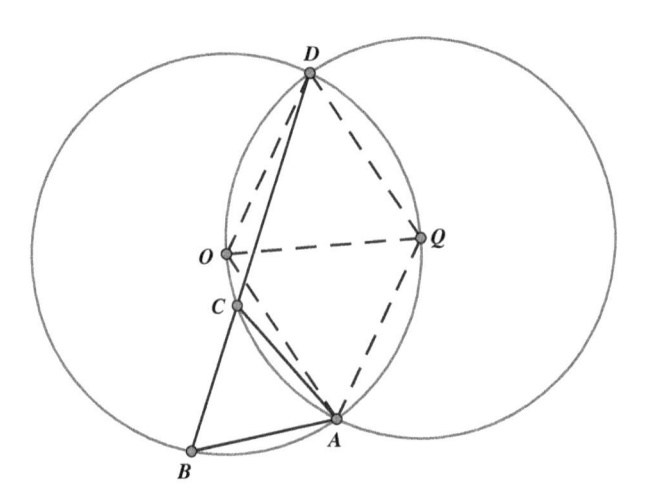

Figure 7-2

In circle Q, we have $\angle ACD = \angle AOD = 120°$. However, $\angle ACD$ is the exterior angle of triangle ABC, and therefore, $\angle ACB = 60°$. Thus, the third angle of triangle ABC, namely $\angle A$, must also equal 60°, and triangle ABC is equilateral.

Circle Wonders   247

The Eyeball Theorem

This unusual theorem involves two non-intersecting circles producing a figure that reminds one of two eyes looking at one another; thus, the name Eyeball Theorem. In Figure 7-3a, we have two non-intersecting circles O and P with tangents emanating from the center of one circle to the other and intersecting circle O and points A and B and intersecting circle P and points C and D.
   Amazingly, we can find that $AB = CD$.

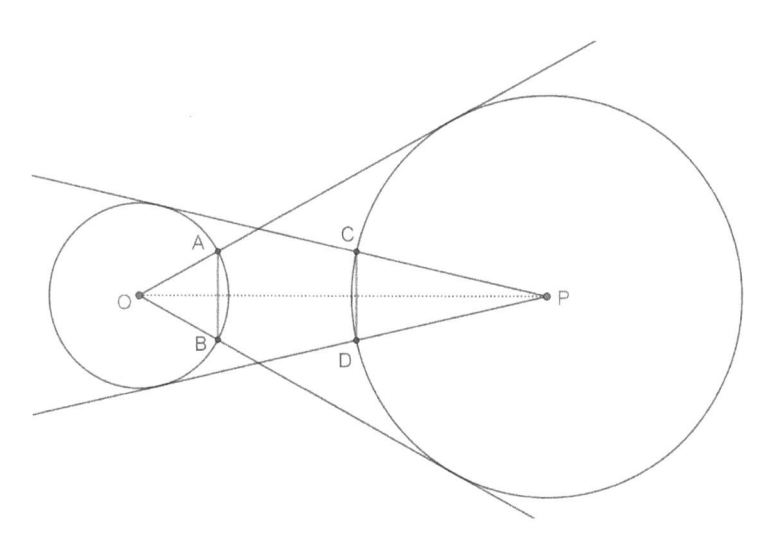

Figure 7-3a

Proof

To simplify this proof we will work with the upper-half of the figure since if we can prove $AE = CF$, then we can certainly double that and get $AB = CD$. To simplify matters further, let $AE = x$, $CF = y$ and we represent the radii of the two circles O and P as r and R, respectively. From similar triangles OAE and OPH, we get $\frac{x}{r} = \frac{R}{d}$, where $OP = d$, which can be written as $x = \frac{rR}{d}$. Similarly, we have similar right triangles CPF and OPG, which gives us $\frac{y}{R} = \frac{r}{d}$, or $y = \frac{rR}{d}$. This implies that $x = y$, and therefore $2x = 2y$, which is essentially $AB = CD$.

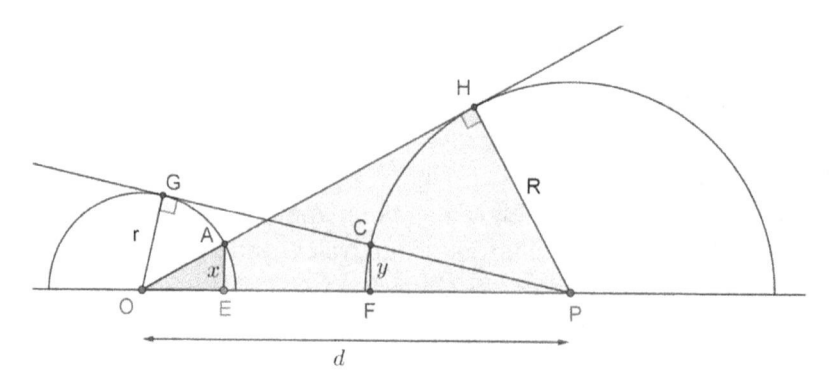

248 A Journey Through the Wonders of Plane Geometry

Figure 7-3b

A Surprising Equality Produced by a Circle and a Square

In Figure 7-4, the circle with center O contains vertex A of square $ABCD$ and also contains the midpoints M and N of sides AD and AB, respectively. Unexpectedly, we find that the tangent from vertex C to circle O at point T has the same length as the side of the square.

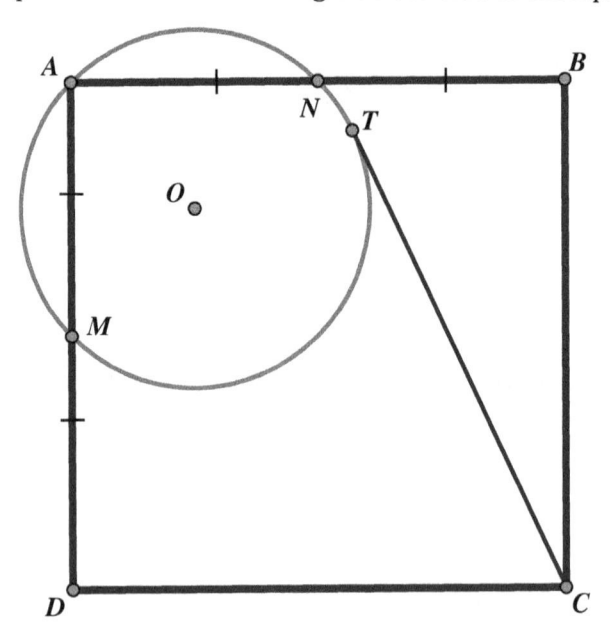

Figure 7-4

Proof

In Figure 7-5, we draw a diagonal AC and line MN. Triangle MAN is an isosceles right triangle, and MN is the diameter of the circle containing points M, A, and N. The midpoint of MN would be the center of the circle. Since MP is parallel to DC and since M is the midpoint of AD, we know that P is the midpoint of AC. Recognizing that AC is a secant and CT is a tangent to the circle with center O, due to the relationship that the tangent is the mean proportional between the entire secant and its external segment, we get $CT^2 = AC \cdot PC$, and $PC = \frac{1}{2}AC$, so that $AC \cdot \frac{1}{2}AC = \frac{1}{2}AC^2$. Applying the Pythagorean theorem to right triangle ABC, we get $AC^2 = AB^2 + BC^2 = 2AB^2$. Thus, we have $\frac{1}{2}AC^2 = AB^2$, which then tells us that $CT^2 = AB^2$. Therefore, $CT = AB$, which is what we set out to prove.

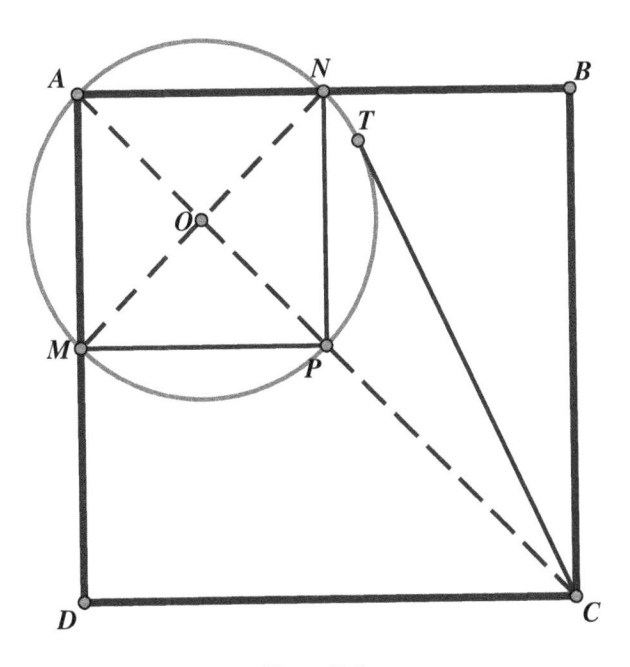

Figure 7-5

The circle also relates to regular polygons other than the equilateral triangle and the square, as we can see with the following regular hexagon.

The Regular Hexagon and the Points on Its Circumcircle

Consider the point P on the circumscribed circle of hexagon $ABCDEF$, as shown in Figure 7-6. Interestingly, we find that $PE + PF = PA + PB + PC + PD$.

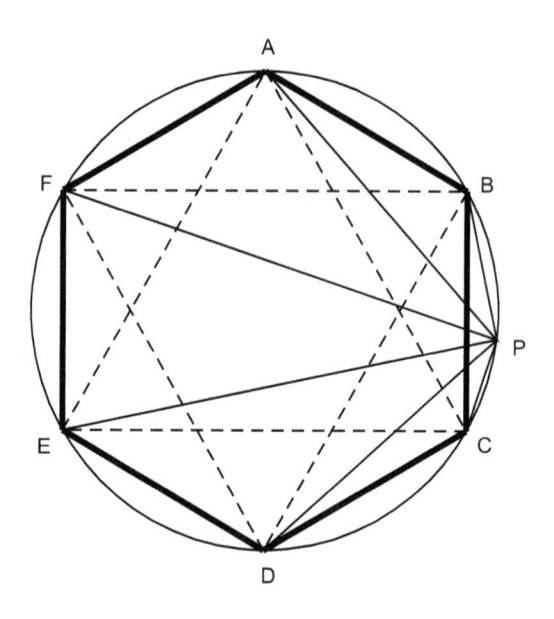

Figure 7-6

Proof

We need to revert to a wonderful theorem (see page 194) discovered by Claudius Ptolemy (ca. 100–170), which states that, for a cyclic quadrilateral, the sum of the products of the opposite sides is equal to the product of the diagonals. Consider the cyclic quadrilateral $APCE$ in Figure 7-6. According to Ptolemy's theorem, $PC \cdot AE + PA \cdot EC = PE \cdot AC$. However, since triangle AEC is equilateral, we have $AE = EC = AC$. Therefore, $PC + PA = PE$. Now consider quadrilateral $FBPD$, in which, by Ptolemy's theorem, we have $PB \cdot DF + PD \cdot BF = PE \cdot AC$. However, since triangle BDF is equilateral, $DF = BF = AC$, and we have

$PB + PD = PE$. By addition, we get $PC + PA + PB + PD = PF + PE$. In a more orderly fashion, we get $PE + PF = PA + PB + PC + PD$, which shows how a circumcircle of a regular hexagon provides an unexpected relationship.

Four Inscribed Circles can be Used to Create a Rectangle

Embedded within cyclic quadrilateral *ABCD* are four circles, each of which is tangent to two sides of the quadrilateral and a diagonal. As we see in Figure 7-7, each of the circles with centers *P*, *Q*, *R*, and *S* is tangent to two sides of the cyclic quadrilateral *ABCD* and a diagonal. Surprisingly, these four centers determine a rectangle *PQRS*. Furthermore, the lines connecting the midpoints of the opposite arcs determined by the cyclic quadrilateral, namely *E*, *F*, *G*, and *H*, are parallel to the sides of the rectangle. That is, $HF \parallel RQ$, and $EG \parallel PQ$.

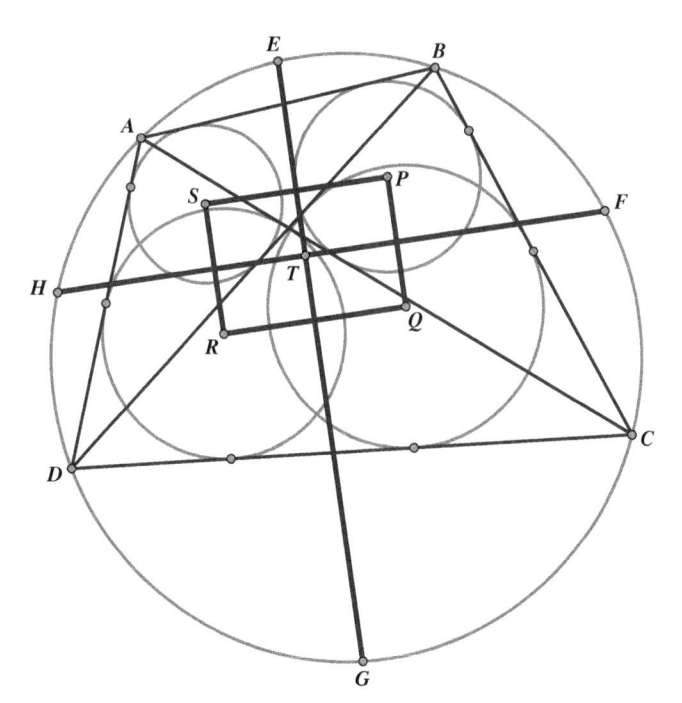

Figure 7-7

Proof

We begin the proof by drawing the following lines: *BP*, *AS*, *AP*, *PF*, *BS*, *SH*, *CP*, *PE*, *DS*, and *SE*. In Figure 7-8, *FH* intersects *EG* at point *T*. Since two chords intersecting in a circle form an angle measured by one-half the sum of the intercepted arcs, $\angle FTG = \frac{1}{2}\left(\widehat{EH} + \widehat{FG}\right) = \frac{1}{2}\left(\widehat{AB} + \widehat{AD} + \widehat{DC} + \widehat{BC}\right) = 90°$, which allows us to conclude that $FH \perp EG$. Since the inscribed circles' centers lie on the angle bisectors of the triangle, we know that we have straight lines *APF*, *BSH*, *CPE*, and *DSE*, which bisect angles $\angle BAC$, $\angle ABD$, $\angle BCA$, and $\angle BDA$, respectively. Also, *AS* and *BP* bisect $\angle ABC$ and $\angle BAD$, respectively. Two angles measured by $\widehat{DGC}$ are equal, so that we have $\angle CAD = \angle CBD$. Therefore, $\angle CBA - \angle DBA = \angle BAD - \angle BAC$. Taking half of each of these quantities gives us $\frac{1}{2}(\angle CBA - \angle DBA) = \frac{1}{2}(\angle BAD - \angle BAC)$, which is to say that

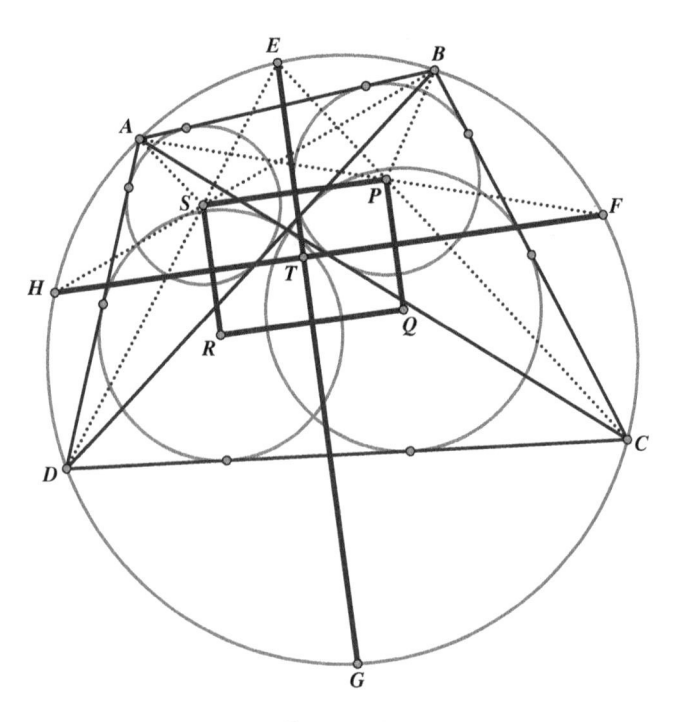

Figure 7-8

∠PBA − ∠SBA = ∠BAS − ∠BAP. Thus, we have ∠PBS = ∠PAS, which allows us to conclude that quadrilateral ASPB is cyclic, since the two angles are measured by the same arc $\overset{\frown}{PS}$. Furthermore, we now have ∠PAB = ∠PSB. This gives us ∠PAB (or ∠FAB) = ∠BHF. Therefore, ∠PSB = BHF, and PS∥FH. Analogously, we find that QR∥FH, and that PQ and RS are both parallel to GE. This establishes that PQRS is a rectangle with sides parallel to EG and FH, which are perpendicular to one another.

Circles Generating Three Equal Lines with a Triangle

We now consider the inscribed and circumscribed circles of the triangle and how they can affect the triangle. In Figure 7-9, a circle with center *I* is inscribed in triangle ABC, and O is the center of the circumscribed circle of triangle ABC. When BI is extended to meet the circumscribed circle at point P, we determine three equal lines, namely AP = PC = PI.

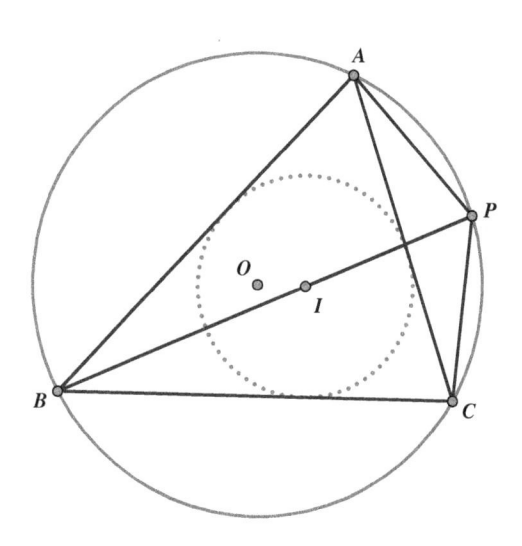

Figure 7-9

Proof

We begin by recognizing that *PB* is the bisector of ∠*ABC* since it contains the center of the inscribed circle. Therefore, $\widehat{AP} = \widehat{CP}$, whereupon it follows that *AP* = *CP*. Similarly, since *AI* bisects ∠*BAC*, we know that ∠*BAI* = ∠*CAI*, as shown in Figure 7-10. The angle *AIP* is the exterior angle of triangle *ABI*, therefore, ∠*AIP* = ∠*ABP* + ∠*BAI*. We can clearly see that ∠*PAI* = ∠*PAC* + ∠*CAI*. Furthermore, we know that ∠*ABP* = ∠*PBC* = ∠*PAC*, as they are angles inscribed in the same arc $\widehat{AP}$. Thus, we have ∠*AIP* = ∠*BAI* + ∠*ABP* = ∠*CAI* + ∠*PAC* = ∠*PAI* and it follows that triangle *API* is isosceles with *AP* = *PI*. We can conclude *AP* = *PC* = *PI*.

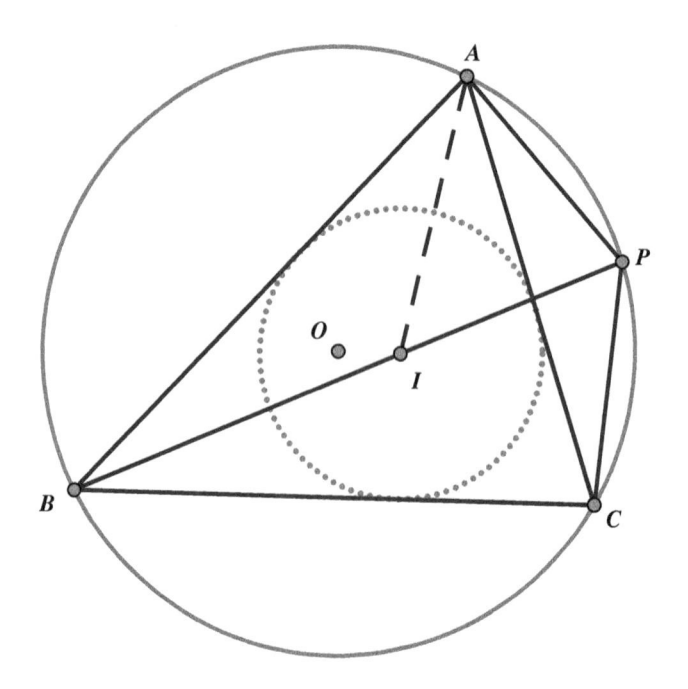

Figure 7-10

Tangent Circles Generate an Unexpected Tangent Circle

When the common external tangent line of two externally tangent circles is the diameter of a third circle, the third circle is tangent to the line of centers of the initial two tangent circles. We see this demonstrated in Figure 7-11, where AB is the common external tangent to the two equal tangent circles with centers O and Q that share point P. Quite unexpectedly, the circle whose diameter is AB shares the common point of tangency P with the two initial tangent circles.

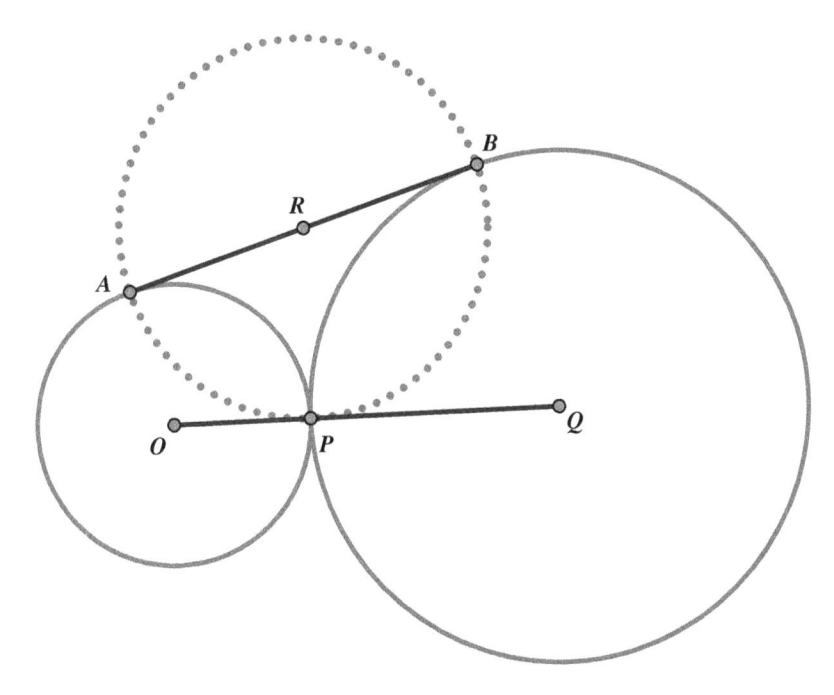

Figure 7-11

Proof

Begin by drawing the lines *AP*, *BP*, and *RP*, where *R* is the midpoint of *AB*, the common tangent to tangent circles *O* and *Q*, as we show in Figure 7-12. The radii *AO* and *BQ* are each perpendicular to the common tangent *AB*. Therefore, angles *OAB* and *QBA* are right angles, since the sum of the angles of a quadrilateral is 360°, and in quadrilateral *ABQO* we have ∠*AOQ* + ∠*BQO* = 180°. These two central angles give us $\widehat{AP} + \widehat{BP} = 180°$, which, in turn, indicates that ∠*BAP* + ∠*ABP* = $\frac{1}{2}\left(\widehat{AP} + \widehat{BP}\right) = \frac{1}{2}180° = 90°$, since each of the angles formed by a tangent and a chord is one-half the intercepted arc. Because angle *APB* is a right angle, it must lie on the circle with diameter *AB*. We now draw *RP* perpendicular to *OPQ*, which makes *RP* a tangent to circles *O* and *Q*. Since we have *AR* = *RP* and *BR* = *RP*, as they are tangents from common points to the respective circles, *RP* must be a radius of the circle with center *R* and therefore is a circle tangent to line *OPQ* at point *P*.

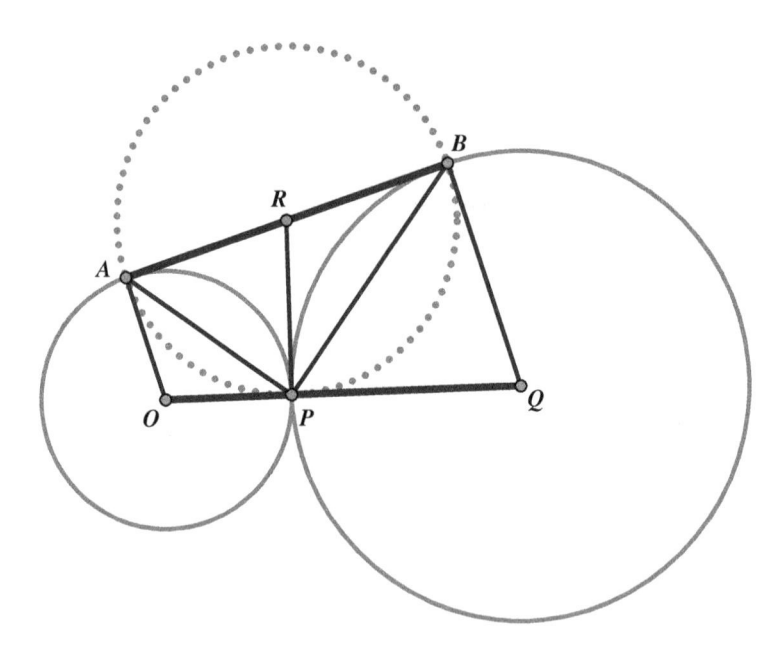

Figure 7-12

Intersecting Circles Curiously Generate Equal Lines

Intersecting circles can certainly bring some strange results. Consider two intersecting circles, where one contains the center of the second circle. These two intersecting circles of two different sizes can produce lines equal in length in a most surprising fashion.

An example of this can be seen in Figure 7-13, where the circle with center A contains the center of circle B. We then draw two chords, CE and DF, in the circle with center B, such that each chord will intersect on circle A at the same point, P. The unexpected result is that $CE = DF$.

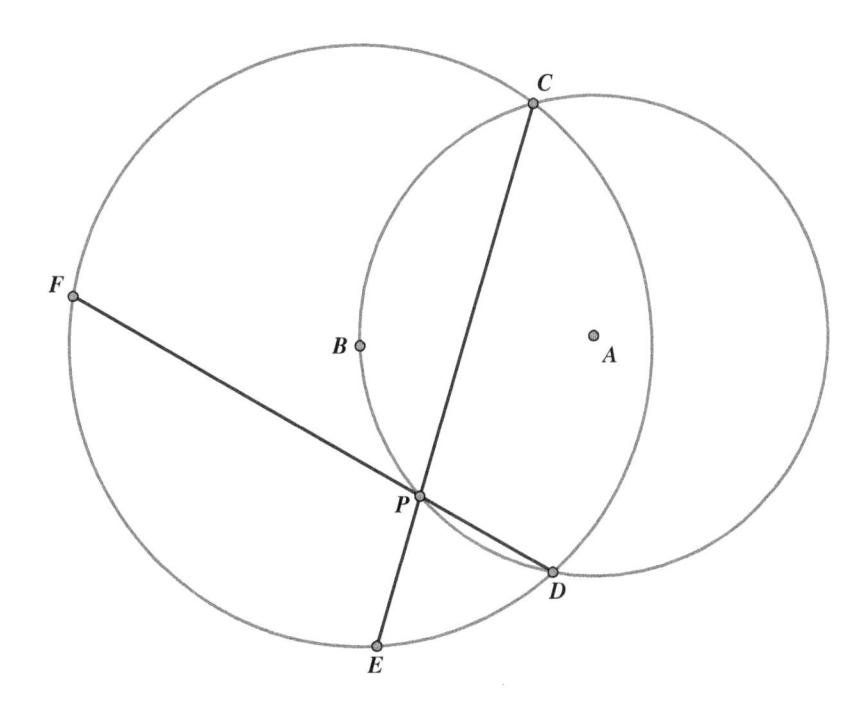

Figure 7-13

258 *A Journey Through the Wonders of Plane Geometry*

Proof

We begin by drawing diameters *CBS* and *DBR* of circle *B*, as shown in Figure 7-14. Since angle *RFD* and angle *SEC* are both inscribed in a semicircle, they are both right angles. Furthermore, in circle *A*, we have $\angle BCP = \frac{1}{2}\widehat{BP}$ and $\angle BDP = \frac{1}{2}\widehat{BP}$. Therefore, $\angle BCP = \angle BDP$. This enables us to establish $\triangle RFD \cong \triangle SEC$, and thus we have $CE = DF$.

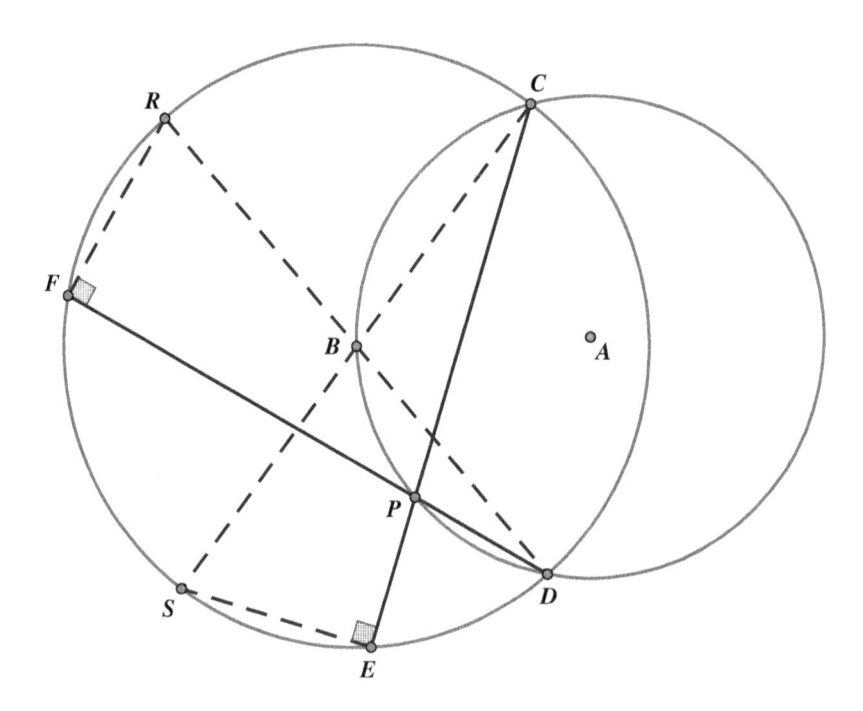

Figure 7-14

New Equal Circles Generated by Three Intersecting Equal Circles

When three equal size circles pass through a common point, the circle through their centers is of equal size to the others. Furthermore, the circle formed by mutual points of intersection is also equal in size to the original three circles. As we can see in Figure 7-15, the circle containing the points of intersection of the original three circles *A*, *B*, and *C* is the same size as the original circles. Also, the circle containing the centers of the three original equal size circles *R*, *Q*, and *S* is also the same size as the original circles.

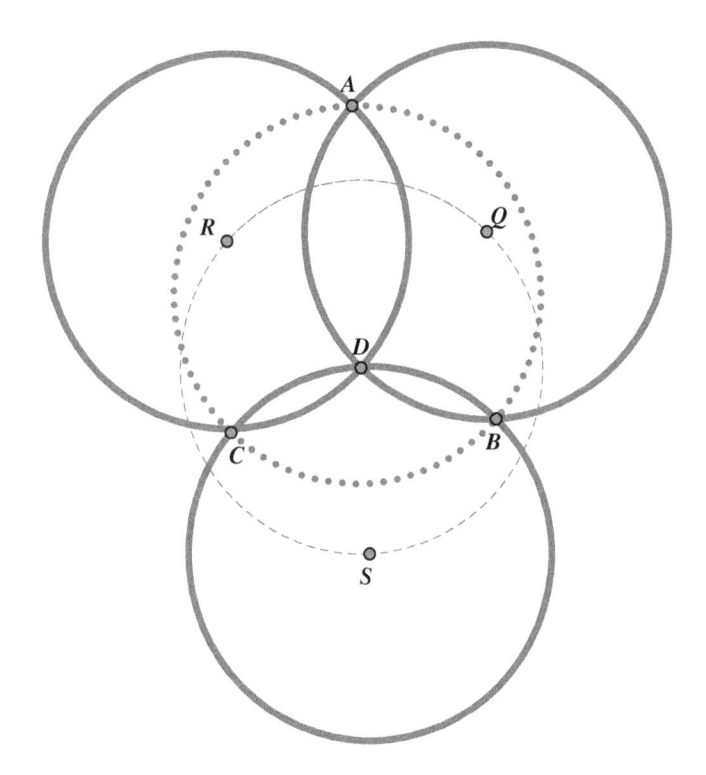

Figure 7-15

Proof

Let us denote the radius of the equal initial circles with r. Due to $QD = RD = SD = r$, the circle through Q, R, and S also has radius r and is centered at D, the common point of the three initial circles. Next, we seek to prove that also the circle through A, B, and C has r as its radius. In Figure 7-16, all dashed lines are of length r, establishing three rhombuses: $ARDQ$, $BSDQ$, and $DSCR$. This enables us to conclude $CR = BQ = r$ and $CR \| BQ$, which tells us that $CBQR$ is a parallelogram. Hence, $CB = RQ$, and similarly we get $AB = RS$ and $AC = QS$. Thus, $\triangle ABC \cong \triangle QRS$ and the circumradius of $\triangle ABC$ must be r, too.

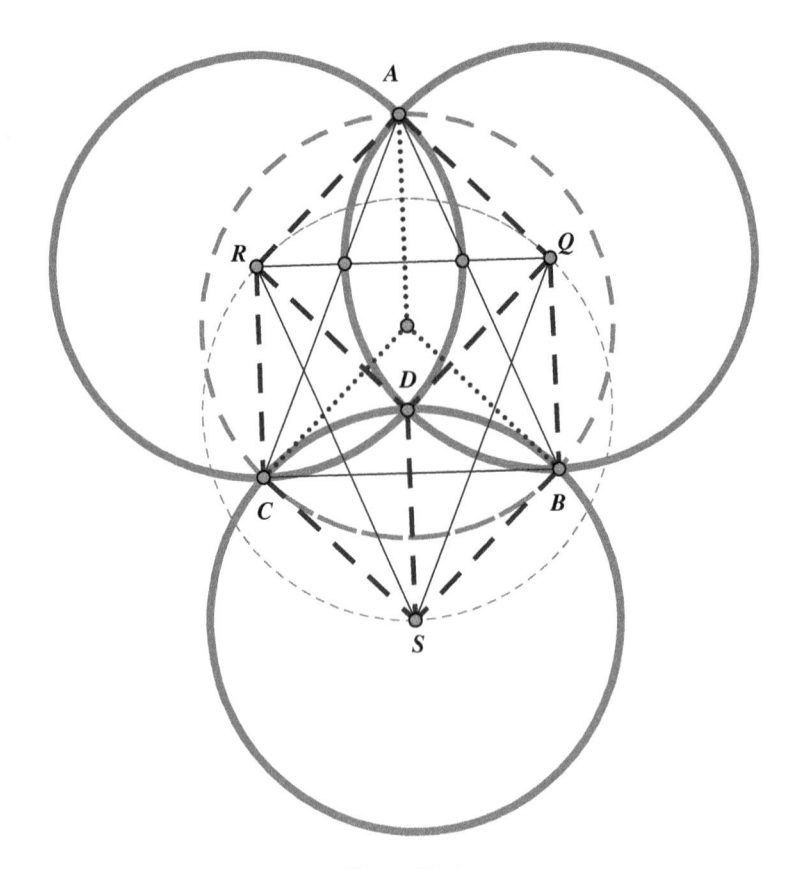

Figure 7-16

Congruent Quadrilaterals Generated by Three Intersecting Circles

In Figure 7-17, we have once again three equal circles centered at Q, R, and S that meet at points A, B, C, and D (all of them passing through D). We also have a fourth circle (dotted in Figure 7-17) containing the three intersection points A, B, and C of the three equal circles; in Figure 7-16, we established that the fourth circle is equal to the three given equal circles. When we join the four points of intersection of the three original equal circles, namely, points A, C, B, and D, we create a quadrilateral congruent to the quadrilateral formed by joining the centers of the four equal circles, namely, points S, Q, R, and P (P is the center of the fourth circle through A, B, and C). Furthermore, quite amazingly, in either of these congruent figures, the line joining any two points is perpendicular to the line joining the other two points.

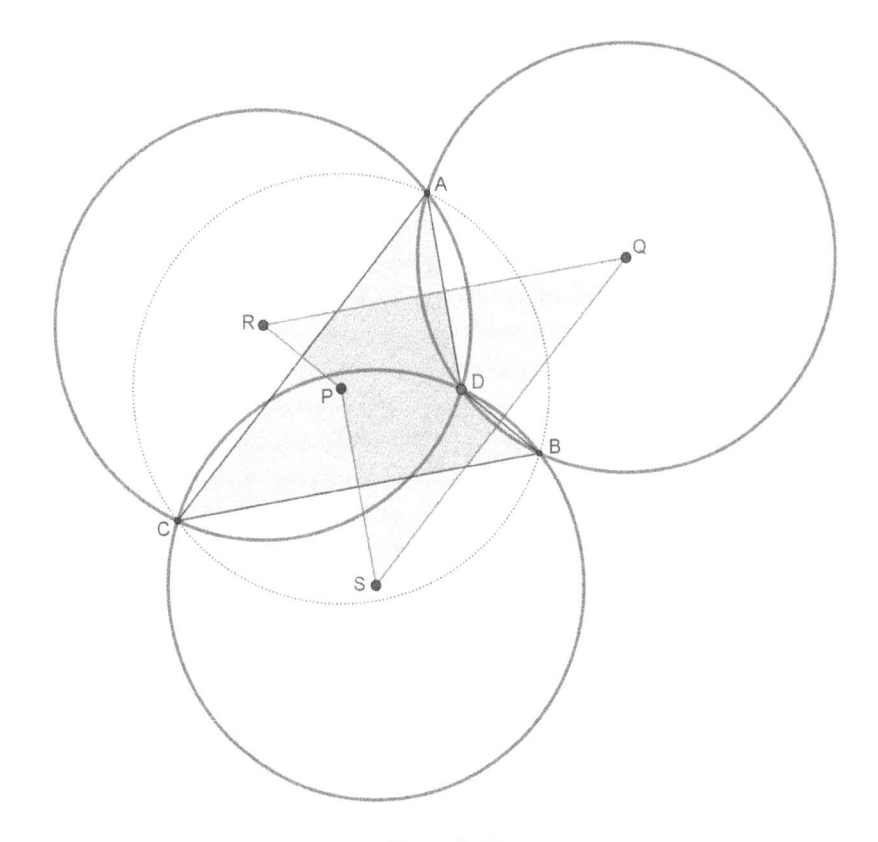

Figure 7-17

Proof

To prove the congruence of these two quadrilaterals, one merely needs to recognize the various parallelograms, such as parallelogram *RQBC*, shown in Figure 7-18. We begin by constructing the circle through *Q*, *R*, *S* centered at *D* and with the same radius as the other four circles. We notice that we have three rhombuses *CSBP*, *SBQD*, and *ARDQ* since all four sides are equal, as they are radii of a circle. We must prove $\angle RAQ = \angle BSC$ to conclude that $\triangle BSC \cong \triangle RAQ$, so that $CB = RQ$ and *RQBC* is a parallelogram. Follow along with the following angle equalities:

$$\angle RAQ = \angle RDQ = 360° - \angle RDS - \angle QDS = 360° - (180° - \angle CSD)$$
$$- (180° - \angle BSD) = \angle CSD + \angle BSD = \angle BSC$$

We then have $CB = RQ$ and $CB \| RQ$. Similarly, the other sides of the quadrilaterals *ACBD* and *SQRP* are pairwise equal and parallel, which

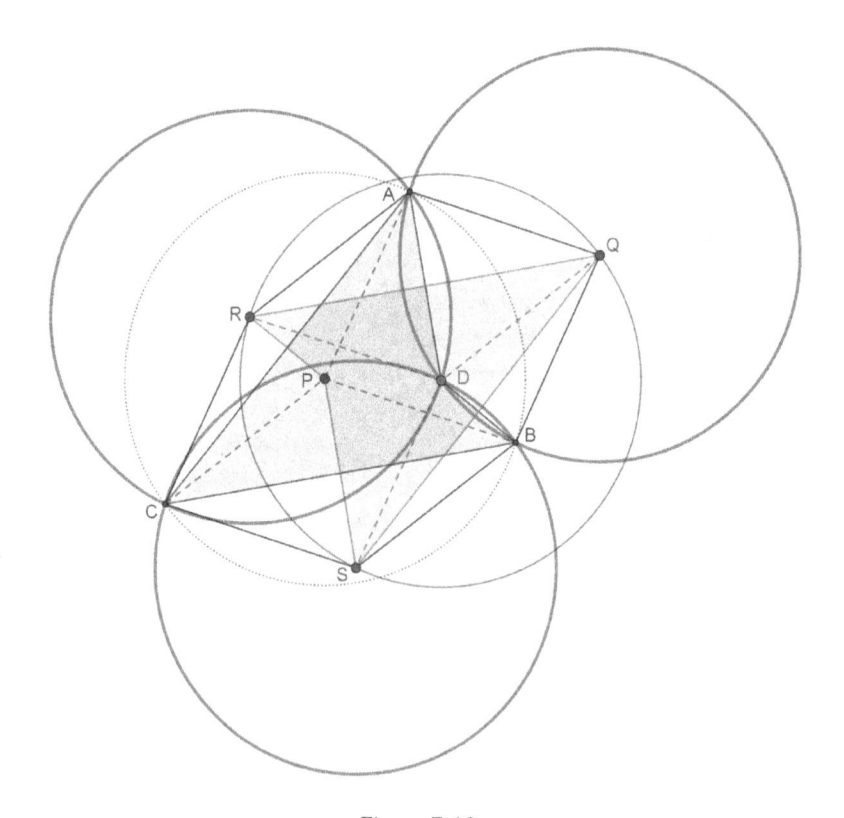

Figure 7-18

establishes the congruence of the quadrilaterals. Keep in mind that the parallelism of the sides guarantees the pairwise equality of the angles in the quadrilaterals.

Miquel's Theorem

Try this experiment. Draw any convenient triangle and select a point on each side. Now construct three circles, each containing two of these points and the vertex determined by the two sides on which these points lie. Although you can do this on paper with the aid of a pair of compasses, it is particularly nice to do this with Geometer's Sketchpad or GeoGebra. What relationship do you notice about these three circles? Your observation should lead you to a theorem published in 1838 by the French mathematician Auguste Miquel (1816–1851). *Miquel's theorem* states that if a point is selected on each side of a triangle, then the circles determined by each vertex and the points on the adjacent sides pass through a common point. This theorem may be viewed in two ways. The expected form is shown in Figure 7-19. However, when two of the selected points are on the extensions of the sides, the theorem still holds, as we can see in Figure 7-21.

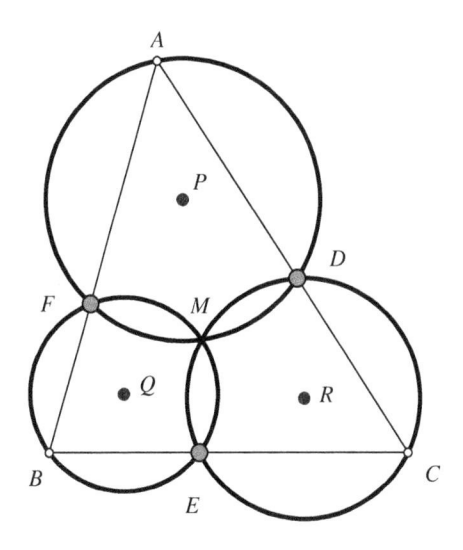

Figure 7-19

264 *A Journey Through the Wonders of Plane Geometry*

A Triangle Generating Intersecting Circles

Three intersecting circles can always produce some astonishing results. For example, consider Figure 7-19, where the three circles contain a point on two adjacent sides of the triangle *ABC* as well as the vertex between them. These three circles then share a common point *M*. This point is known as the *Miquel point*, also named after Auguste Miquel.

Proof

Consider the situation where point *M* is inside triangle *ABC*, as shown in Figure 7-20. Points *D*, *E*, and *F* are any points on the sides of *AC*, *BC*, and *AB*, respectively, of triangle *ABC*. We will consider the circles *Q* and *R*, determined by the points *F*, *B*, *E* and *D*, *C*, *E*, respectively, to meet at point *M*. We will now draw line segments *MF*, *ME*, and *MD*.

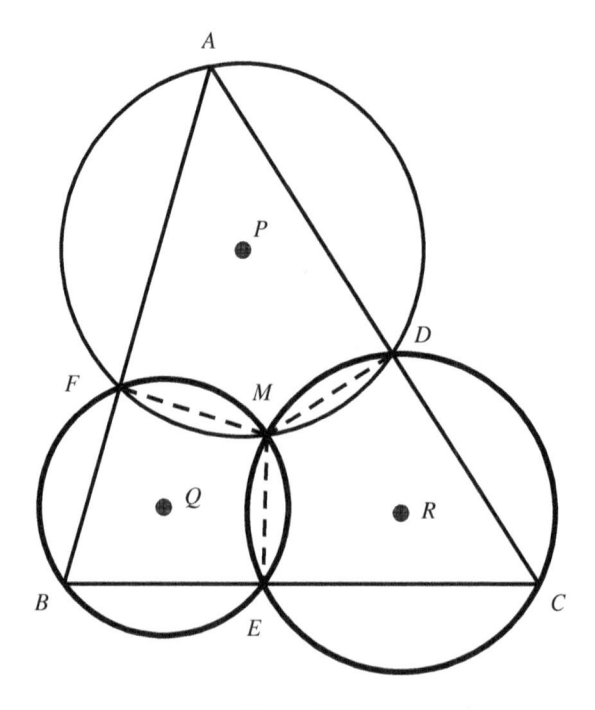

Figure 7-20

Because quadrilateral *BFME* is cyclic, and the opposite angles of a cyclic quadrilateral are supplementary, $\angle FME = 180° - \angle B$. Similarly, quadrilateral *CDME* is also cyclic; therefore, $\angle DME = 180° - \angle C$. When we add these two angles, we get $\angle FME + \angle DME = 360° - (\angle B + \angle C)$, or written another way, $\angle B + \angle C = 360° - \angle FME + \angle DME = \angle DMF$. However, in triangle *ABC*, we notice that $\angle B + \angle C = 180° - \angle A$. It follows that $\angle DMF = 180° - \angle A$, which allows us to conclude that quadrilateral *AFMD* is cyclic. Therefore, the circle containing points *D*, *A*, and *F* contains a point common to the other two circles, namely, point *M*.

We could also configure the situation with circles determined by points on the extensions of the sides of the triangle *ABC*, where the configuration would look like that shown in Figure 7-21.

In each case, you will notice that each of the circles contains two points on the sides of triangle *ABC* (or their extensions) and a vertex between them. Also, the three circles contain a common point *M*.

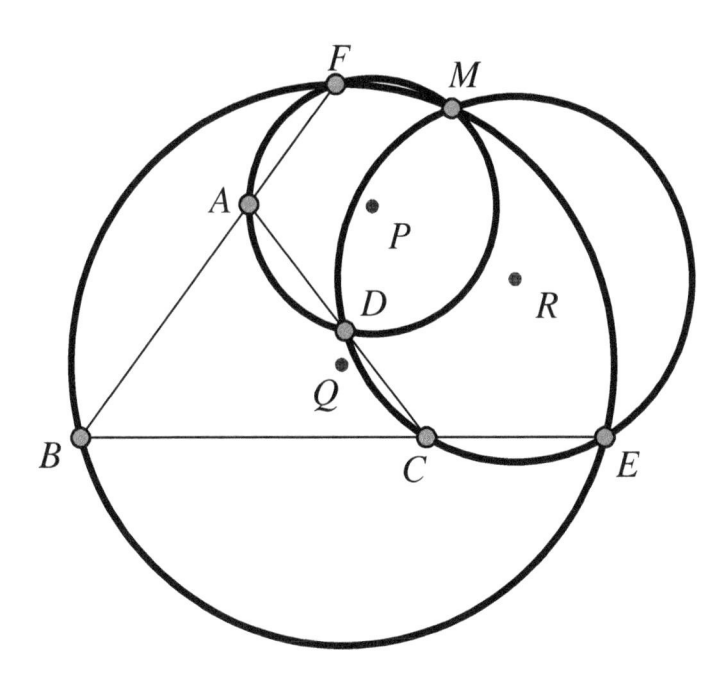

Figure 7-21

More on the Miquel Relationship

In Figure 7-22, we see the complete quadrilateral, where we find four triangles △ABD, △BFE, △CDE, and △ACF. When we draw the circumscribed circles about each of these triangles, unexpectedly, we find that they all contain a common point M, which is the Miquel point. Furthermore, the centers of these four circles, P, Q, R, and S, all lie on the same circle.

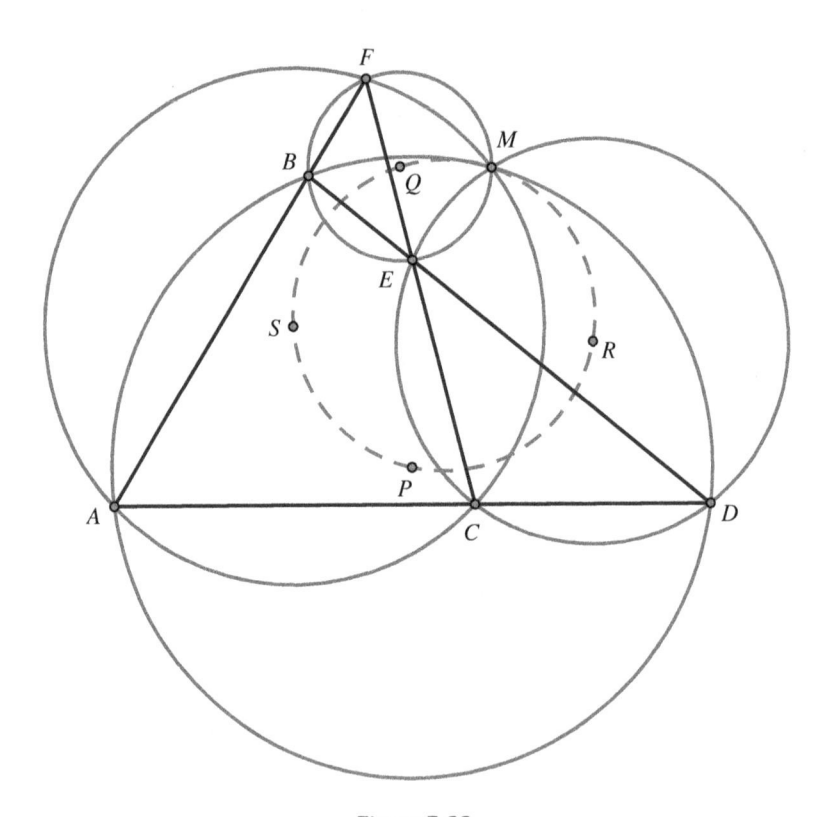

Figure 7-22

Proof

Let *M* be the second intersection point of the circumcircles of △*ACF* and △*ECD*, and let *FM* meet the circumcircle of △*ECD* at point *H*, as shown in Figure 7-23. Since *ACMF* and *CDHM* are cyclic quadrilaterals, we have ∠*ACM* = ∠*MHD* = 180° − ∠*BFM*; thus, *HD*||*AF*. Also, *EDHM* is a cyclic quadrilateral, so that ∠*MEB* = ∠*MHD* = 180° − ∠*BFM*. From this, we can conclude that *BEMF* is a cyclic quadrilateral, which means that *M* also lies on the circumcircle of △*BEF*. Similarly, *M* also lies on the circumcircle of △*ADB*.

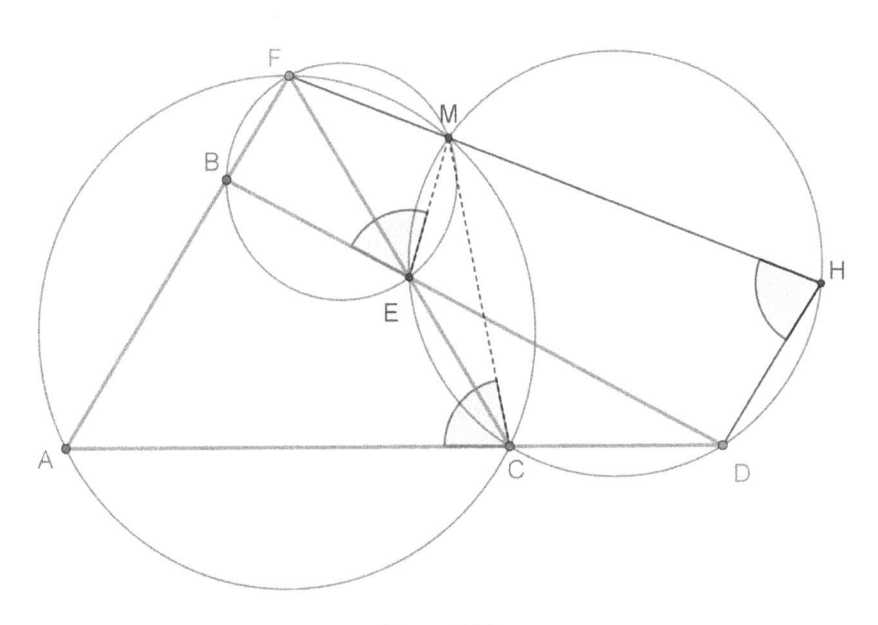

Figure 7-23

Now we can establish a proof that *P*, *Q*, *R*, and *S* are concyclic with *M*. First, we shall consider the points *P*, *Q*, *S*, *M*, as shown in Figure 7-24. The two quadrilaterals *PESM* and *QCSM* are kites with a common side *MS*, which will prove to be a common chord in the circle through *P*, *Q*, *S*, and *M*. Because inscribed angles are half the corresponding center angles, we have $\angle MPS = \frac{1}{2} \angle MPE = \angle MFE = \angle MFC = \frac{1}{2} \angle MQC = \angle MQS$. This enables us to conclude that *M*, *P*, *Q*, and *S* are concyclic, and we similarly prove that *M*, *Q*, *R*, and *S* are concyclic to yield our claimed result that all five points *M*, *P*, *Q*, *R*, and *S* are concyclic.

Without a proof, suppose we were to take five intersecting lines instead of the four we used in Figure 7-24. If we were to consider these lines by taking them four a time, we would end up with five Miquel points, which lie on the same circle: the *Miquel circle*. Furthermore, each of the sets of four lines creates a circle, as we have seen above, which contains the four circumcenters. To take this step a bit further, we notice that the five circles all pass through a common point.

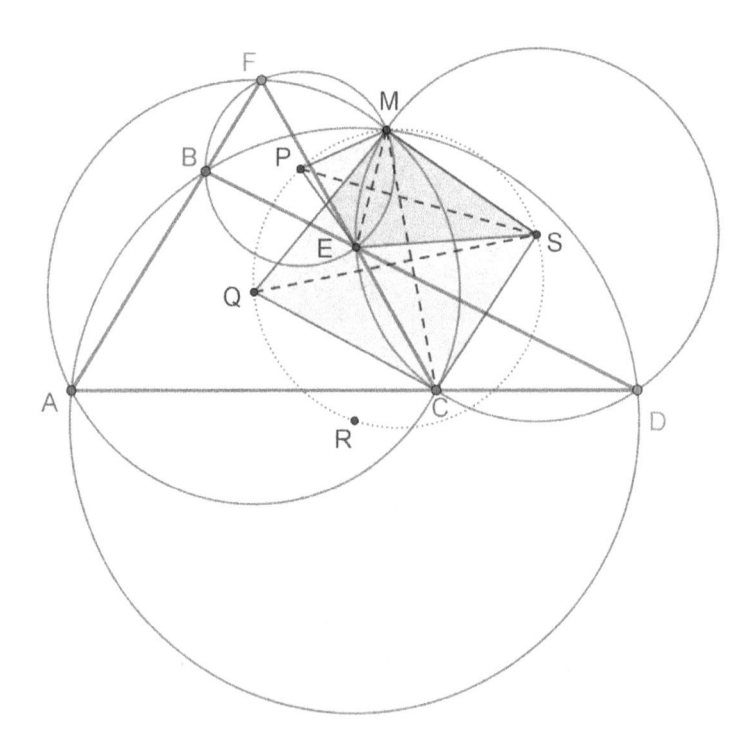

Figure 7-24

Another Application of Miquel's Theorem

We say that a triangle is inscribed in a second triangle if each of the vertices of the first triangle lies on the sides (or their extensions) of the second triangle. Thus, we state the following relationship: Two triangles inscribed in the same triangle that have a common Miquel point are similar. That is, point M in Figure 7-25 is a Miquel point for $\triangle DEF$ and $\triangle D'E'F'$, which are inscribed in $\triangle ABC$. With these restrictions, these two triangles are similar.

Proof

Consider $\triangle DEF$ and $\triangle D'E'F'$, which have the same Miquel point M, as we show in Figure 7-25. From our first application of Miquel's theorem, we find that $\angle MFB \cong \angle MDA$ and $\angle MF'A \cong \angle MD'C$. Therefore, $\triangle MF'F \sim \triangle MD'D$. Similarly, $\triangle MD'D \sim \triangle ME'E$.

Thus, $\angle FMF' \cong \angle DMD' \cong \angle EME'$. By addition, $\angle F'MD' \cong \angle FMD$, $\angle F'ME' \cong \angle FME$, and $\angle E'MD' \cong \angle EMD$. Also, because of the above similar triangles, $\frac{MF}{MF'} = \frac{MD}{MD'} = \frac{ME}{ME'}$. Since two triangles are similar if two pairs of corresponding sides are proportional and the included angles congruent, we get the following:

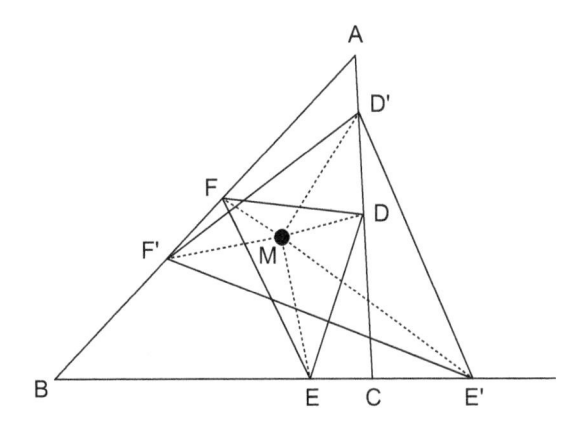

$$\triangle F'MD \sim \triangle FMD, \triangle F'ME' \sim \triangle FME, \text{ and } \triangle E'MD' \sim \triangle EMD.$$

Figure 7-25

Therefore, $\frac{F'D'}{FD} = \frac{F'M}{FM}$, and $\frac{F'E'}{FE} = \frac{F'M}{FM}$. Thus, $\frac{F'D'}{FD} = \frac{F'E'}{FE}$. Similarly, $\frac{ED'}{ED} = \frac{F'E'}{FE}$.

This proves $\triangle DEF \sim \triangle D'E'F'$ because the corresponding sides are proportional.

More on the Application of Miquel's Theorem

The centers of Miquel circles of a given triangle determine a triangle similar to the given triangle, as shown in Figure 7-26.

Proof

For each pair of circles, we draw common chords FM, EM, and DM. Furthermore, PQ intersects circle Q at point N, and RQ intersects circle Q at point L, as shown in Figure 7-26. Since the line connecting centers of two circles is the perpendicular bisector of their common chord, PQ is the perpendicular bisector of FM

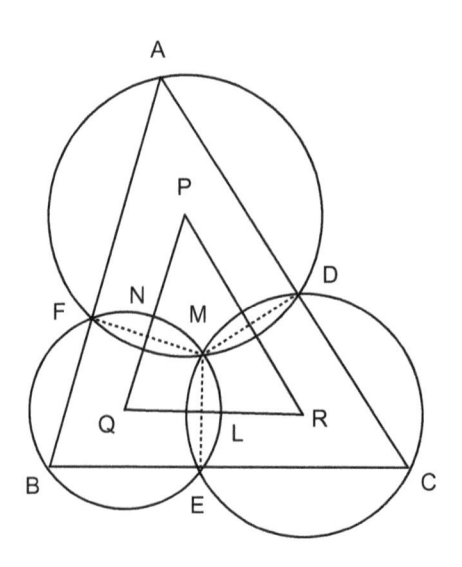

Figure 7-26

so that $\widehat{FN} = \widehat{NM}$. Similarly, QR bisects $\widehat{EM}$ so that $\widehat{ML} = \widehat{LE}$. Now $\angle NQL = \widehat{NM} + \widehat{ML} = \frac{1}{2}\widehat{FE}$, and $\angle FBE = \frac{1}{2}\widehat{FE}$. Therefore, $\angle NQL = \angle FBE$. In a similar fashion, we can prove that $\angle QPR = \angle BAC$. Thus, $\triangle PQR \sim \triangle ABC$.

Miquel's Theorem Extended

Another variation on Miquel's theorem involves taking any circle and selecting four points A, B, C, and D on the circle, and then drawing a circle through each pair of consecutively placed points. We mark the other points at which each pair of consecutive circles intersects as A', B', C', D'. These latter points also lie on a circle, shown in Figure 7-27 with dashed lines.

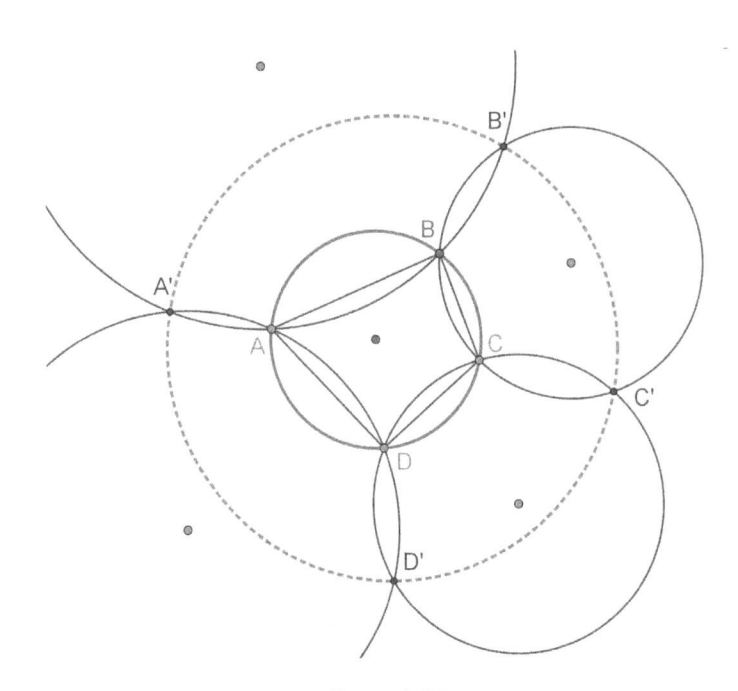

Figure 7-27

272 A Journey Through the Wonders of Plane Geometry

Proof

In Figure 7-27, we have four cyclic quadrilaterals, $A'ABB'$, $B'BCC'$, $C'CDD'$, $D'DAA'$, and we need to show that $\angle A'D'C' + \angle A'B'C' = 180°$, which is done by the following:

$\angle A'D'C' + \angle A'B'C' = \underbrace{\angle A'D'D + \angle DD'C'}_{\angle A'D'C'} + \underbrace{\angle A'B'B + \angle BB'C'}_{\angle A'B'C'}$

$= (180° - \angle A'AD) + (180° - \angle DCC') + (180° - \angle A'AB) + (180° - \angle BCC')$

$= 4 \cdot 180° - \underbrace{(\angle A'AD + \angle A'AB)}_{360° - \angle BAD} - \underbrace{(\angle DCC' + \angle BCC')}_{360° - \angle DCB}$

$= \angle BAD + \angle DCB = 180°$.

Three Intersecting Circles Generate Similar Triangles and More

The three intersecting circles with centers A, B, and C all share the common point of intersection O, as shown in Figure 7-28. Triangle ABC is formed by joining the centers of the three circles. The lines joining the common point of intersection O with each of the circles' centers intersect each circle at points D, E, and F to determine a second triangle DEF. The two triangles ABC and DEF have their corresponding sides parallel, making the triangles similar. The unexpected result is that the sides of triangle DEF contain the intersection points P, Q, and R of the pairs of circles.

Proof

We begin by drawing the common chord OP, shown in Figure 7-29. Line AB, which joins the centers of the two circles, is perpendicular to OP. Also, AB is the bisector of OP. Next, we have $\angle PDO = \frac{1}{2}\widehat{PMO} = \widehat{MO}$. However, $\angle MAO = \widehat{MO}$; therefore, $\angle PDO = \angle MAO$, which allows us to conclude $PD \parallel AB$. Similarly, we have $PE \parallel AB$, which determines that DPE is a straight line. In a similar fashion, we can show that DRF and FQE are also straight lines. Therefore, we have similar triangles, and the sides of triangle DEF contain the intersection points P, Q, and R.

Circle Wonders **273**

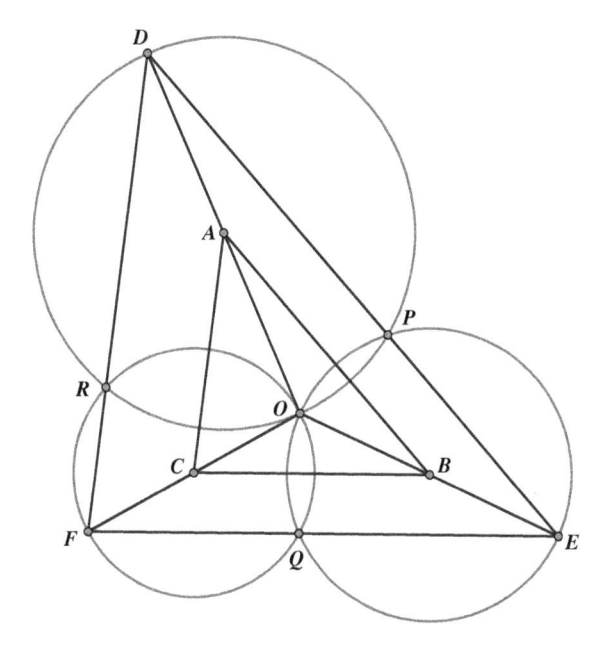

Figure 7-28

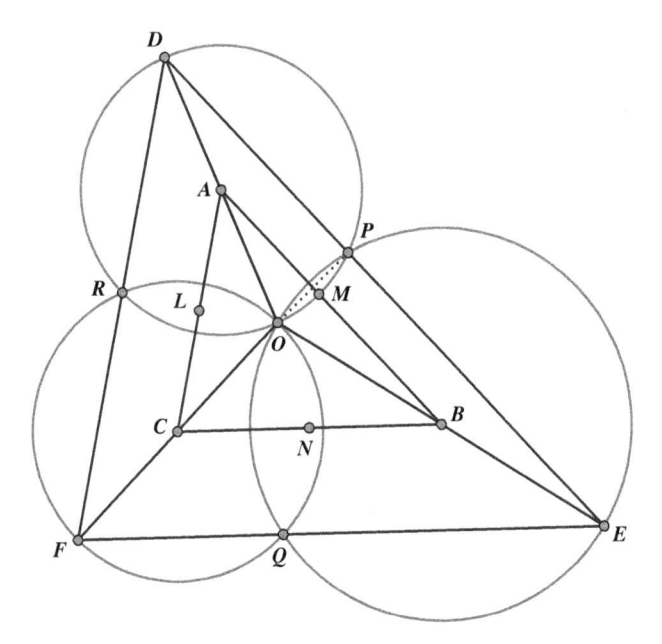

Figure 7-29

Another way of looking at this problem is noticing that △ABC and △DEF are homothetic (see Chapter 11) with center O and factor 2, as are the triangles △LMN and △PQR.

A Surprising Relationship Generated by an Inscribed and Escribed Circle of a Triangle

In Figure 7-30, triangle *ABC* has an inscribed circle with center *I* that is tangent to side *AB* at point *D*. The escribed circle of triangle *ABC* with center *O* is tangent to side *AB* at point *E*. A curious relationship evolves, namely, $AD \cdot DB = AE \cdot EB$,[1] which equals the area of a rectangle whose sides are the radii of the two circles.

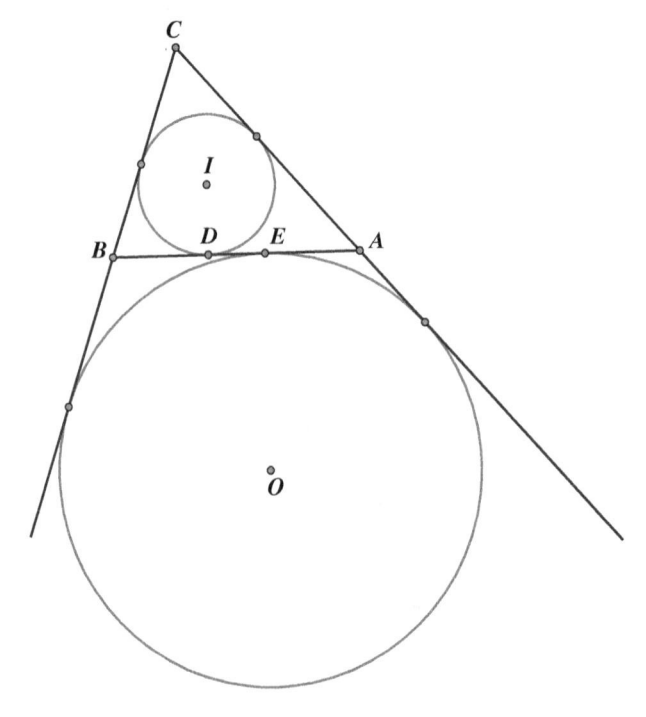

Figure 7-30

---

[1] Even a stronger result holds (see Figure 11-13b): $AD = EB$ and $AE = DB$.

Proof

In Figure 7-31, point I is the center of the inscribed circle tangent to side AB of triangle ABC at point D, point O is the center of the escribed circle tangent to side AB at point E. We then draw the following lines: AI, BI, AO, and BO. The extension of ID intersects this circle at point F. We then draw FO. The bisectors of the interior and exterior angles at point A are AI and AO, respectively. Similarly, the bisectors of the interior and exterior angles at point B are BI and BO, respectively. Therefore, $\angle IAO = IBO = 90°$, whereupon quadrilateral $AIBO$ is cyclic, with IO as the diameter of its circumscribed circle. Therefore, $\angle IFO = 90°$, and similarly, $\angle IGO = 90°$. We also have rectangle $DEOF$ so that $EO = DF$, and therefore, $AD \cdot BD = ID \cdot DF = ID \cdot EO$. Similarly, we have $AE \cdot EB = ID \cdot EO$, which is the area of a rectangle with sides ID and EO.

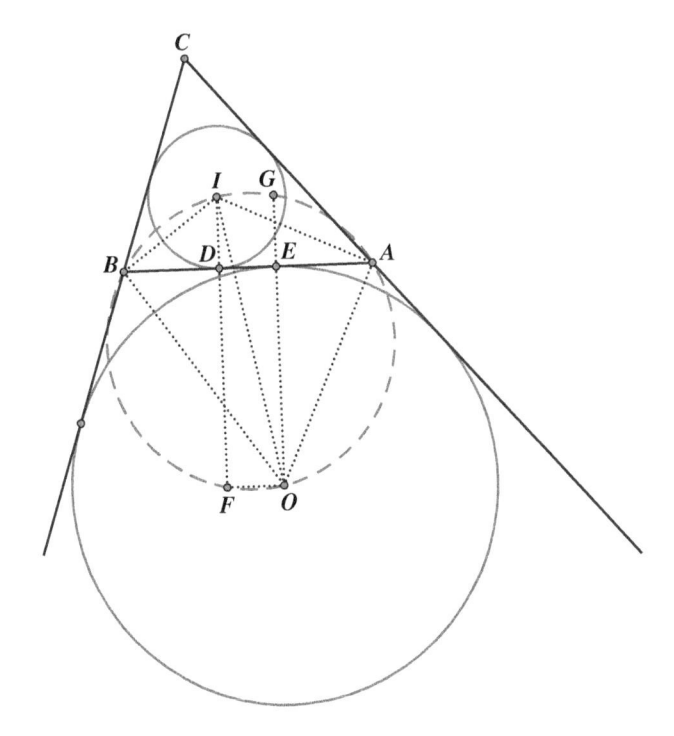

Figure 7-31

Three Related Circumscribed Circles

In Figure 7-32, point P lies above three collinear points A, B, and C, which determine three triangles, PAB, PBC, and PAC, whose circumscribed circles have centers at points R, S, and T. From this relatively simple configuration, we find that these three centers are concyclic with the point P.

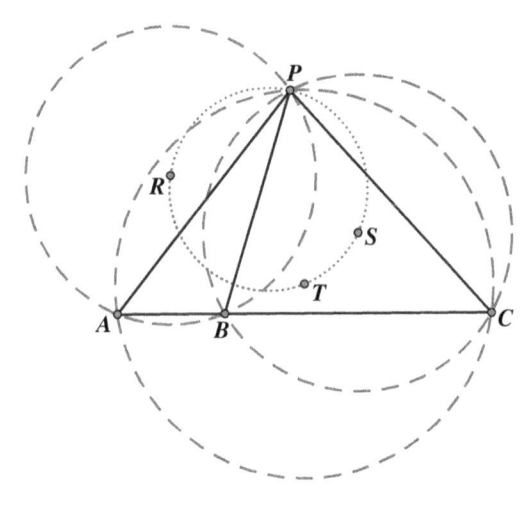

Figure 7-32

Proof

Our objective here is to show that quadrilateral PRTS in Figure 7-33 is a cyclic quadrilateral. To do that, we need to show that one pair of opposite angles are supplementary; if we can show that ∠PRT = ∠PSQ, where Q is the intersection point of TS with PC, we will have accomplished that task. Since $TR \perp PA$, we also know that $\angle PRT = \frac{1}{2}\angle PRA$, and similarly, $TSQ \perp PC$, and we thus have $\angle PSQ = \frac{1}{2}\angle PSC$. Furthermore, in circle S, we have ∠PSC = 2∠PBC. Considering cyclic quadrilateral PUAB, we find that ∠PBC = ∠PUA. In circle R, $\angle PUA = \frac{1}{2}\angle PRA$. Therefore, ∠PRT = ∠PSQ, and since ∠PST is supplementary to ∠PSQ, we can conclude that the angles ∠PRT and ∠PST are supplementary. Therefore, quadrilateral PRTS as a cyclic quadrilateral.

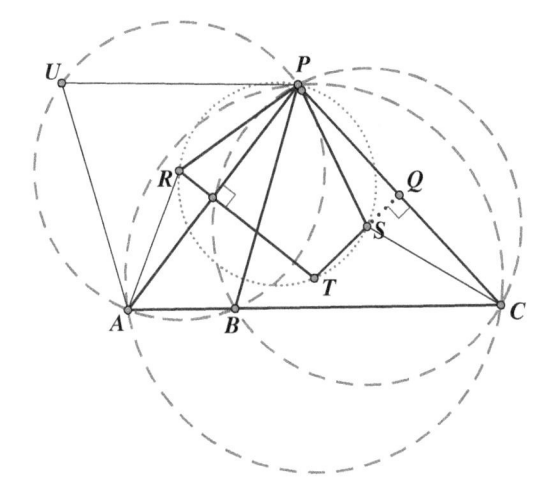

Figure 7-33

The Pythagorean Theorem Extended

One of the most popular relationships in geometry, and the one that most people recall from school, is the Pythagorean theorem. Simply stated, this theorem says that the sum of the squares of the legs of a right triangle is equal to the square of the hypotenuse. This can also be stated as the sum of *the areas of* the squares on the legs of a right triangle is equal to *the area of* the square on the hypotenuse. The "squares" can be replaced by any similar figures drawn on the sides of a right triangle, and we can state that the sum of *the areas of* the *similar polygons* on the legs of a right triangle is equal to *the area of* the *similar polygon* on the hypotenuse.[2] Furthermore, this can then be restated for the specific case of semicircles (which are, of course,

---

[2] Let F_a and F_b be two similar figures erected on the legs a and b, respectively, of a right triangle, and F_c the corresponding similar figure on hypotenuse c. Since the ratio of areas of similar figures is the squared ratio of corresponding lengths (see Chapter 11), we have area $F_a = \frac{a^2}{c^2}$ area F_c and area $F_b = \frac{b^2}{c^2}$ area F_c, which yields area F_a + area $F_b = \frac{a^2+b^2}{c^2}$ area F_c. From the standard version of the Pythagorean theorem, we know $\frac{a^2+b^2}{c^2} = 1$; thus, area F_a + area F_b = area F_c.

similar): the sum of *the areas of* the semicircles on the legs of a right triangle is equal to *the area of* the semicircle on the hypotenuse. In Figure 7-34, the areas of the semicircles relate as follows:

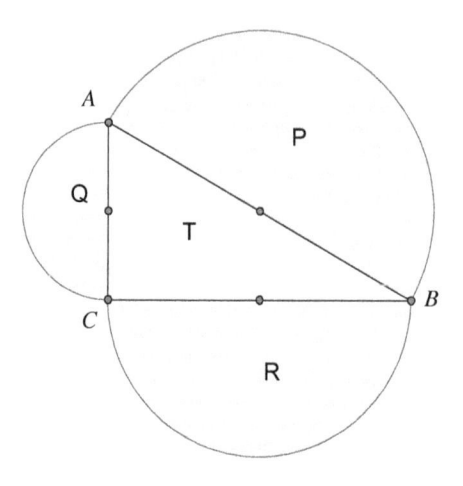

$$\text{area } P = \text{area } Q + \text{area } R.$$

Figure 7-34

Suppose we now flip semicircle *P* over the rest of the figure (using *AB* as its axis). We would get the figure shown in Figure 7-35.

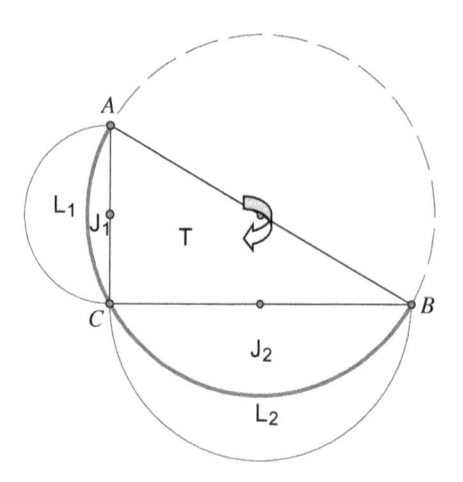

Figure 7-35

Circle Wonders **279**

Let us now focus on the lunes (shaded) formed by the two semicircles, shown in Figure 7-36. We label them L_1 and L_2.

Earlier in Figure 7-34, we established that area P = area Q + area R. In Figure 7-36, that same relationship can be written as follows:

area J_1 + area J_2 + area T = area L_1 + area J_1 + area L_2 + area J_2.

If we subtract area J_1 + area J_2 from both sides, the astonishing result is area T = area L_1 + area L_2.

That is, we have a rectilinear figure (the triangle) equal to some non-rectilinear figures (the lunes). This is quite unusual since the measures of circular figures seem to always involve π, while rectangular (or straight-line figures) do not.

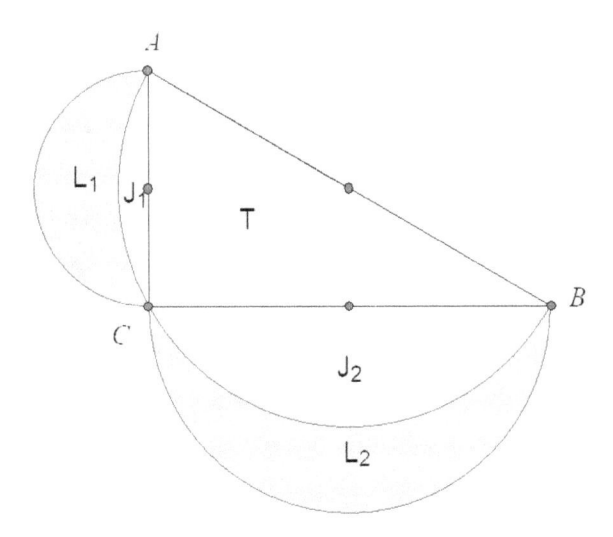

Figure 7-36

Equal Areas Formed by Semicircles

On the three sides of right triangle ABC, semicircles are drawn so that unusual area relationships evolve. In Figure 7-37, the areas within the bold curves is such that $X + Y + \triangle ABC = S$.

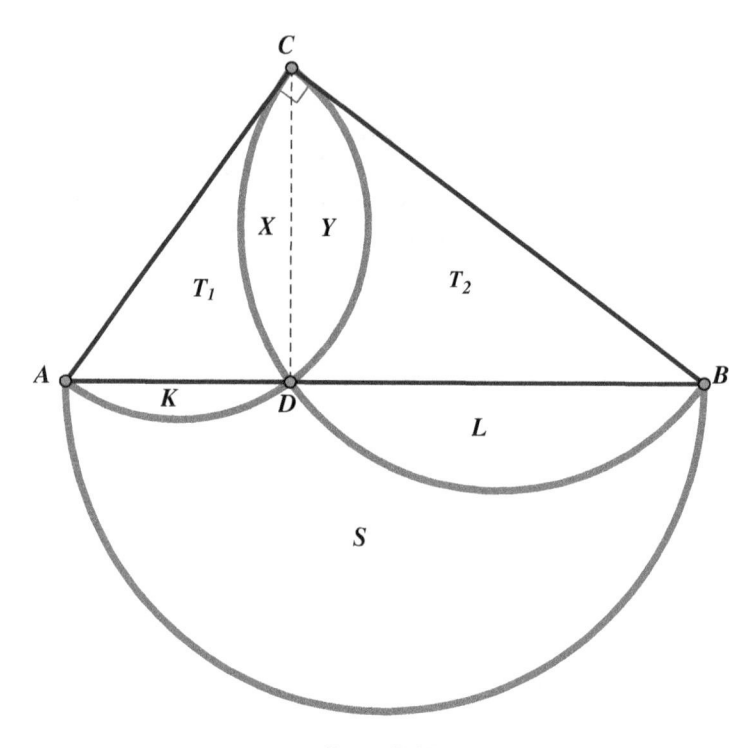

Figure 7-37

Proof

Before we begin, remember that the area of triangle $ADC = T_1$ and the area of triangle $BDC = T_2$, as can be seen in Figure 7-37. Also keep in mind that $T_1 + T_2 = \triangle ABC$. The area of semicircle ADC equals $K + Y + T_1$, and the area of semicircle BDC equals $L + X + T_2$. Since by the generalized Pythagorean theorem the sum of the areas of the semicircles on the legs of a right triangle is equal to the area of the semicircle on hypotenuse, which is $K + L + S$, we have the following relationship: $(K + Y + T_1) + (L + X + T_2) = K + L + S$. Therefore, $X + Y + \triangle ABC = S$.

Surprises of Equal Areas Formed by Semicircles

In Figure 7-38, we have a further unexpected equality of areas from these overlapping semicircles, where $M + N + R + P = \triangle ABC$.

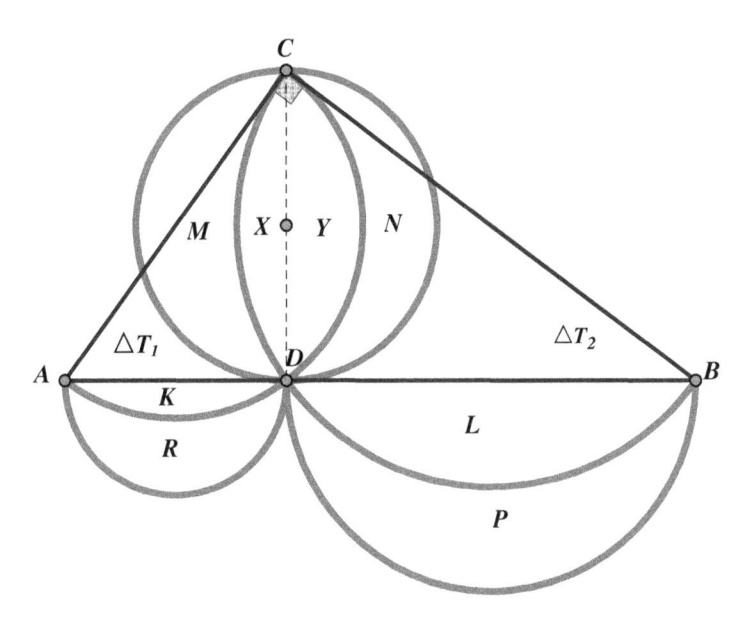

Figure 7-38

Proof

In Figure 7-38, as an application of the Lune of Hippocrates, we see $R + N = \triangle T_1 = \triangle ADC$ and $P + M = \triangle T_2 = \triangle BDC$. Thus, by addition, we get $M + N + R + P = \triangle T_1 + \triangle T_2 = \triangle ABC$.

The Semicircles Forming the Arbelos

The semicircle has an important role among circle problems. This is seen with the *arbelos*, or *shoemaker's knife*, which is shown Figure 7-39 and formed by three semicircles where the sum of the diameters of the two smaller semicircles is equal to that of the larger semicircle.

It is not intuitively obvious, however, that the sum of the two smaller semicircular arcs is equal in length to the larger semicircular arc. As we can see in Figure 7-40, where the radii of the three semicircles are $AD = r_1$, $BE = r_2$, and $AO = R$, the sum of the two smaller semicircular arc lengths is $\pi r_1 + \pi r_2 = \pi(r_1 + r_2) = \pi R$, which is the arc length of the larger semicircle.

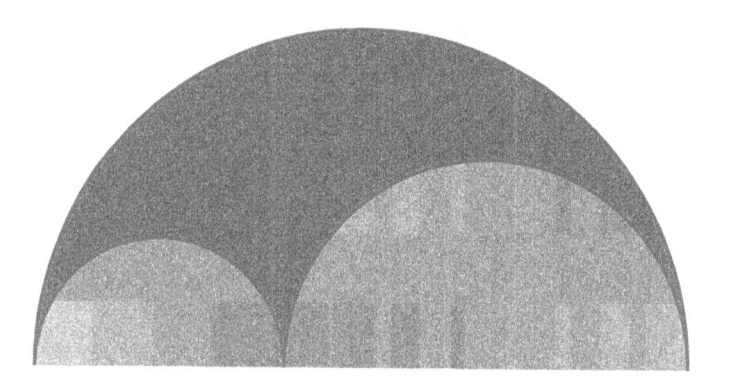

Figure 7-39

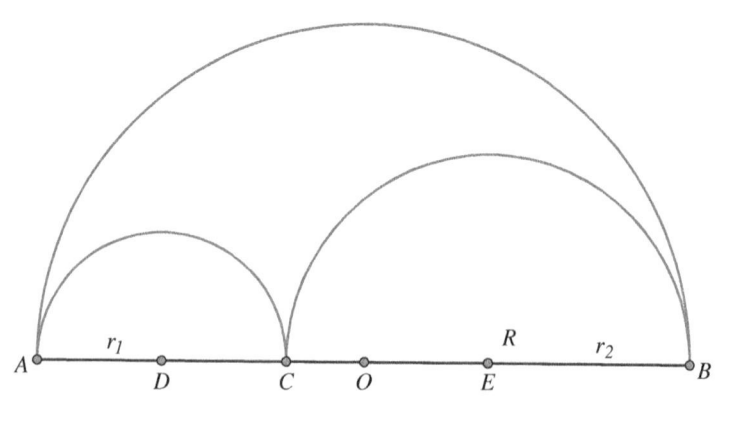

Figure 7-40

The Area of the Arbelos

We construct a perpendicular line to the line segment AB through point C to intersect the larger semicircle at point H, as shown in Figure 7-41. The area of the circle with diameter CH equals the area of the arbelos.

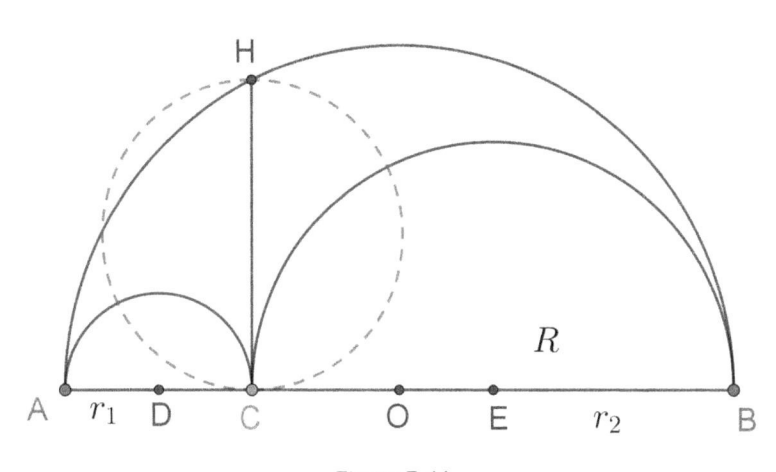

Figure 7-41

Proof

The area of the arbelos can be found by taking the area of the largest semicircle and subtracting from it the areas of the two smaller semicircles, shown in Figure 7-41, as follows: $\frac{\pi R^2}{2} - \left(\frac{\pi r_1^2}{2} + \frac{\pi r_2^2}{2}\right) = \frac{\pi}{2}\left(R^2 - r_1^2 - r_2^2\right)$. However, $R = r_1 + r_2$; therefore, $\frac{\pi}{2}\left(R^2 - r_1^2 - r_2^2\right) = \frac{\pi}{2}\left((r_1 + r_2)^2 - r_1^2 - r_2^2\right) = \pi r_1 r_2$.

HC is on the one hand the diameter of the circle, and on the other hand a mean proportional between the two segments along the hypotenuse, namely, AC and BC, which means $HC^2 = 2r_1 \cdot 2r_2 = 4r_1 r_2$. Consequently, the radius of the circle is then $\sqrt{r_1 r_2}$, and its area $\pi r_1 r_2$, which is the same area as the arbelos.

A Rectangle in the Arbelos Configuration

In the arbelos configuration, one can see many more astonishing properties. This time we focus on a special rectangle formed by points H and C and the two intersection points F and G of the circle with diameter HC and the semicircles, as seen in Figure 7-42. Furthermore, HA passes through point F and HB passes through point G.

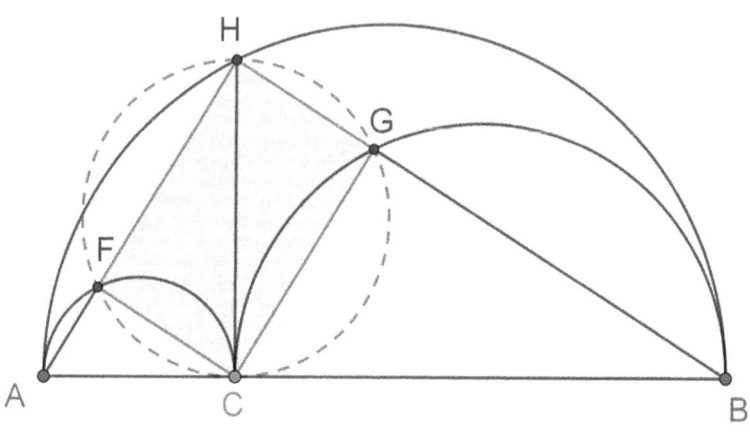

Figure 7-42

Proof

Since a triangle inscribed in a semicircle is a right angle, the quadrilateral $FCGH$ has two right angles at F and G. Similarly, $\angle AFC$ and $\angle CGB$ are two more right angles in Figure 7-42 and are inscribed in the two smaller semicircles. This proves the last mentioned claim that HA passes through F and HB passes through G. Using the three angles of 90° in quadrilateral $FCGH$ in Figure 7-42, another right angle, $\angle AHB = 90°$, is established, which enables us to conclude that $FCGH$ has four right angles and is thus a rectangle. As a bonus, we get $FG = HC$ since rectangles have equal diagonals.

A Common Tangent

F and G are not only opposite vertices of a rectangle. We also find that FG is a common tangent to both semicircles of the arbelos, as can be seen in Figure 7-43, revealing another unexpected phenomenon in the arbelos configuration.

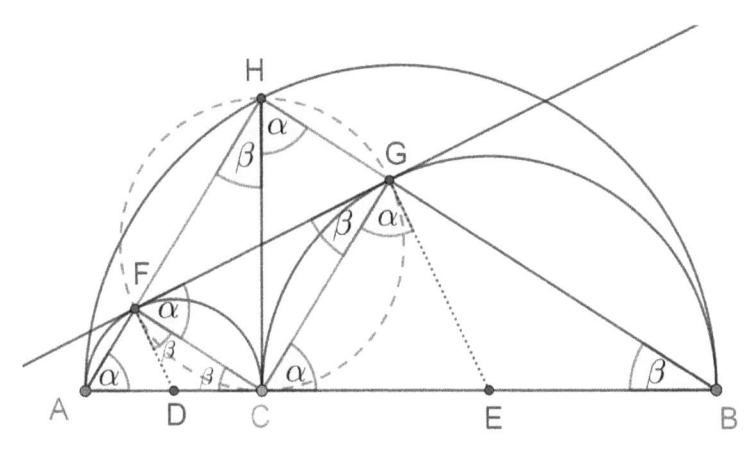

Figure 7-43

Proof

In Figure 7-43, we let α, β be the two acute angles at A and B in the right triangle ABH so that $\alpha + \beta = 90°$. Angle BCG and angle GHC are both inscribed in arc GC so that $\angle BCG = \angle GHC = \alpha$. We also have isosceles triangle GEC, so that $\angle ECG = \angle CGE = \alpha$. Also, since angle GHC and angle CFG are both one-half the measure of arc $\overset{\frown}{GC}$, we have $\angle GHC = \angle CFG = \alpha$. Similarly, $\angle FHC = \angle FGC = \beta$. Also, $\angle FHC$ and $\angle DFC$ are both one half the measure of arc $\overset{\frown}{FC}$, and therefore, $\angle FHC = \angle FGC = \beta$. Since we already know that $\alpha + \beta = 90°$, we have $FD \perp FG$ and $GE \perp FG$. Since in both cases the radius is perpendicular to the line FG, we can conclude that FG is tangent to both semicircles, as claimed. Furthermore, we can see $SF = SC = SG$ in two different ways: on the one hand, these line segments are half diagonals in a rectangle, and on the other hand, they are equal tangent segments to the semicircles.

Another Arbelos Surprise

There are many surprising relationships relating to the arbelos. In Figure 7-44, the midpoints P and U of the two smaller semicircular arcs, and the midpoint Q of the reflection (in AB) of the larger semicircular arc, along with point C, form a quadrilateral whose area (shaded) is equal to the sum of the squares of the radii of the two smaller semicircles.

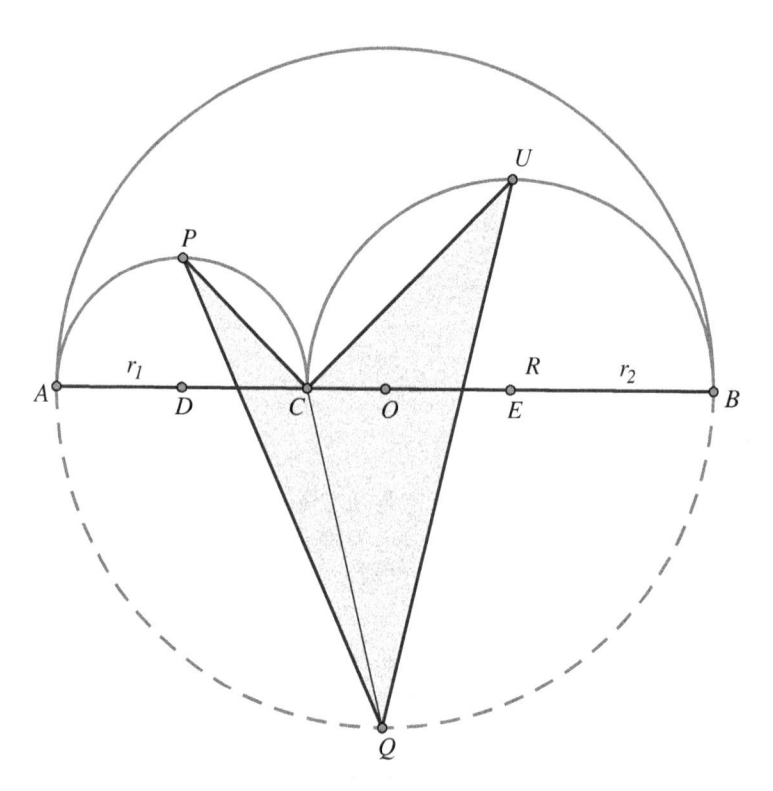

Figure 7-44

Proof

In Figure 7-45, $\angle APC$ is inscribed in a semicircle and is therefore a right angle. Also, because P and Q are midpoints of their respective semicircular arcs, $\angle PAC = 45° = \angle QAC$, thereby making $\angle PAQ$ a right angle. A rectangle is completed by constructing a perpendicular, QT, to the extension of PC. When two triangles share a common base and have equal altitudes, they have equal areas. That is the case with $\triangle PCQ$ and $\triangle PCA$, which are consequently equal in area. Similarly, $\triangle UCQ$ is equal in area to $\triangle UCB$, since QB is parallel to CU. Thus, area$\triangle PCQ$ = area$\triangle APC = \frac{1}{4}(2r_1)^2 = r_1^2$. In a similar fashion, it can be shown that area$\triangle UCQ = \frac{1}{4}(2r_2)^2 = r_2^2$. Consequently, the area of the shaded quadrilateral is equal to $r_1^2 + r_2^2$, which we wanted to prove.

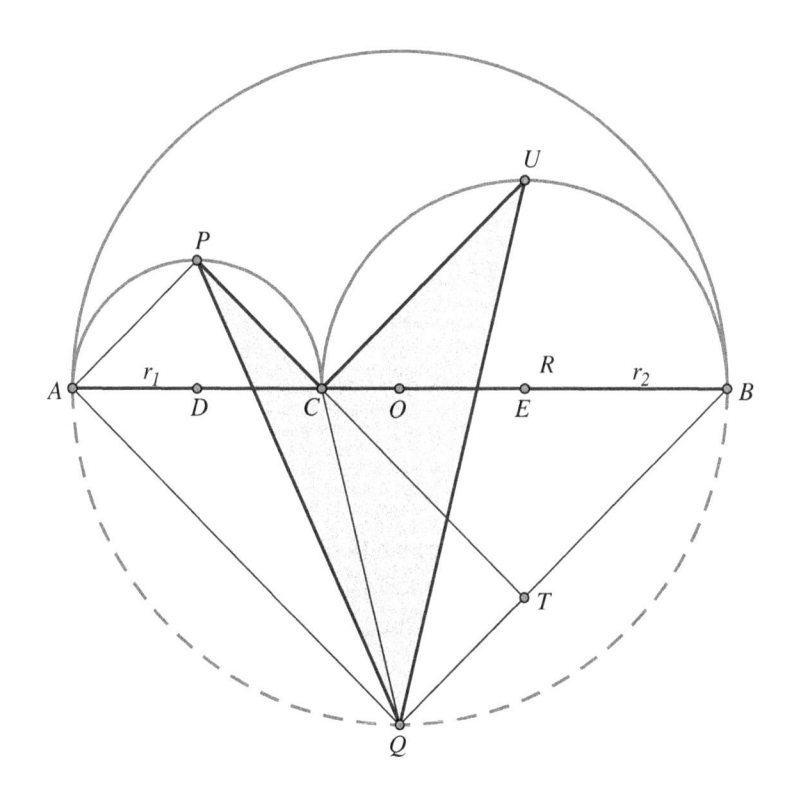

Figure 7-45

Some Entertainment with the Arbelos

Let's consider an entertaining equality of areas. In Figure 7-46, two semicircles (with centers D and E) sit along the diameter of a larger semicircle. One of the semicircles overlaps the larger semicircle, and the other is below it. We then draw a tangent to the small semicircle at point T from the external point A. With AT as a diameter, construct a circle with center R. The unexpected result is that the area of the darker shaded region is equal to the area of the circle with center R.

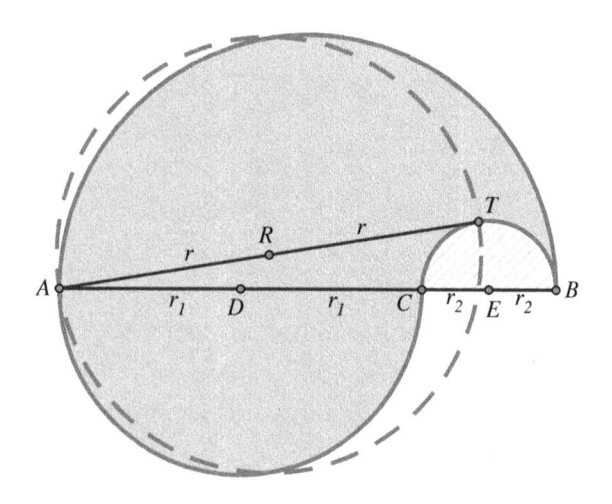

Figure 7-46

Proof

Using the line segments as marked in Figure 7-46, the area of the lower semicircle is $\frac{1}{2}\pi r_1^2$. The area of the upper portion of the darker shaded region is the area of the large semicircle minus the area of the smallest semicircle: $\frac{1}{2}\pi(r_1+r_2)^2 - \frac{1}{2}\pi r_2^2 = \frac{1}{2}\pi(r_1^2+2r_1r_2+r_2^2) - \frac{1}{2}\pi r_2^2 = \frac{1}{2}\pi(r_1^2+2r_1r_2)$. Therefore, the area of the entire darker shaded region is $\frac{1}{2}\pi r_1^2 + \frac{1}{2}\pi(r_1^2+2r_1r_2) = \frac{1}{2}\pi(2r_1^2+2r_1r_2) = \pi(r_1^2+r_1r_2)$. To find the area of the dashed circle with diameter AT, we apply the Pythagorean theorem to $\triangle ATE$: $(2r)^2 + r_2^2 = (2r_1 + r_2)^2 = 4r_1^2 + 4r_1r_2 + r_2^2$, which can be written as $4r^2 = 4r_1^2 + 4r_1r_2$, or $r^2 = r_1^2 + r_1r_2$. Thus, the area of the circle with radius r is simply $\pi r^2 = \pi(r_1^2 + r_1r_2)$, which is also the area of the darker shaded region.

Chapter 8

Polygons and Polygrams

Books on plane geometry, when it comes to linear features, typically focus on triangles and quadrilaterals. In this chapter, we will focus on general polygons.

The Sum of the Interior Angles of a Polygon

Beginning with the notion that the sum of the measures of the angles of a triangle is 180°, one can usually conjecture that the sum of the measures of the angles of a quadrilateral is 360°, since the sum of the measures of the angles of a rectangle (the special type of quadrilateral) is 4 · 90° = 360°. How can one carry this knowledge to find the sum of the interior angles of *any* polygon?

Here, we learn how to "triangulate" the area inside the polygon. That is, partition the region inside the polygon into triangular regions with no overlap and no region omitted. We offer three possible triangulations in Figures 8-1, 8-2, and 8-3. For each, we will derive the sum of the measures of the interior angles of the polygon.

In Figure 8-1, where a vertex of the given polygon is selected and the diagonals are drawn from there to each of the other vertices, it is clear that the sum of the measures of the interior angles of the polygon is the same as the sum of the measures of the angles of the triangles. Using this drawing scheme, the number of triangles will also be 2 less than the number of sides of the polygon.

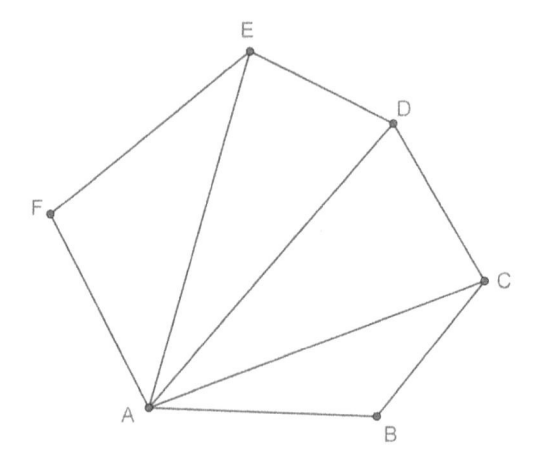

292 *A Journey Through the Wonders of Plane Geometry*

Figure 8-1

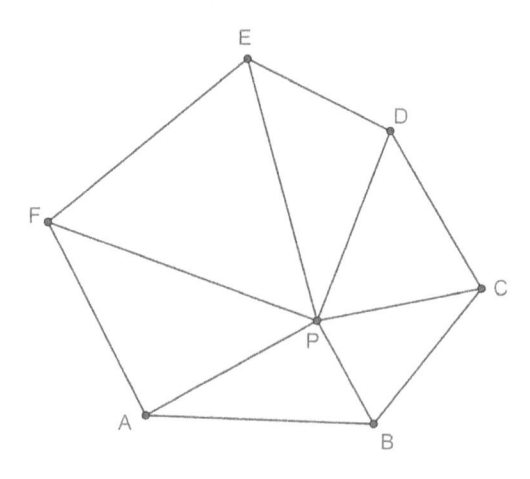

Figure 8-2

Therefore, the angle sum will be 180° times the number of triangles, which for an *n*-sided polygon is $(n-2) \cdot 180°$.

In Figure 8-2, a point is chosen in the interior of the polygon. From that point, segments are drawn to each of the vertices of the polygon. There are *n* triangles for an *n*-sided polygon. The sum of the measures of the interior angles of the polygon is the sum of the measures of the angles of the triangles minus the measures of the angles at point *P*, which is 360°. Therefore, the sum of the measures of the interior angles of an *n*-sided polygon is $n \cdot 180° - 360° = (n-2) \cdot 180°$.

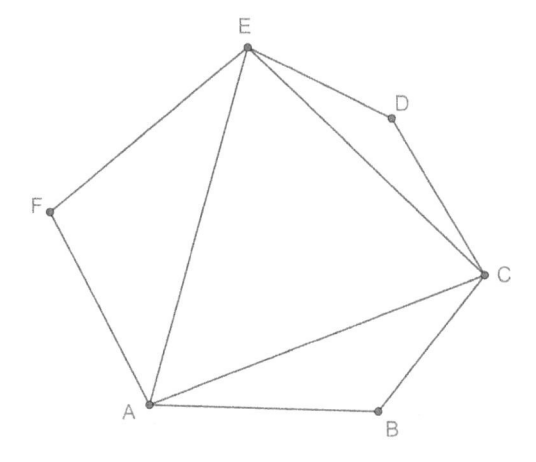

Figure 8-3

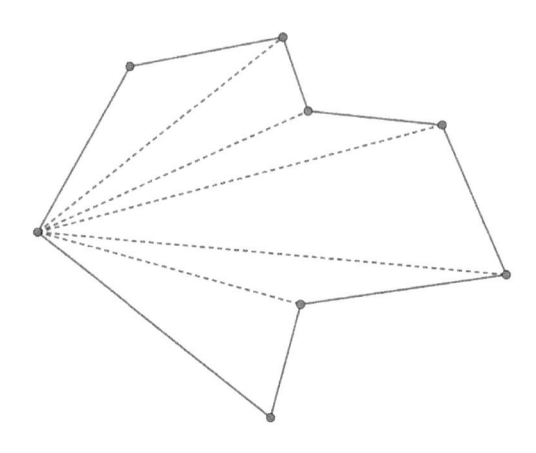

Figure 8-4

In Figure 8-3, the region is triangulated in such a way that diagonals are drawn in such a way that no diagonals intersect any other ones (except at an endpoint). Again, the number of triangles needed to triangulate the region is two less than the number of sides of the polygon. Hence, the result is once again $(n-2) \cdot 180°$.

A motivated reader is encouraged to triangulate the interior region of a polygon in any suitable way and then reach the appropriate interior angle-measure sum. This also works for concave polygons, as we can see in Figure 8-4.

Exterior Angles

To begin, we will consider the most basic polygons: triangles. The sum of the interior angles of a triangle equals 180°, while the sum of the exterior angles is 360°. One can see the latter in two different ways, which can easily be generalized to polygons.

1. Let α, β, γ be the interior angles of a triangle and α', β', γ' its exterior angles. Then one can calculate $\alpha' + \beta' + \gamma' = (180° - \alpha) + (180° - \beta) + (180° - \gamma) = 3 \cdot 180° - (\alpha + \beta + \gamma)$. Since $\alpha + \beta + \gamma = 180°$, we get $\alpha' + \beta' + \gamma' = 2 \cdot 180° = 360°$.

2. A more general approach uses a sketch and a different point of view. We will display the situation for a quadrilateral, and we will see that this approach immediately generalizes to arbitrary *simple*[1] *n*-gons, regardless of the number of sides (*n*) and if it is convex or concave.

Let us imagine that one goes around the convex quadrilateral counterclockwise one vertex at a time, starting at D (shown in Figure 8-5). At each vertex, one must rotate exactly by the exterior angle (counterclockwise, in sum $\alpha' + \beta' + \gamma' + \delta'$). Finally, one will have completed exactly one turn, because one has gone exactly once around the quadrilateral. Thus, we have $\alpha' + \beta' + \gamma' + \delta' = 360°$. This would be the same with every other *convex n*-sided polygon. The sum of the exterior angles is always 360°. In case of a *concave n*-gon, starting at G (shown in Figure 8-6), a vertex with a reflex angle v, the exterior angle would be negative (here it is important to think of *oriented*[2] angles: $v' = 180° - \underset{>180°}{v} < 0$) and one would have to turn v' in the other direction (clockwise, at points D and G in Figure 8-6). Nevertheless, the same result holds: the sum of the exterior angles equals 360° (Note: Negative exterior angles at reflex interior angles) because one has gone around the polygon counterclockwise exactly once.

---

[1] This means that the boundary does not cross itself and that the polygon has no holes.
[2] These are angles with signs, positive counterclockwise, negative clockwise. Sometimes these angles are also called "directed angles".

Polygons and Polygrams **295**

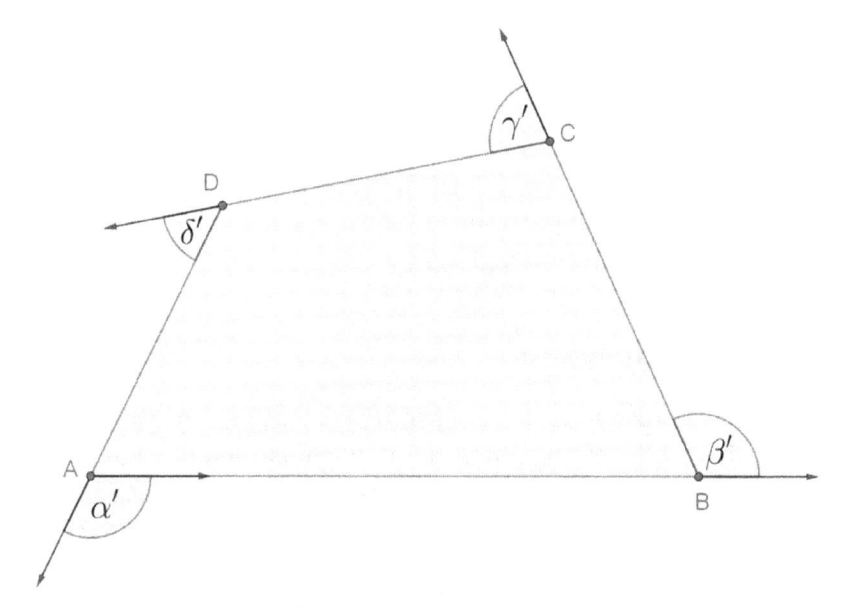

Figure 8-5

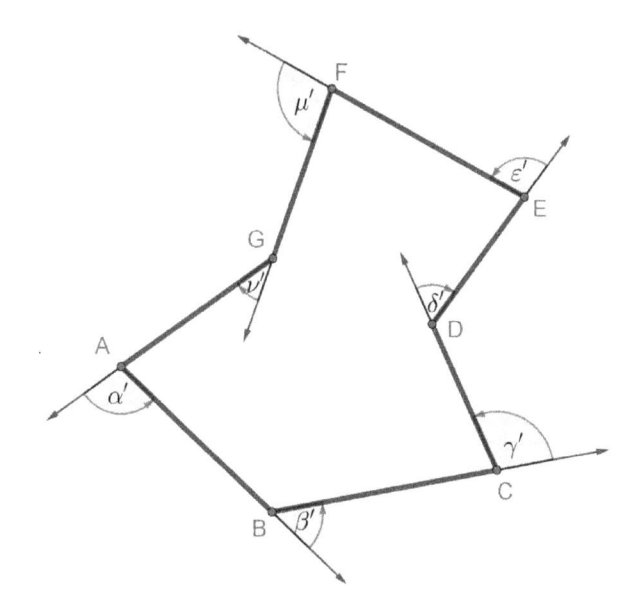

Figure 8-6

So, in a sense, the *sum of the exterior angles* of a simple *n*-gon turns out to be the easier object (as it is 360° for all *n*) than the *sum of its interior angles*.

The corresponding formula for the *sum of the interior angles* of a *n*-gon, $(n - 2) \cdot 180°$, is obtained by triangulating the polygon and noticing that it can be partitioned into $n - 2$ triangles, each having 180° as sum of its interior angles, as we saw above.

Angles in Polygrams

We will now consider some astonishing properties concerning angle sums of the angles of *polygrams*. First, we must define what a polygram is. The most famous example is the *pentagram*, which has the same number of vertices as a convex pentagon (see Figure 8-7), but connects alternating vertices, as shown in Figure 8-8.

The common notation of such polygrams is $\{n/m\}$. In the above case, the notation $\{5/2\}$ would indicate that we have $n = 5$ vertices, and $m = 2$ indicates that we join every second point, counterclockwise, starting at *A* and continuing with *B*, *C*, *D*, and *E*, yielding the pentagram *ABCDE*. It is clear that $m = 3$ yields the same result, which can also be seen in Figure 8-8 (*AEDCB* is the same as *ABCDE*, just in the other direction).

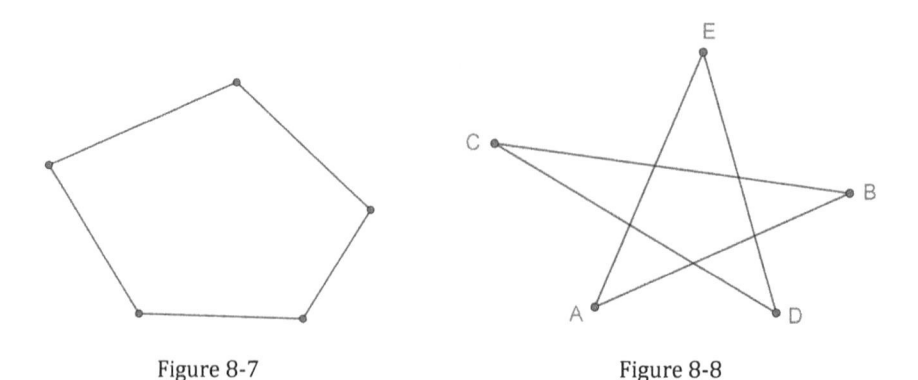

Figure 8-7

Figure 8-8

We will restrict our focus to cases of $\{n/m\}$ where $m < \frac{n}{2}$ and $\gcd(n, m) = 1$, because $\{n/m\}$ is the same as $\{n/n - m\}$. In an n-gon, going m points further *counterclockwise* yields the same as going $n - m$ points further *clockwise*; thus, $\{n/m\}$ is the same as $\{n/n - m\}$, just in the other direction. If we had $\gcd(n, m) = d > 1$, then the polygram would be divided into d polygons (this is called a *compound polygon*), as we can see for example in $\{6/2\} = 2\{3/1\}$ (Figure 8-9), a hexagon divided into two triangles. Another example would be $\{12/3\} = 3\{4/1\}$, a dodecagon divided into three quadrilaterals (Figure 8-10). As mentioned earlier, such cases we will not consider.

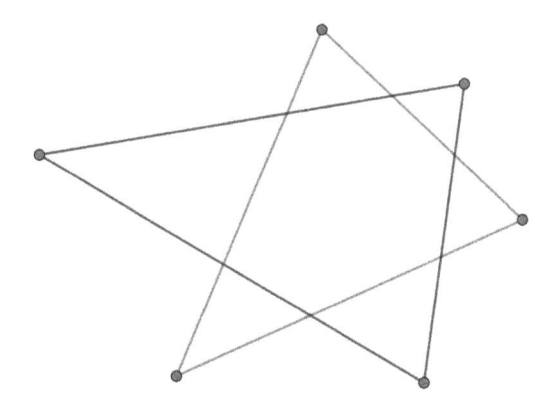

Figure 8-9

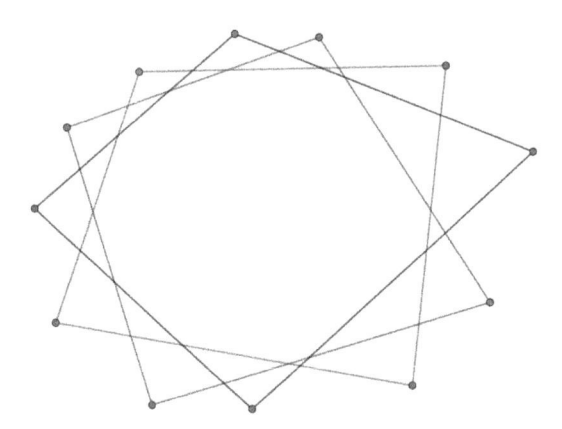

Figure 8-10

298   A Journey Through the Wonders of Plane Geometry

The interesting polygrams with $n \leq 10$ sides are (see Figure 8-11 with the *regular* versions):

- Pentagram – {5/2}
- Heptagram – {7/2} and {7/3}
- Octagram – {8/3}
- Nonagram – {9/2} and {9/4}
- Decagram – {10/3}

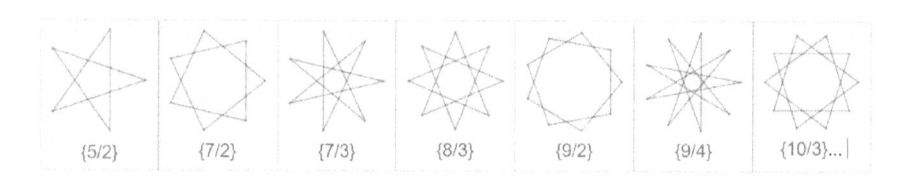

| {5/2} | {7/2} | {7/3} | {8/3} | {9/2} | {9/4} | {10/3}... |

Figure 8-11

For a *regular* pentagram, it is not difficult to determine the sum of the angle measures at the five points (vertices) as 180°. But what about pentagrams that are *not regular*, like in Figure 8-8? Dynamic geometry will yield the conjectured insight that in all cases of a pentagram, the angle measure sum at the vertices is 180°. There are many different proofs for that phenomenon, and the simplest one is sort of a "proof without words" (Figure 8-12). Just creating parallels to each of the sides at one vertex immediately yields the claimed result.[3]

Furthermore, this easy and elucidating technique can be used to show that the angle sum at the vertices is 180° in the (not necessarily regular) cases of the heptagram {7/3} and the nonagram {9/4}. Additionally, it can also be used to determine the angle sum in the other mentioned cases, the only difference being that the resulting angle is not 180° but a multiple of it (360°, 540°, 720°, 900°). In Figure 8-13, we apply the technique to the case {8/3} and the angle sum 360°.

---

[3] This scheme can be seen in W. Zeuge. *Nützliche und schöne Geometrie*. SpringerSpektrum, Berlin, 2021.

Polygons and Polygrams **299**

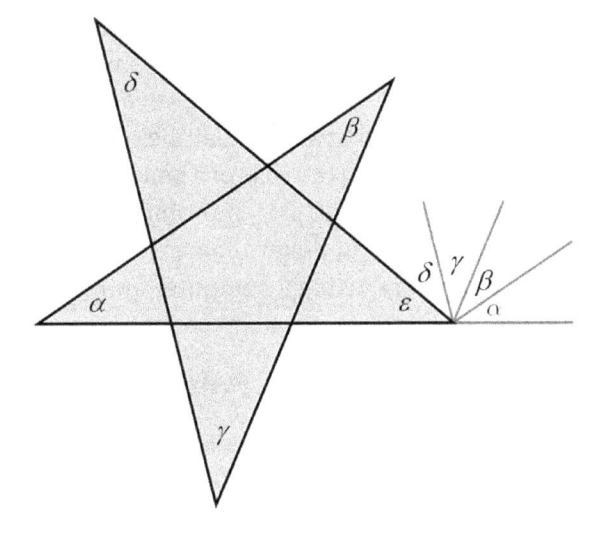

Figure 8-12

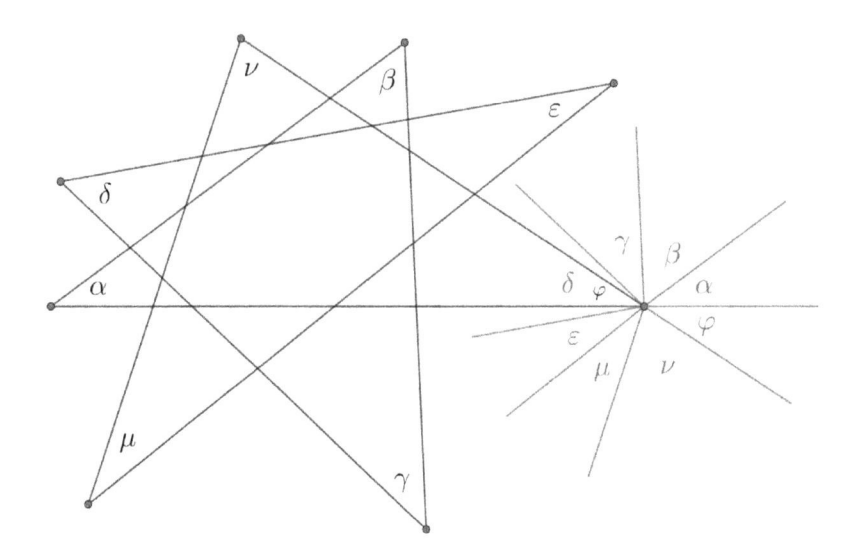

Figure 8-13

For the other heptagram {7/2}, we get an angle sum of 540°, the decagram {10/3} yields an angle sum of 720°, and the other nonagram {9/2} has angle sum of 900°. It is impressive to see the power of the simple method "translating" lines parallel through one single point. Most other proofs, even in the case of a pentagram, involve much more algebra and geometry. At https://puzzling.stackexchange.com/questions/17681/five-angles-in-a-star one can find another short explanation, without algebra, using a computer animation.

Lengths of Line Segments in Polygrams

There is another curiosity concerning polygrams. Let us start again with the pentagram {5/2} and denote the line segments and angles as in Figure 8-14. It turns out[4] that $a_1 b_1 c_1 d_1 e_1 = a_2 b_2 c_2 d_2 e_2$ always holds true, even in irregular pentagrams.

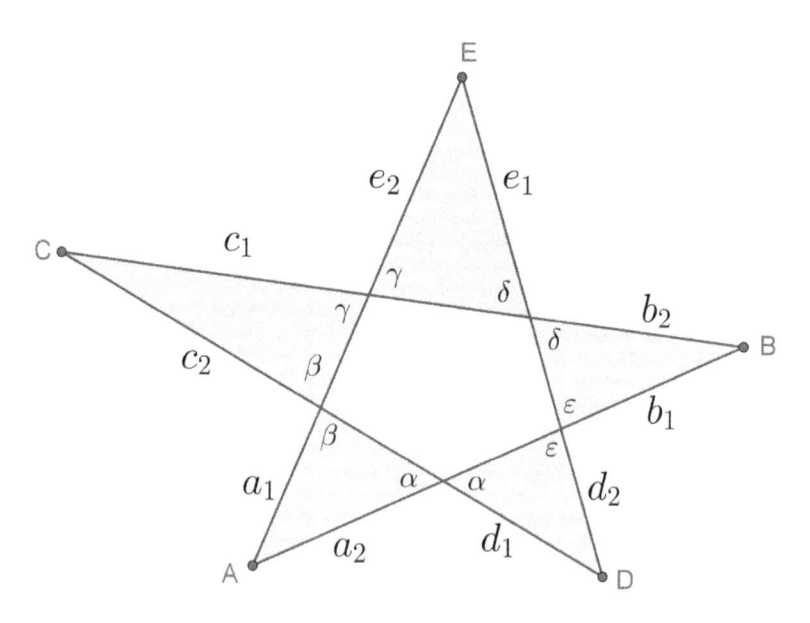

Figure 8-14

---

[4] See e.g. C. S. Lee. Polishing the star. *College Mathematics Journal* **29**, 144–145, (1998).

Proof

Applying the law of sines in each of the shaded triangles in Figure 8-14 yields.

$$\frac{a_1}{a_2} = \frac{\sin\alpha}{\sin\beta}, \quad \frac{b_1}{b_2} = \frac{\sin\delta}{\sin\varepsilon}, \quad \frac{c_1}{c_2} = \frac{\sin\beta}{\sin\gamma}, \quad \frac{d_1}{d_2} = \frac{\sin\varepsilon}{\sin\alpha}, \quad \frac{e_1}{e_2} = \frac{\sin\gamma}{\sin\delta}.$$

Multiplying these five equations together gives us $\frac{a_1 b_1 c_1 d_1 e_1}{a_2 b_2 c_2 d_2 e_2} = 1$, which is what we sought to prove. This immediately generalizes to other polygrams {(2n + 1)/2}, such as {7/2} or {9/2}. Once more, we see in a striking way that polygrams (not only regular polygrams!) have some unexpected "regularities."

Determining the Area of a Regular Dodecagon With Circumradius 1

Most regular polygons inscribed in a unit circle have an irrational area, such as $\frac{3}{2}\sqrt{3}$ for the regular hexagon. It is easy to see that a square inscribed a unit circle has an area of $\sqrt{2} \cdot \sqrt{2} = 2$, an integer. One can also see the area of a regular dodecagon inscribed a unit circle is exactly 3, which can be obtained by using the well-known trigonometric area formula for the area of triangles: $\frac{ab}{2}\sin\gamma$, where the lengths of the sides are a, b and the included angle measure is γ. Applied to the regular dodecagon, this gives its area as

$$12 \cdot \frac{1 \cdot 1}{2} \cdot \underbrace{\sin 30°}_{\frac{1}{2}} = 3.$$

In order for the area of a regular n-gon inscribed a unit circle to be a rational number, $\sin\frac{360°}{n}$ must be a rational number, which is the case for $n = 4$ and $n = 12$. According to Niven's theorem,[5] there are no other such cases.

---

[5] Ivan Morton Niven (1915–999), American-Canadian mathematician.

An Unexpected Dissection of a Regular Dodecagon

In a regular dodecagon, the "central" rectangle containing two opposite sides, which in Figure 8-15 is shaded, partitions the dodecagon into thirds.

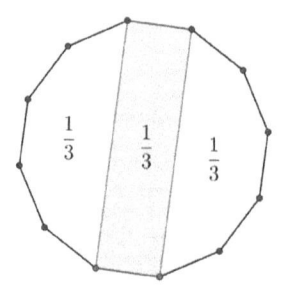

Figure 8-15

Proof

Let a be the side length of the regular dodecagon and r its inradius. The area of the whole dodecagon is $12 \cdot \frac{ar}{2} = 6ar$, which is obtained by partitioning the dodecagon into 12 triangles formed by the radii of the dodecagon. The area of the shaded rectangle is $2ar$, which is one-third of the whole area. Thus, by symmetry, the other two regions must each also be one-third of the area of the dodecagon.

An Isosceles Triangle in a Regular Octagon

In a regular octagon, an isosceles triangle is shaded, as shown in Figure 8-16. The challenge is to find the fraction of the shaded and

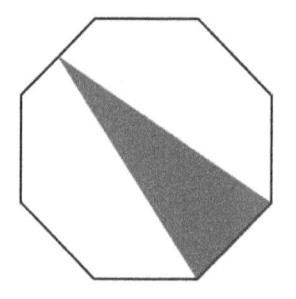

Figure 8-16

unshaded areas of the area of the entire octagon without actually calculating the areas.

Solution

If the apex of the isosceles triangle were in the center of the octagon, the triangle obviously would have $\frac{1}{8}$ of the octagon. Since the shaded triangle in Figure 8-16 has the same base (a side of the octagon) and twice the altitude, it must have twice the area, or $\frac{1}{4}$ of the octagon. Hence, the fraction of the area of each unshaded pentagon is $\frac{1-\frac{1}{4}}{2} = \frac{3}{8}$.

An Astonishing Property of Regular Heptagons

Let A, B, C, and D be consecutive vertices of a regular heptagon, as shown in Figure 8-17. Let P denote the intersection point of AC and BD. It follows that $AB + AP = AD$.

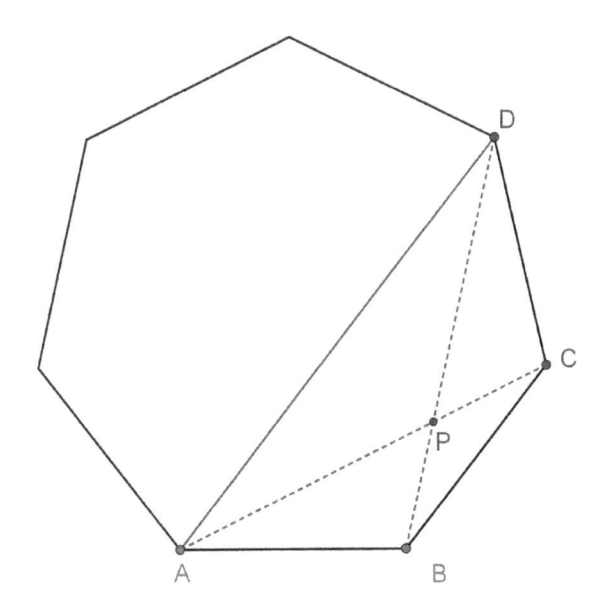

Figure 8-17

Proof

Let $\alpha = \frac{1}{2} \cdot \frac{360°}{7} = \frac{180°}{7}$. By the inscribed angle theorem $\alpha = \angle ADB = \angle BDC = \angle CAD$, as shown in Figure 8-18. From the angle sum in $\triangle ACD$, we can conclude $\angle ACD = 180° - 3\alpha = 180° - 3\left(\frac{180°}{7}\right) = 180°\left(1 - \frac{3}{7}\right) = 180°\left(\frac{4}{7}\right) = 4\left(\frac{180°}{7}\right) = 4\alpha$; thus, the exterior angle $\angle CPD = 2\alpha$. We rotate $\triangle PCD$ about P counterclockwise so that $D' = A$ (note: in isosceles triangle APD, we have $AP = DP$). Then, C' lies on AD, and we have on the one hand $AB = CD = C'D'$, and on the other hand $\angle PC'D = 3\alpha = \angle DPC'$. This enables us to conclude that $\triangle C'PD$ is isosceles, proving the claimed result $AB + AP = AC' + DP = AC' + C'D = AD$.

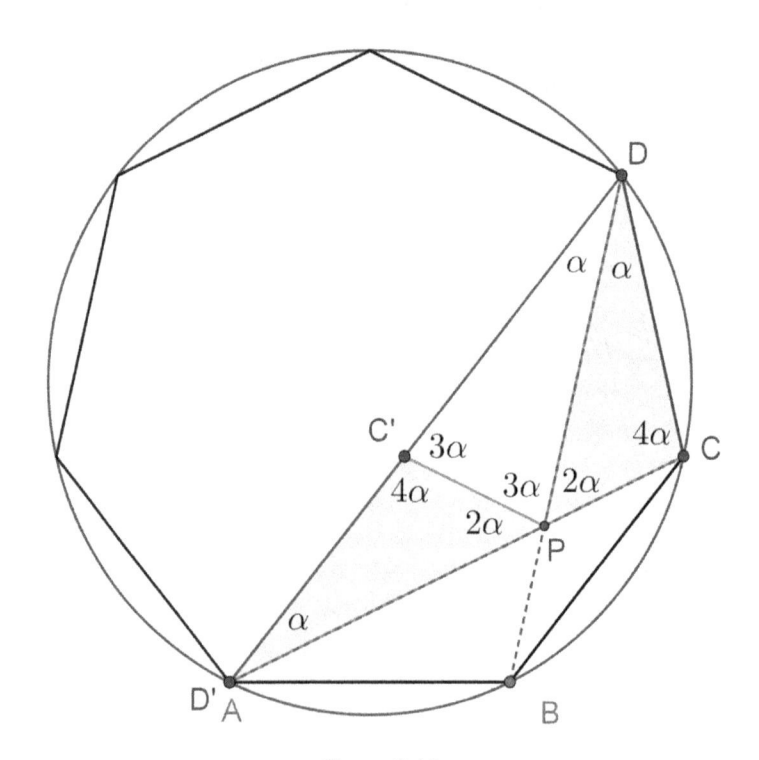

Figure 8-18

An Octagon with Fixed Area Ratio Within a Parallelogram

If each vertex of a parallelogram is joined with midpoints of the two opposite sides, as shown in Figure 8-19, then these lines form a convex octagon. In the general case, this octagon is not regular, but it has some striking properties: (1) the opposite sides are parallel and equal; (2) its area is always $\frac{1}{6}$ of the initial parallelogram.

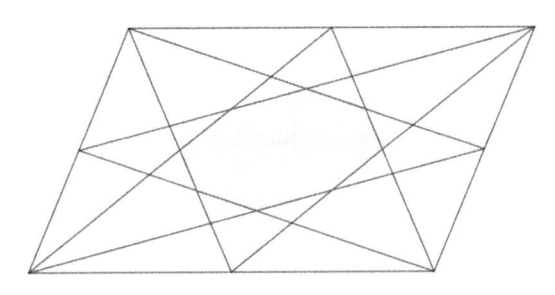

Figure 8-19

Proof

Property (1) is clear by the half-turn symmetry of the parallelogram. Because of the parallelogram's half-turn symmetry, the opposite sides of the mentioned octagon also have this half-turn symmetry; thus, they are equal and parallel. The proof for property (2) requires some further figures (Figures 8-20, 8-21 and 8-22). In Figure 8-20, the

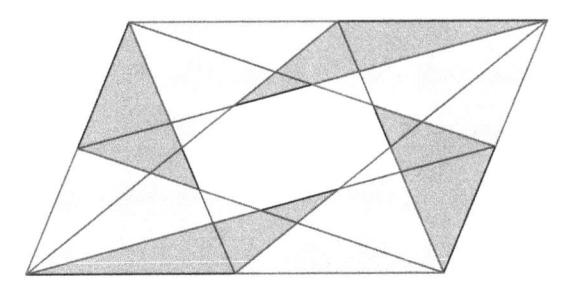

Figure 8-20

octagon, shaded in Figure 8-19, is unshaded, and two groups of triangles are shaded light gray and dark gray. These two groups are drawn again separately in Figure 8-21 and Figure 8-22. Let us assume the initial parallelogram has an area of 1. The four light gray triangles of Figure 8-22 have an area of $\frac{1}{8}$ each (half a side of the parallelogram as base, and half the height of the parallelogram as altitude), and they all add up to $\frac{1}{2}$.

Now consider the four dark gray triangles of Figure 8-22. The diagonals of the parallelogram divide it into pairs of congruent triangles with area $\frac{1}{2}$. Since the medians of every triangle dissect it into 6 smaller triangles with equal area, we know that each dark gray triangle has area $\frac{1}{6} \cdot \frac{1}{2} = \frac{1}{12}$. Hence, four of them sum up to $\frac{1}{3}$, and for the octagon, an area of $1 - \frac{1}{2} - \frac{1}{3} = \frac{1}{6}$ remains.

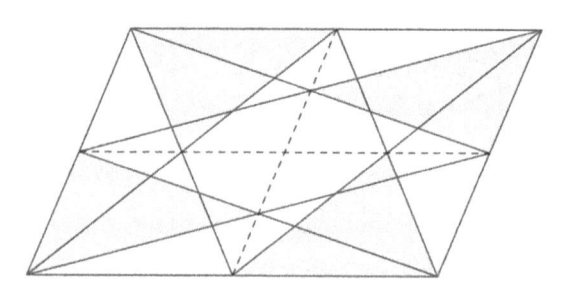

Figure 8-21

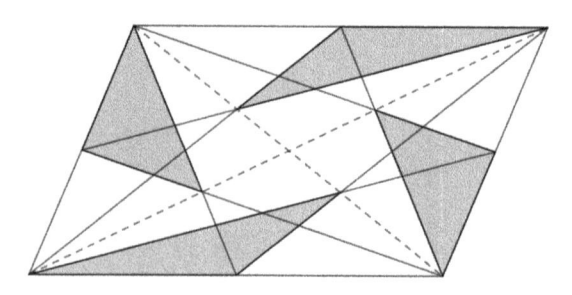

Figure 8-22

Similar Octagons Inscribed in Congruent Squares

In Figures 8-23 and 8-24, two octagons are inscribed in congruent squares. In the octagon of Figure 8-23, the vertices are joined with the midpoints of the opposite sides. In the octagon of Figure 8-24, the

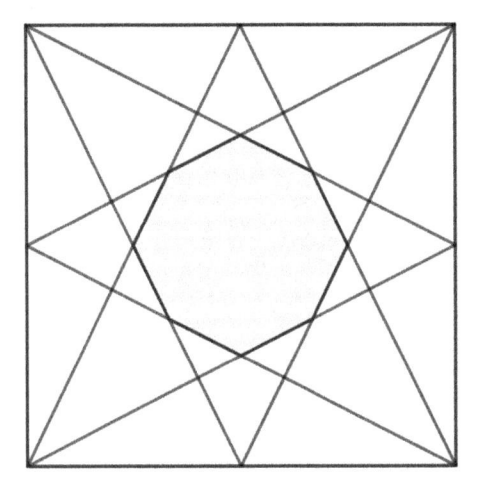

Figure 8-23

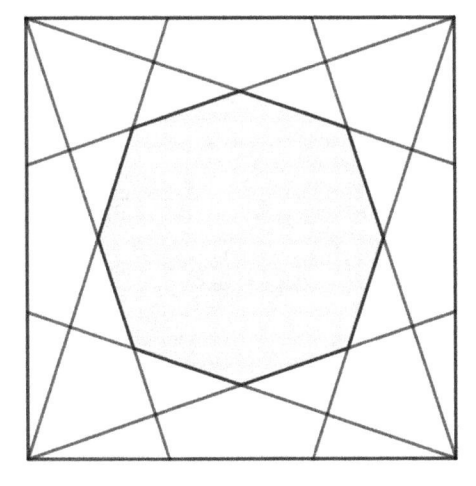

Figure 8-24

vertices are joined with the trisection points of the opposite sides. Unexpectedly, the two octagons are *similar*. Furthermore, the ratio of their areas is 1:2.

For another approach to this phenomenon (based on the tangent function), see Alsina & Nelsen, 2023.[6]

A short and elucidating proof requires a phenomenon that is easy to see and will help us a lot. In Figure 8-25, the diagonal *AC* is trisected by the points of intersection *P* and *Q* with *DG* and *DF*. Since *E* is the midpoint of *AD* and *EB*∥*DF*, we know *AP* = *PQ*. Similarly, with *F* instead of *E*, *PQ* = *QC*, and altogether *AP* = *PQ* = *QC*.

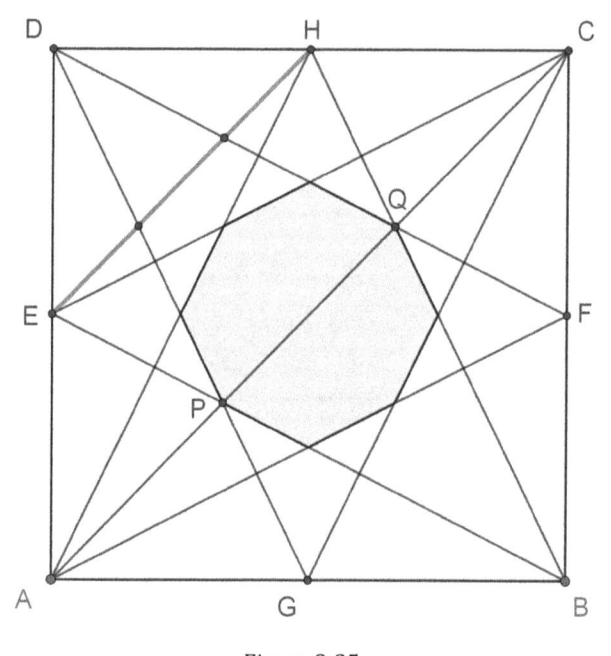

Figure 8-25

---

[6] C. Alsina & R. B. Nelsen R. B. *A Panoply of Polygons*. AMS /MAA. Dolciani Mathematical Expositions, vol 58, p. 140 and 233f (2023).

Since the triangles $\triangle DAC$ and $\triangle DEH$ (Figure 8-25) are homothetic with center D and factor $\frac{1}{2}$ (see Chapter 11), this implies that EH is also trisected by DG and DF.

Now we are prepared to prove the similarity of the octagons in Figure 8-23 and Figure 8-24. With the help of Figure 8-26 and the above relationships, we can see that the octagon of Figure 8-23 is also an octagon of the type of Figure 8-24 with respect to the square with vertices at the midpoints of the initial square. Since the two squares have the similarity ratio 1:$\sqrt{2}$, the ratio for the areas of the octagons of Figure 8-23 and Figure 8-24 is 1:2.

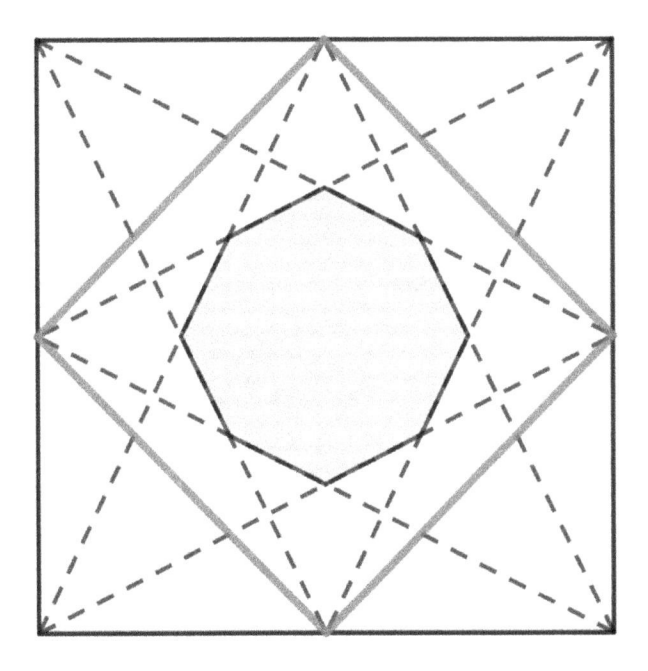

Figure 8-26

Regular n-gons Inscribed in a Unit Circle: Sums of Squared Chords

In this section, we are interested in polygon diagonals, more generally, in chords and their unexpected properties concerning the sum of their squares.

Let us start with a concrete example, consider a regular octagon ($n = 8$) inscribed in a unit circle and an arbitrary point P on this unit circle, as shown in Figure 8-27. We then draw all the chords $s_1, \ldots, s_8$

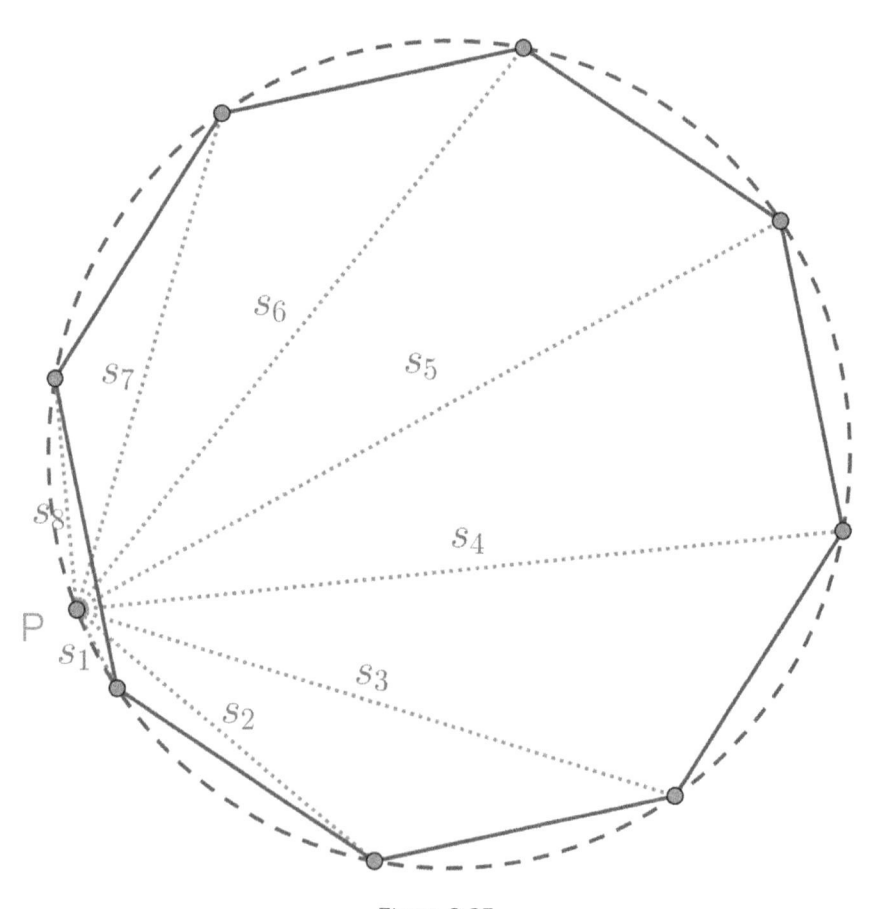

Figure 8-27

from point P to the vertices of the octagon and calculate the sum of their squares: $s_1^2 + \cdots + s_8^2$. With dynamic geometry you will observe that this sum equals 16, independent of the position of P on the unit circle. However, a proof that this is in fact true is necessary.

Proof

Since $n = 8$ is an even number, each vertex of the octagon has a diametrical opposite vertex (diameter = 2). By applying the Pythagorean theorem to those right triangles formed with the diameter as the hypotenuse, such as $s_2^2 + s_6^2 = 4$, we can get the equations $s_1^2 + s_5^2 = 4, s_2^2 + s_6^2 = 4, s_3^2 + s_7^2 = 4, s_4^2 + s_8^2 = 4$, the sum of which is 16. If P happens to be one of the vertices, then one of the chords $s_1, \ldots, s_8$ vanishes, but nothing else changes.

This amazing relationship can be generalized to other regular n-gons with even $n = 2k$.

This unexpected relation $s_1^2 + \cdots + s_n^2 = 2n$ even holds true for odd n, but then the proof is not as simple, as it could employ complex numbers. Motivated readers are invited to explore this situation with dynamic geometry, say with $n = 5$ or $n = 7$.

Regular n-gons Inscribed in a Unit Circle: Products of Chords

In this section, we will discuss an even more astonishing phenomenon regarding chords in a regular n-gon inscribed in a unit circle. In contrast to the above section, the starting point P now must be one of the *vertices* of the regular n-gon.

Let us start with the easiest case, $n = 4$. Given a square inscribed in the unit circle, shown in Figure 8-28, take an arbitrary vertex point P and join it with the other three vertices. Take the product $s_1 \cdot s_2 \cdot s_3$ of the resulting chords and the result will be 4. This is easy to see because $s_1 = s_3 = \sqrt{2}$ and $s_2 = 2$ (diameter) so that $s_1 \cdot s_2 \cdot s_3 = \sqrt{2} \cdot 2 \cdot \sqrt{2} = 4$.

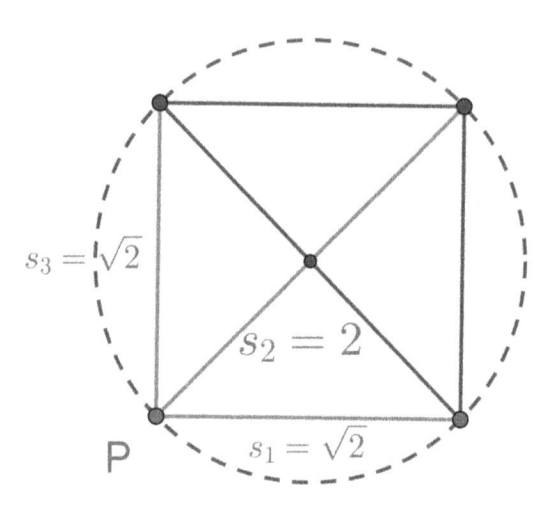

Figure 8-28

Now we will explore the case $n = 6$, the regular hexagon, shown in Figure 8-29. In this case, we have $s_1 = s_5 = 1$ and $s_3 = 2$ (diameter). The chords $s_2 = s_4$ are twice the altitude of an equilateral triangle with side 1, which means that each has length $\sqrt{3}$. Therefore, $s_1 \cdot s_2 \cdot s_3 \cdot s_4 \cdot s_5 = 6$.

Let us now examine two regular polygon cases for odd n. The easiest is $n = 3$. The Pythagorean theorem makes clear that the side length of an equilateral triangle inscribed within the unit circle is $\sqrt{3}$, which in this case equals s_1 and s_2; thus, in the case $n = 3$, we have the relation $s_1 \cdot s_2 = 3$.

The next odd number is $n = 5$. The side length of a regular pentagon inscribed in a unit circle is $s_1 = \sqrt{\frac{5-\sqrt{5}}{2}} = s_4$ (explanation below), shown in Figure 8-30. The corresponding diagonals $s_3 = s_4$

Polygons and Polygrams **313**

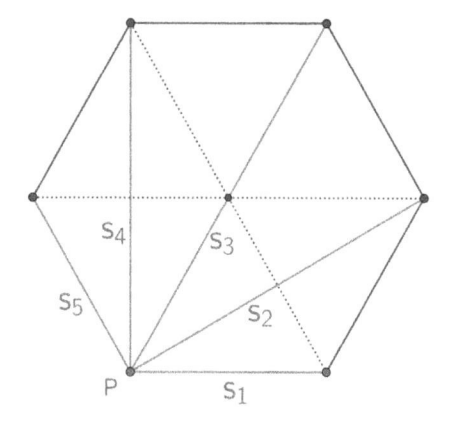

Figure 8-29

we get by multiplication of this length by the *Golden Ratio* $\varphi = \frac{1+\sqrt{5}}{2}$ (see Chapter 1). The product $s_1 \cdot s_2 \cdot s_3 \cdot s_4$ thus yields

$$\underbrace{\sqrt{\frac{5-\sqrt{5}}{2}}}_{s_1} \cdot \underbrace{\left(\frac{\sqrt{5-\sqrt{5}}}{2} \cdot \frac{1+\sqrt{5}}{2}\right)}_{s_2} \cdot \underbrace{\left(\frac{\sqrt{5-\sqrt{5}}}{2} \cdot \frac{1+\sqrt{5}}{2}\right)}_{s_3} \cdot \underbrace{\sqrt{\frac{5-\sqrt{5}}{2}}}_{s_4}$$

$$= \left(\frac{5-\sqrt{5}}{2} \cdot \frac{1+\sqrt{5}}{2}\right)^2 = \left(\frac{5+4\sqrt{5}-5}{4}\right)^2 = 5.$$

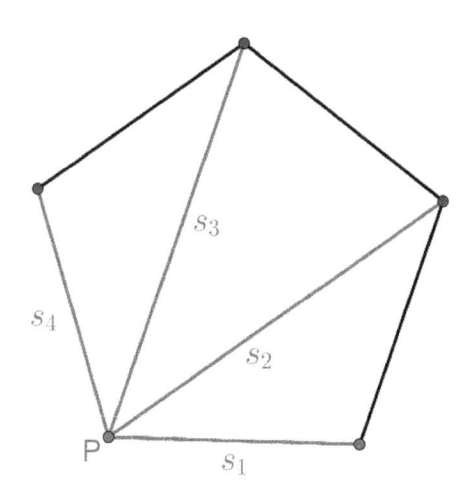

Figure 8-30

To determine the side length $\sqrt{\frac{5-\sqrt{5}}{2}}$ of a regular pentagon inscribed in a unit circle, we see in Figure 8-31 that this side length $a = 2\sin 36°$, and in Figure 8-32 we see that $\cos 36° = \frac{\varphi}{2}$. Remember that $AC = a \cdot \varphi$, as we had in Chapter 1; thus, we can use the characteristic equation $\varphi^2 = \varphi + 1$ for the Golden Ratio φ:

$$a = 2\sin 36° = 2\sqrt{1 - \cos^2 36°} = 2\sqrt{1 - \left(\frac{\varphi}{2}\right)^2} = \sqrt{4 - \frac{\varphi^2}{\varphi+1}} = \sqrt{4 - (\varphi+1)}$$

$$= \sqrt{3 - \varphi} = \sqrt{3 - \frac{1+\sqrt{5}}{2}} = \sqrt{\frac{5-\sqrt{5}}{2}}$$

This enables us to make the following conjecture:

For arbitrary $n \geq 3$ in a regular n-gon inscribed a unit circle, the relation $s_1 \cdots s_{n-1} = n$ holds, where s_i ($i = 1, \ldots, n-1$) are all chords joining one vertex with all the others.

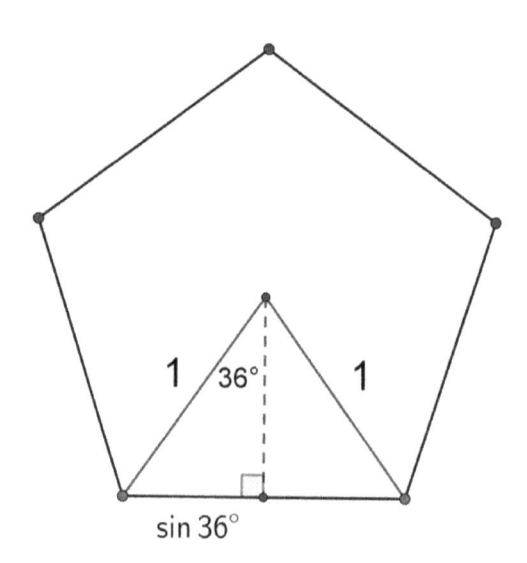

Figure 8-31

Polygons and Polygrams **315**

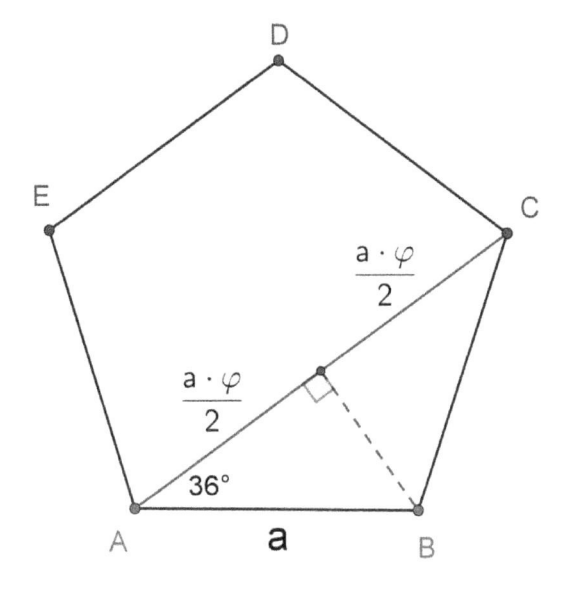

Figure 8-32

This conjecture is correct, but the general proof can hardly be given geometrically. One could use complex numbers for a short and elegant proof, but that is beyond the scope of this book. Instead, the reader is invited to do the calculations for other concrete values of n (e.g., $n = 8, 10, 12$), or to research the corresponding values and thereby confirm the conjecture for these n.

This theorem has interesting and unexpected consequences. Here is one example: Imagine the huge number $n = 1000$; according to the theorem (conjecture) above $s_1 \cdots s_{999} = 1000$, where the s_i ($i = 1, \ldots, 999$) denote the corresponding chords of the regular 1000-gon. For $n = 500$ we have $t_1 \cdots t_{499} = 500$, where the t_j ($j = 1, \ldots, 499$) denote the corresponding chords of the regular 500-gon. For every $j = 1, \ldots, 499$, we have $t_j = s_{2j}$ (note: taking every second vertex in a 1000-gon yields a 500-gon). Thus, we know the product of the s_i even when index $i = 2j$ is 500: $s_2 \cdot s_4 \cdots s_{996} \cdot s_{998} = 500$. Hence, the product of the s_i with odd index must be 2: $s_1 \cdot s_3 \cdots s_{997} \cdot s_{999} = 2$, which is smaller (than 500) by the factor 250.

This is due to the very small values of $s_1 = s_{999} \approx \frac{\pi}{500} \approx \frac{1}{160}$, the side length of the regular 1000-gon. If one takes one of them out of the product, then the number of the odd indexed chords is equal to the number of the even indexed chords, and one gets approximately 320 for the product, much closer to 500.

It is obvious that the analogous phenomenon holds true for every even number $n = 2k$. The product of the even-indexed chords is k, and the product of the odd-indexed chords is always 2, independent of k, even for arbitrarily huge even numbers like 1 million ($s_1 \cdot s_3 \cdots s_{999,997} \cdot s_{999,999} = 2$) or 1 trillion. This is a genuinely unexpected consequence.

The generalization even goes one step further. Let $n = j \cdot k$ with $s_1 \cdot s_2 \cdots s_{jk-1} = j \cdot k$. If we take in this $(j \cdot k)$-gon every k-th vertex, we have a j-gon; thus, $s_k \cdot s_{2k} \cdots s_{(j-1)k} = j$, which enables us to conclude that the product of all the other s_i (where i is not divisible by k) must be k. In other words, trying to eliminate the value k by leaving out of the product all the s_i whose index is a multiple of k takes the product exactly to k (so "k is back again"). What a curiosity!

Lattice Polygons and Pick's Theorem

A polygon that has integer coordinates for all its vertices (lattice points) is called a *lattice polygon*. Let i be the number of lattice points that are in the interior of the polygon, and let b be the number of lattice points on its boundary (including vertices as well as lattice points along the sides of the polygon). The area A of this polygon is $A = i + \frac{b}{2} - 1$.

This is *Pick's theorem*, for simple polygons (these are polygons that do not intersect themselves and have no holes) and it was published in 1899 by the Austrian mathematician Georg Alexander Pick (1859–1942). A consequence of this theorem is that the area of a lattice polygon is always an integer or .5. If asked "Do you think there exists a lattice polygon with an area of, say, 3456.8?" many people, ignorant of this theorem, would probably answer "Yes, why not?" Bear in mind that the lattice polygon can be arbitrarily complicated, with hundreds or even thousands of vertices, some concave (meaning a reflex interior angle at this vertex), but the correct answer to the question is "No, it cannot exist!"

We will now prove this theorem step by step. These steps again show the importance of problem solving strategies such as considering special cases and working backwards.

Step 1: We will prove that Pick's theorem holds for lattice rectangles with sides on lines parallel to the axes, and for lattice right triangles with legs on lines parallel to the axes.

These are two special cases and will prove useful to the whole theorem. Let us start with the case of lattice rectangles with sides on lines parallel to the axes, as shown in Figure 8-33. Let the natural numbers a, and c be the length and the width of the rectangle. The area is $A = ac$.

When counting the lattice points of a side, we count on every side one of the vertices, and it is clear that $b = 2a + 2c$. From Figure 8-33, we have $i = (a - 1)(c - 1)$. If we insert that into the claimed formula, we get $i + \frac{b}{2} - 1 = (a-1)(c-1) + (a+c) - 1 = ac$, and this is the area A.

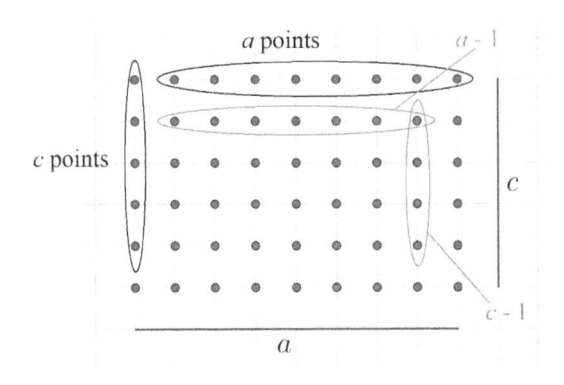

Figure 8-33   Lattice rectangle with sides parallel to the axes

318 A Journey Through the Wonders of Plane Geometry

Now consider a lattice right triangle with legs on lines parallel to the axes, as we show in Figure 8-34.

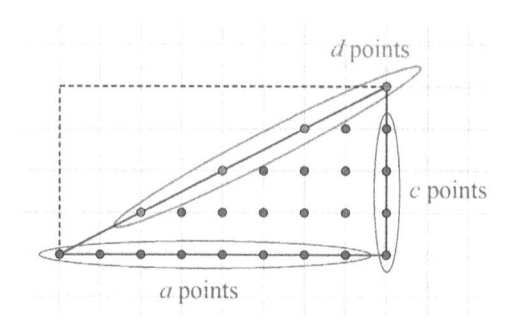

Figure 8-34   Lattice triangle with legs parallel to the axes

Here, we have $b = a + c + d$, where d denotes the number of lattice points on the diagonal (hypotenuse), with one vertex included, $i = \overbrace{\frac{(a-1)(c-1)}{2}}^{\text{interior of the rectangle}} - \overbrace{(d-1)}^{\text{hypotenuse (without vertices)}}$. Inserting $b = a + c + d$ and $i = \frac{(a-1)(c-1)-(d-1)}{2}$ yields $i + \frac{b}{2} - 1 = \frac{(a-1)(c-1)-(d-1)+a+c+d}{2} - 1 = \frac{ac}{2}$, which is the area.

One could consider many other special cases. Instead, the next steps, 2 and 3, *work backwards* to answer the question: What would we have to know in order to prove the theorem generally? Answer: We would have to know that the formula holds for arbitrary lattice triangles, and that the validity of Pick's formula for an arbitrary lattice polygon can be derived by its validity for the triangles that triangulate the polygon.

Step 2: We prove that Pick's formula is *additive* in the following sense: If two lattice polygons $P_{1,2}$ are put together so that they share one common side, and if Pick's formula holds for $P_{1,2}$, then it holds also for the bigger polygon $P = P_1 \cup P_2$, where the common side of $P_{1,2}$ is deleted, as can be seen in Figure 8-35.

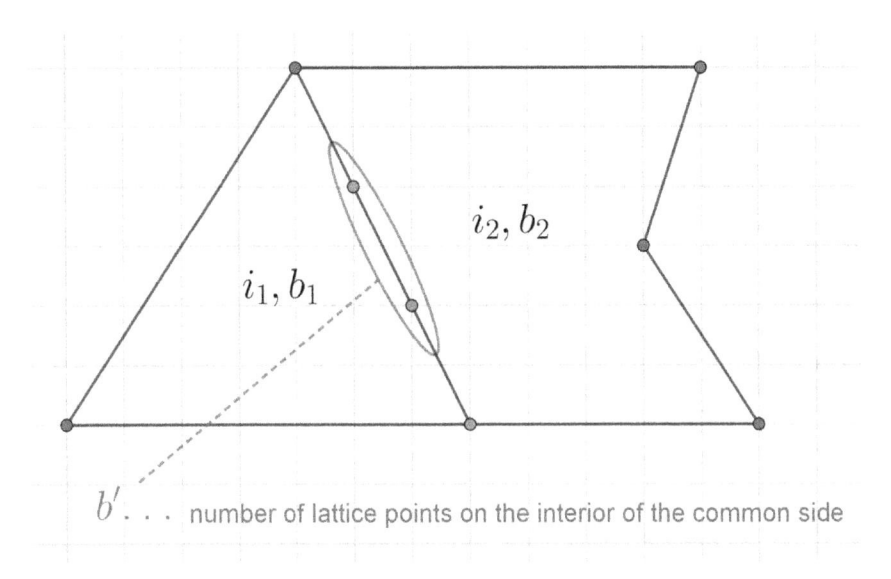

Figure 8-35 *Additivity* of Pick's formula

Proof

From the precondition, we know that Pick's formula holds for $P_{1,2}$, that means $A_1 = i_1 + \frac{b_1}{2} - 1$ and $A_2 = i_2 + \frac{b_2}{2} - 1$. We also know $A = A_1 + A_2$. Thus, we must prove that in the big polygon $i + \frac{b}{2} - 1 = \left(i_1 + \frac{b_1}{2} - 1\right) + \left(i_2 + \frac{b_2}{2} - 1\right)$.

How can we express i, b in the big polygon? We have $i = i_1 + i_2 + b'$ and $b = b_1 + b_2 - 2b' - 2$, where b' is counted twice in $b_1 + b_2$, but it must not be counted at all for b; the two vertices at the ends of the common side are counted twice in $b_1 + b_2$, but we must count them only once. And again, inserting these values yields $i + \frac{b}{2} - 1 = (i_1 + i_2 + b') + \left(\frac{b_1}{2} + \frac{b_2}{2} - b' - 1\right) - 1 = \left(i_1 + \frac{b_1}{2} - 1\right) + \left(i_2 + \frac{b_2}{2} - 1\right)$, as claimed.

Step 3: We prove that Pick's formula holds for all lattice triangles.

Proof

We want to put the lattice rectangle R from step 1 around the lattice triangle T. Here, three cases are possible (depending on the position of point C, seen in Figures 8-36a, 8-36b, and 8-36c):

(1) One needs three added triangles from step 1 (Figure 8-36a)

(2) One needs two added triangles from step 1 (Figure 8-36b)

(3) One needs three added triangles and one added rectangle from step 1 (Figure 8-36c)

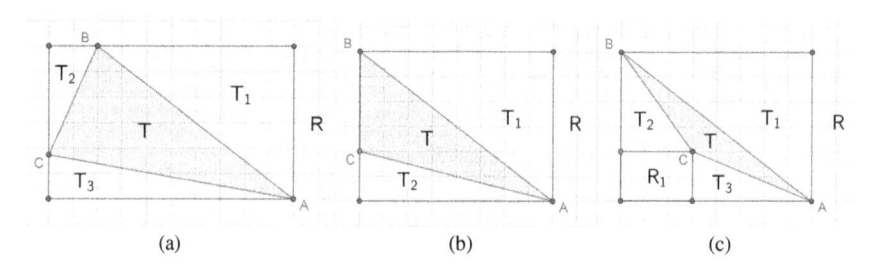

Figure 8-36a–c Three cases of putting a lattice rectangle around a lattice triangle

We will use the most complex case (3) – the others work analogously. Let $P(V)$ denote the number resulting from Pick's formula applied to a polygon V. We know from step 1 that $P(R)$, $P(R_1)$, $P(T_1)$, $P(T_2)$, $P(T_3)$ are the corresponding areas. The additivity of Pick's formula (step 2) gives us $P(R) = P(R_1) + P(T_1) + P(T_2) + P(T_3) + P(T)$. Thus, $P(T)$ must also be the area of T since it complements the areas of R_1, T_1, T_2, T_3 to the area of lattice rectangle R.

Step 4: We use steps 1, 2, 3 for proving Pick's theorem.

Proof

Every lattice polygon can be triangulated by lattice triangles, and from step 3 we know that Pick's formula holds for all these triangles. Thus, due to step 2 (the formula is *additive*), it holds for all lattice polygons.

Chapter 9
Geometric Surprises

There are many ways we can be surprised and impressed by the wonders of geometry. There are situations where relationships occur that are unexpected, some of which we have seen in prior chapters, such as the collinearity of points and concurrency of lines. There are also special relationships that often carry the name of the mathematician who discovered them and relate to familiar figures such as triangles, quadrilaterals, and circles. The placement and relationships of shapes, and not just their internal properties, will lead to rather counterintuitive results. Many of these named theorems provide powerful ways to understand, admire, and solve some unusual geometric problems. We begin by introducing theorems discovered centuries ago that can serve as invaluable tools in developing and justifying some amazing geometric relationships.

Desargues' Surprising Theorem

In his book *Manière universelle de M. Desargues, pour pratiquer la perspective*, the French mathematician Gérard Desargues (1591–1661) presents a theorem about an unusual geometric relationship that in the nineteenth century became fundamental to projective geometry. It involves placing any two triangles in a position that will enable the three lines joining corresponding vertices to be concurrent. Remarkably, when this is achieved, the extensions of the pairs of corresponding sides intersect in three collinear points. As we can see in Figure 9-1, $\triangle A_1B_1C_1$ and $\triangle A_2B_2C_2$ are situated so that the lines joining the corresponding vertices, A_1A_2, B_1B_2, and C_1C_2, are concurrent, and the extensions of the pairs of corresponding sides intersect in three collinear points; this is Desargues' theorem. More specifically, lines B_2C_2 and B_1C_1 meet at A'; lines A_2C_2 and A_1C_1 meet at B'; and lines B_2A_2 and B_1A_1 meet at C'.

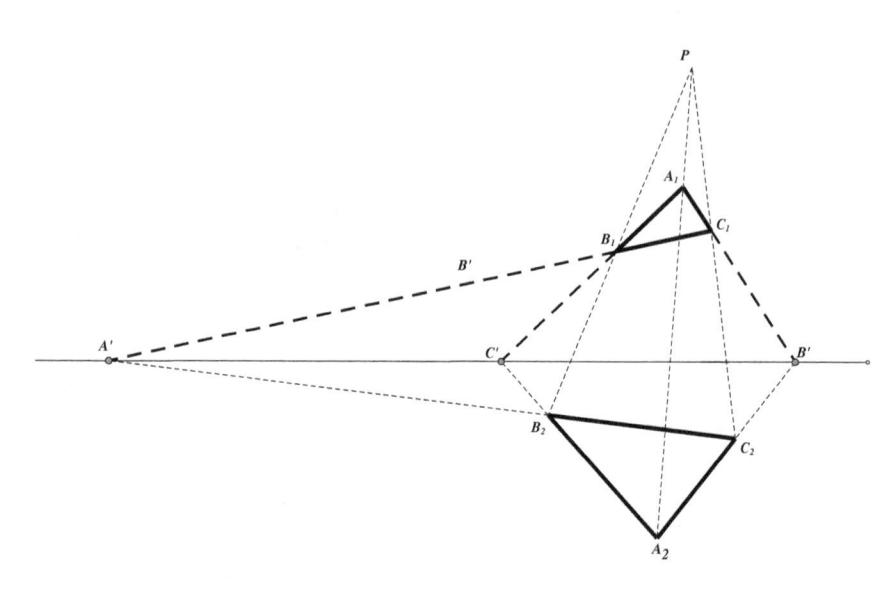

Figure 9-1

Proof

We shall prove *Desargues' theorem* by using Menelaus' theorem. Consider $A'B_1C_1$ to be a transversal of $\triangle PB_2C_2$, as seen in Figure 9-1. Therefore, by Menelaus' theorem:

$$\frac{PB_1}{B_2B_1} \cdot \frac{B_2A'}{C_2A'} \cdot \frac{C_2C_1}{PC_1} = 1. \tag{I}$$

Similarly, considering $C'B_1A_1$ as a transversal of $\triangle PB_2A_2$, by Menelaus' theorem:

$$\frac{PA_1}{A_2A_1} \cdot \frac{A_2C'}{B_2C'} \cdot \frac{B_2B_1}{PB_1} = 1. \qquad \text{(II)}$$

Now taking $B'A_1C_1$ as a transversal of $\triangle PA_2C_2$, once again, by Menelaus' theorem:

$$\frac{PC_1}{C_2C_1} \cdot \frac{C_2B'}{A_2B'} \cdot \frac{A_2A_1}{PA_1} = 1. \qquad \text{(III)}$$

By multiplying (I), (II), and (III), we get $\frac{B_2A'}{C_2A'} \cdot \frac{A_2C'}{B_2C'} \cdot \frac{C_2B'}{A_2B'} = 1$. Thus, applying Menelaus' theorem to $\triangle A_2B_2C_2$ yields points A', B', and C' collinear. It should be noted that the converse of Desargues' theorem is also true.

Pascal's Theorem

Blaise Pascal (1623–1662), a contemporary of Desargues, is regarded today as one of the true geniuses in the history of mathematics. Although eccentricities kept him from achieving his true potential, he is considered one of the originators of the formalized study of probability (an outgrowth of his correspondences with Pierre de Fermat) and he made many important contributions to other branches of mathematics. Of interest here is one of his contributions to geometry.

In 1640, at the age of sixteen, Pascal published a one-page paper entitled *Essay pour les coniques*. It contained a theorem that Pascal referred to as *mysterium hexagrammicum*. The work impressed René Descartes, who could not believe it was the work of a boy. The theorem states that the intersections of the opposite sides of a hexagon inscribed in a conic section are collinear. For our purposes, we shall consider only the case where the conic section is a circle, and the hexagon has *no* pair of opposite sides parallel. The theorem states that if a hexagon, with no pair of opposite sides parallel, is inscribed

324 *A Journey Through the Wonders of Plane Geometry*

in a circle, then the intersections of the opposite sides are collinear. This is illustrated in Figure 9-2.

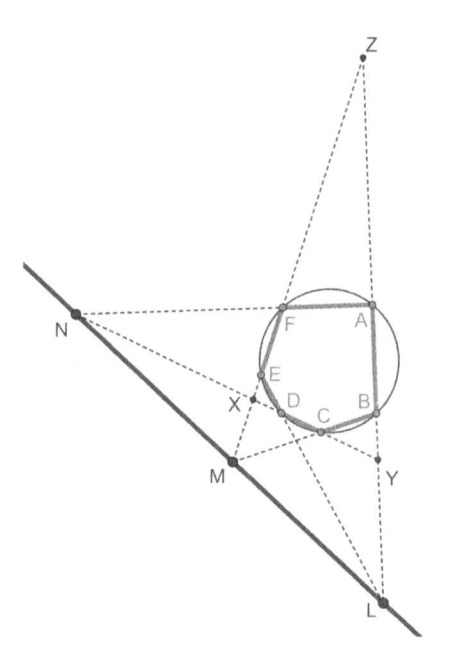

Figure 9-2

Proof

Hexagon *ABCDEF*, shown in Figure 9-2, is inscribed in a circle. The pairs of opposite sides *AB* and *DE* intersect at point *L*, sides *CB* and *EF* intersect at point *M*, and sides *CD* and *AF* intersect at point *N*. Furthermore, *EF* intersects *CN* at point *X*, *DC* intersects *AL* at point *Y*, and *EF* intersects *AB* at point *Z*. Consider *BC* to be a transversal of $\triangle XYZ$. By Menelaus' theorem:

$$\frac{ZB}{YB} \cdot \frac{YC}{XC} \cdot \frac{XM}{ZM} = 1. \tag{I}$$

Taking *AF* to be a transversal of △*XYZ*, by Menelaus' theorem:

$$\frac{ZA}{YA} \cdot \frac{XF}{ZF} \cdot \frac{YN}{XN} = 1, \qquad \text{(II)}$$

Also, since *DE* is a transversal of △*XYZ*, by Menelaus' theorem:

$$\frac{YD}{XD} \cdot \frac{XE}{ZE} \cdot \frac{ZL}{YL} = 1. \qquad \text{(III)}$$

By multiplying (I), (II), and (III), we get

$$\frac{YM}{MZ} \cdot \frac{XN}{NY} \cdot \frac{ZL}{LX} \cdot \frac{(ZB)(ZA)}{(ZE)(ZF)} \cdot \frac{(XF)(XE)}{(XC)(XD)} \cdot \frac{(YD)(YC)}{(YA)(YB)} = 1. \qquad \text{(IV)}$$

When two secant segments are drawn to a circle from an external point, the product of the lengths of one secant and its external segment equals the product of the lengths of the other secant and its external segment. Thus,

$$\frac{(ZB)(ZA)}{(ZE)(ZF)} = 1, \qquad \text{(V)}$$

$$\frac{(XF)(XE)}{(XC)(XD)} = 1, \qquad \text{(VI)}$$

and

$$\frac{(YD)(YC)}{(YA)(YB)} = 1. \qquad \text{(VII)}$$

By substituting (V), (VI), and (VII) into (IV), we get

$$\frac{XM}{ZM} \cdot \frac{YN}{XN} \cdot \frac{ZL}{YL} = 1.$$

Thus, once again by Menelaus' theorem, points *M*, *N*, and *L* must be collinear. Furthermore, this theorem can be extended in the following manner: If a hexagon has its vertices on a circle in any order, then the intersections (if they exist) of the opposite sides are collinear. For an example of this variation, follow the above proof using the diagram in Figure 9-3. Only one minor adjustment needs to be made, and that is the justifications for (V) through (VII), which are now based on the intersections of chords, the product of whose segments are equal. Remember, the same pairs of "opposite sides" are used here as were used earlier.

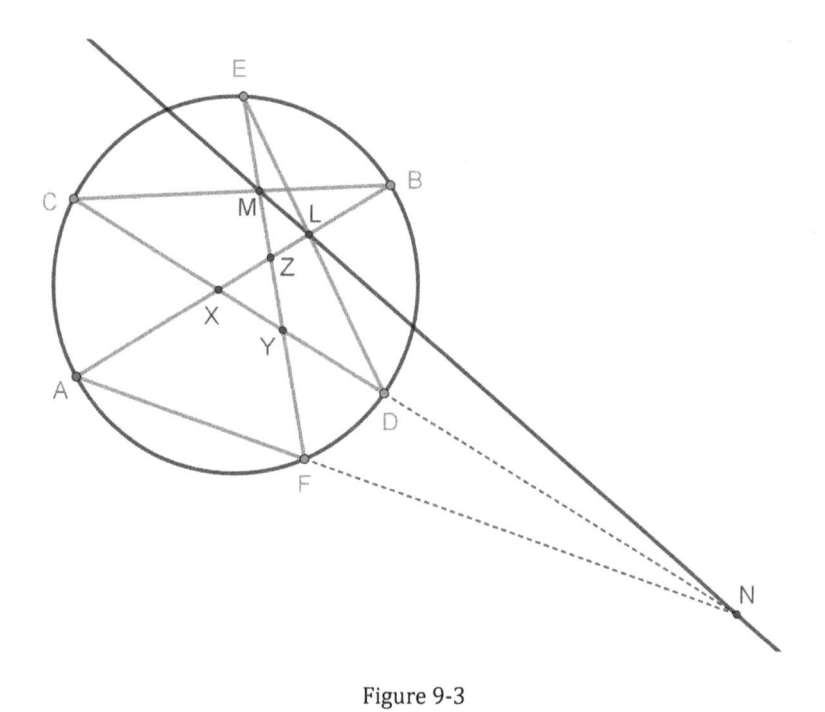

Figure 9-3

Pascal's Theorem Application 1

Pascal's theorem has many applications, but we shall only consider a few here. We begin by considering Figure 9-4, where point P is any point in the interior of $\triangle ABC$. Points M and N are the feet of the perpendiculars from P to AB and AC, respectively. $AK \perp CP$ at K and $AL \perp BP$ at L. Unexpectedly, we find that KM, LN, and BC are concurrent.

Proof

We can easily prove that the points A, K, M, P, N, and L all lie on the circle with diameter AP, shown in Figure 9-4. This can be justified by realizing that right angles AKP and AMP are inscribed in the same semicircle, as is the case for right angles ALP and ANP. Now using the variation to Pascal's theorem, we notice that for inscribed hexagon $AKMPNL$, the pairs of opposite sides intersect as follows: $AM \cap LN = B$, $AN \cap KP = C$, and $KM \cap LN = Q$. By Pascal's theorem, B, C, and Q are collinear, which is to say that KM, LN, and BC are concurrent.

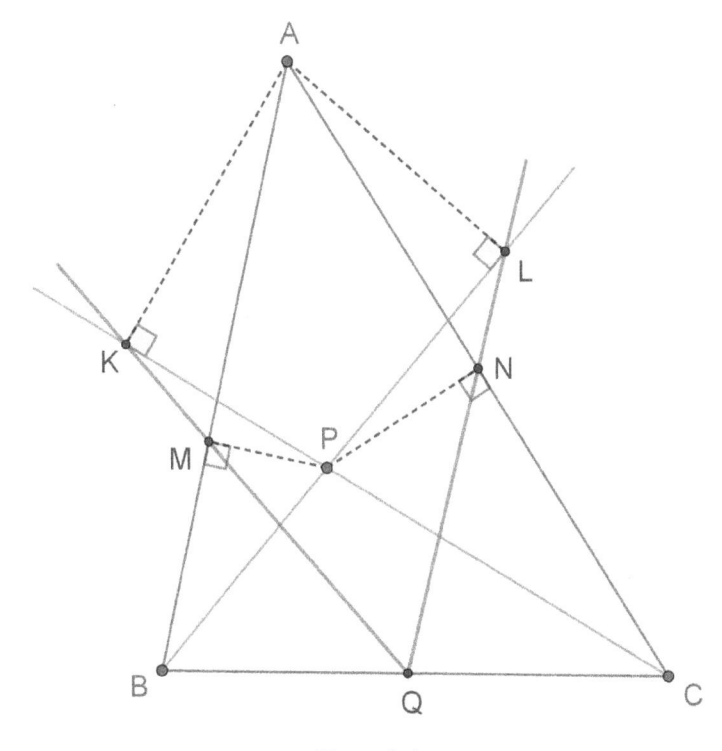

Figure 9-4

Pascal's Theorem Application 2

Another example of the power of Pascal's theorem can be seen in Figure 9-5, where we select any point P not on $\triangle ABC$ and a line ℓ containing P and intersecting sides BC, AB, and AC at points X, Y, and Z, respectively. Let AP, BP, and CP intersect the circumcircle of $\triangle ABC$ at points R, S, and T, respectively. We can now show that RX, SZ, and TY are concurrent at a point Q on the circumcircle of triangle ABC.

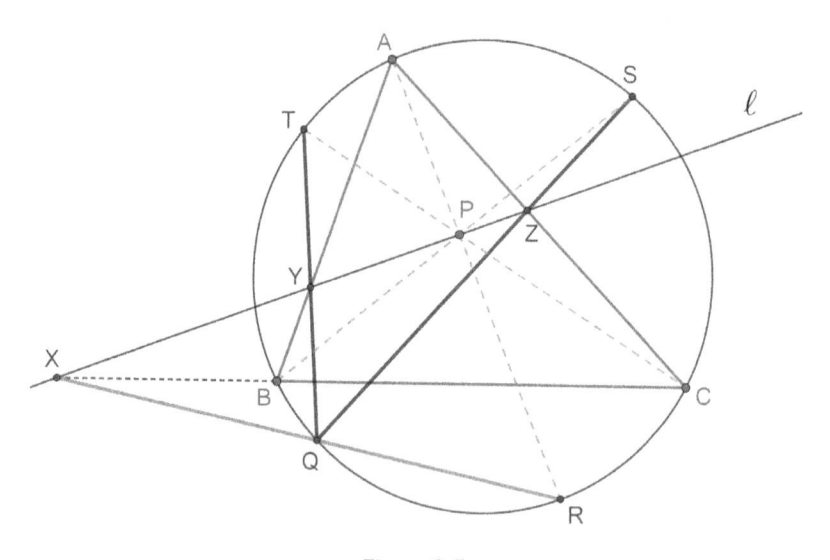

Figure 9-5

Proof

In Figure 9-5, we let RX intersect the circumcircle at Q. Consider hexagon $ARQTCB$ and apply Pascal's theorem to it. We notice that, since AP intersects TC at P and RQ intersects CB at X, then TQ intersects AB at a point on ℓ, which must be Y, since $AB \cap \ell = Y$. Now consider hexagon $ARQSBC$. Similarly, since $AR \cap SB = P$, and $RQ \cap CB = X$, then SQ intersects AC at a point on line ℓ, which must be point Z. Thus RX, SZ, and TY are concurrent.

Stewart's Theorem

Finding the length of "any" Cevian, which we recall is a segment that has one endpoint at a vertex of a given triangle and the other endpoint on the opposite side, sounds like a very difficult conundrum. The famous Scottish geometer Robert Simson (1687–1768), who made major contributions to Euclidean geometry by adapting ancient scriptures into modern formats, presented this unique problem in lectures. He allowed his notes to be used by his prized student Matthew Stewart (1717–1785) in Stewart's famous publication *General Theorems of Considerable Use in the Higher Parts of Mathematics* (Edinburgh, 1746). Simson's generosity was motivated by his desire to see Stewart obtain the professorship of mathematics at the University of Edinburgh. It is ironic that Simson was credited with the theorem on the Simson line, which he did not know, but was not credited with Stewart's theorem, which he developed. We shall still refer to the theorem by Stewart's name, because he was the author of the book in which it first appeared. Stewart's theorem states the following relationship as referred to in Figure 9-6: $a^2 n + b^2 m = c(d^2 + mn)$.

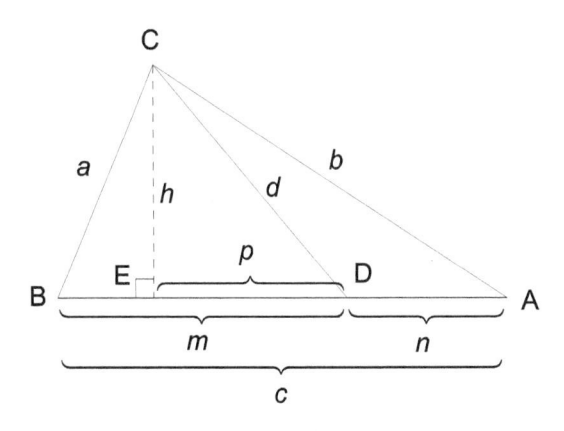

Figure 9-6

Derivations of Stewart's Theorem

1. **Using the Pythagorean Theorem:** This theorem yields a relation between the lengths of the sides of the triangle and the length of a Cevian of the triangle. Referring to Figure 9-6, we have $a^2n + b^2m = c(d^2 + mn)$. In Figure 9-6, for $\triangle ABC$, let $BC = a$, $AC = b$, $AB = c$, $CD = d$. Point D divides AB into two segments: $BD = m$, and $DA = n$. Draw altitude $CE = h$ and let $DE = p$. In order to proceed with the proof of Stewart's theorem, we will first derive two necessary relationships. The first one applies to $\triangle CBD$, where the Pythagorean theorem for $\triangle CBE$ gives us $CB^2 = CE^2 + BE^2$.

Since

$$BE = m - p, \text{ we have } a^2 = h^2 + (m - p)^2. \tag{I}$$

By applying the Pythagorean theorem to $\triangle CDE$, we have $CD^2 = CE^2 + DE^2$, or $d^2 = h^2 + p^2$, or $h^2 = d^2 - p^2$. We replace this value of h^2 in equation (I) to obtain $a^2 = d^2 - p^2 + (m - p)^2 = d^2 - p^2 + m^2 - 2mp + p^2$.

Thus,

$$a^2 = d^2 + m^2 - 2mp. \tag{II}$$

A similar argument is applicable to triangle ACD. Applying the Pythagorean theorem to $\triangle ACE$, we find that $AC^2 = AE^2 + CE^2$.

Since $CE = n + p$, we have $b^2 = h^2 + (n + p)^2$. (III)

Above we had $h^2 = d^2 - p^2$, so we again substitute for h^2 in equation (III) as follows:

$$b^2 = d^2 - p^2 + (h + p)^2 = d^2 - p^2 + n^2 + 2np + p^2 = d^2 + n^2 + 2np.$$

Thus,

$$b^2 = d^2 + n^2 + 2np. \tag{IV}$$

Equations (II) and (IV) give us the formulas we need.
Now multiply equation (II) by n to get

$$a^2n = d^2n + m^2n - 2mnp, \tag{V}$$

and multiply equation (IV) by m to get

$$b^2m = d^2m + mn^2 + 2mnp. \tag{VI}$$

We add (V) and (VI) to get

$$b^2m + c^2n = d^2m + d^2n + m^2n + mn^2 + 2mnp - 2mnp.$$

Then $b^2m + a^2n = d^2(m + n) + mn(m + n)$. Since $m + n = c$, we have $b^2m + a^2n = d^2c + mnc$. This is equivalent to $b^2m + a^2n = c(d^2 + mn)$, which is the relationship we set out to prove.

2. **Using the Law of Cosines:** With the help of the *law of cosines*, the proof is considerably shorter. Once again considering Figure 9-6, we see that ∠BDC and ∠ADC are supplementary angles, so that $\cos\angle ADC = -\cos\angle BDC$. Applying the law of cosines yields $a^2 = m^2 + d^2 - 2md \cos\angle BDC$ in triangle △BDC and $b^2 = n^2 + d^2 + 2nd \cos\angle BDC$ in triangle △ADC. Therefore, we can express $\cos\angle BDC$ in each of these equations to establish $\frac{m^2+d^2-a^2}{2md} = \frac{b^2-n^2-d^2}{2nd}$, or equivalently, $m^2n + d^2n - a^2n = b^2m - n^2m - d^2m$. This can be rewritten as $a^2n + d^2m = d^2\underbrace{(n+m)}_{c} + mn\underbrace{(n+m)}_{c}$, or as $a^2n + b^2m = c(d^2 + mn)$, as we sought to prove.

Stewart's theorem can be applied to a variety of situations, some of which are offered here.

Stewart's Theorem Application 1

Isosceles triangle *ABC*, shown in Figure 9-7, has two congruent sides, $AB = AC = 17$. A line $AD = 16$ is drawn from the vertex to the base.

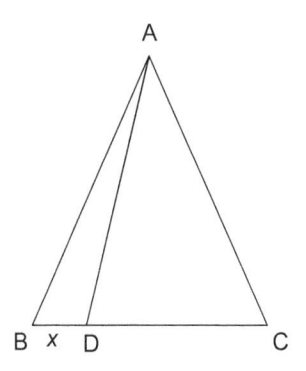

Figure 9-7

If one segment of the base, as cut by this line, exceeds the other by 8, find the measures of the two segments.

Solution: In Figure 9-7, for triangle ABC, we have $AB = AC = 17$, and $AD = 16$. Let $BD = x$ so that $DC = x + 8$. By Stewart's theorem, $(AB)^2(DC) + (AC)^2(BD) = BC[(AD)^2 + (BD)(DC)]$. Therefore, $(17)^2(x+8) + (17)^2(x) = (2x+8)[(16)^2 + x(x+8)]$, and $x = 3$. Thus, $BD = 3$ and $DC = 11$.

Stewart's Theorem Application 2

In the right triangle in Figure 9-8, the sum of the squares of the distances from the vertex of the right angle to the trisection points along the hypotenuse is equal to $\frac{5}{9}$ of the square of the measure of the hypotenuse.

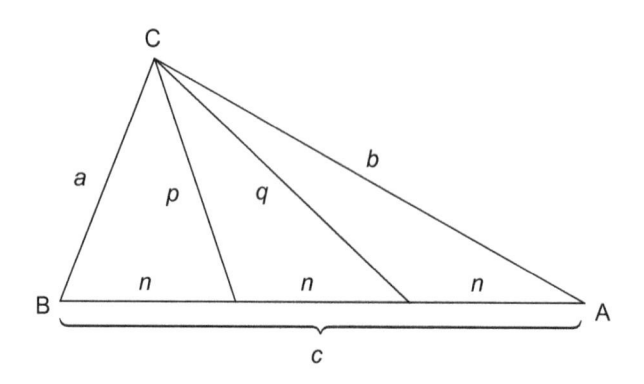

Figure 9-8

Solution: We apply Stewart's theorem to triangle ABC in Figure 9-8 using p as the internal line segment to find that

$$2a^2n + b^2n = c(p^2 + 2n^2). \tag{I}$$

Using q as the internal line segment, we get

$$a^2n + 2b^2n = c(q^2 + 2n^2). \qquad (II)$$

By adding (I) and (II), we get

$$3a^2n + 3b^2n = c(4n^2 + p^2 + q^2).$$

Since $a^2 + b^2 = c^2$, we have

$$3n(c^2) = c(4n^2 + p^2 + q^2).$$

Since $3n = c$, we have

$$c^2 = (2n)^2 + p^2 + q^2.$$

However, $2n = \tfrac{2}{3}c$; therefore,

$$p^2 + q^2 = c^2 - \left(\frac{2}{3}c\right)^2 = \frac{5}{9}c^2.$$

Brianchon's Theorem

In 1806, Charles Julien Brianchon (1785–1864), a twenty-one-year-old student at the École Polytechnique in Paris, published an article in the *Journal de L'École Polytechnique* that was to become one of the fundamental contributions to the study of conic sections in projective geometry. His development leads to a restatement of the then-somewhat-forgotten theorem of Pascal and its extension, upon which Brianchon developed a new theorem that now bears his name. *Brianchon's theorem* states that in any hexagon circumscribed about a conic section, the three diagonals intersect each other at a common point. This bears a curious resemblance to Pascal's theorem. The two theorems are, in fact, *duals* of one another. Before we make the comparisons, we recall the concept of duality in geometry. To best understand this concept, consider the following examples of duality statements:

Statement	Dual statement
1. Two distinct **points** determine a unique **line**.	1. Two distinct **lines** determine a unique **point**.
2. Any **point** contains an infinite number of **lines**.	2. Any **line** contains an infinite number of **points**.
3. Only one **triangle** is determined by three **non-collinear points**.	3. Only one **trilateral** is determined by three **non-concurrent lines**.

This last example of duality implies that there are other related words that need also be changed when forming the dual of a statement. Specifically, notice that *collinear* and *concurrent* are dual words, as are *triangle* and *trilateral*.

This concept can be seen by comparing the following versions of each theorem to appreciate the "dual" connection. Notice the relationship by comparing the bold words.

Pascal's theorem	Brianchon's theorem
The **points of intersection** of the opposite **sides** of a hexagon **inscribed in** a conic section are **collinear**.	The **lines joining** the opposite **vertices** of a hexagon **circumscribed about** a conic section are **concurrent**.

Notice that the two statements above are alike except for the bold printed words, which are duals of one another. As with Pascal's theorem, we shall consider *Brianchon's theorem* only for the conic section that is a circle. That is, if a hexagon is circumscribed about a circle, the lines containing opposite vertices are concurrent, as shown in Figure 9-9.

The simplest proof of this theorem requires a knowledge of concepts from projective geometry. However, we will provide a proof of this theorem using only Euclidean methods.

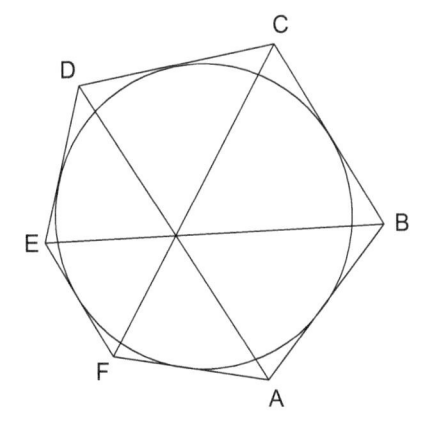

Figure 9-9

Proof

As seen in Figure 9-10, the sides of hexagon *ABCDEF* are tangent to a circle at points *T*, *N*, *L*, *S*, *M*, and *K*. Points *K′*, *L′*, *N′*, *M′*, *S′*, and *T″* are chosen on *FA*, *DC*, *BC*, *FE*, *DE*, and *BA*, respectively, so that *KK′* = *LL′* = *NN′* = *MM′* = *SS′* = *TT″*. Next, construct circle *P* tangent to *BA* and *ED* at

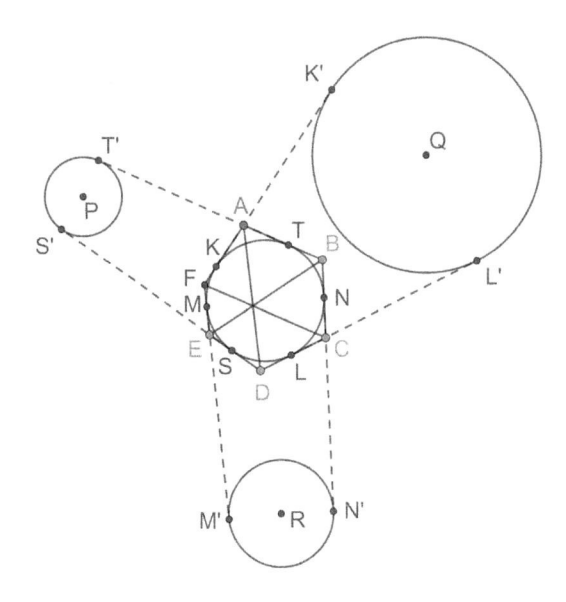

Figure 9-10

points T' and S', respectively. The existence of this circle is easily justified. Similarly construct circle Q tangent to FA and DC at points K' and L', respectively. Then construct circle R tangent to FE and BC at points M' and N', respectively. Since two tangent segments to a circle from an external point have the same length, $FM = FK$. We already know that $MM' = KK'$. Therefore, by addition, $FM' = FK'$.

Similarly, $CL = CN$ and $LL' = NN'$. By subtraction, $CL' = CN'$. We now notice that points F and C are each endpoints of a pair of congruent tangent segments to circles R and Q. Thus, these points determine the radical axis[1] CF of circles R and Q. Using the same technique, we can easily show that AD is the radical axis of circles P and Q, and that BE is the radical axis of circles P and R.

We can prove that the radical axes of three circles with non-collinear centers (taken in pairs) are concurrent. Therefore, CF, AD, and BE are concurrent. We should note that the only way in which these circles would have had collinear centers is if the diagonals had coincided. This is impossible!

An Application of Brianchon's Theorem

Brianchon suggested the following application immediately after the statement of his new theorem. Pentagon $ABCDE$ is circumscribed about a circle, with points of tangency at F, M, N, R, and S. If diagonals AD and BE intersect at P the result is, curiously, that C, P, and F are collinear, as can be seen in Figure 9-11.

Proof

Consider the hexagon in Figure 9-11 circumscribed about a circle with sides AF and EF merged into one line segment. Now, AFE is a side of a circumscribed pentagon with F as one point of tangency. Thus, we can view the pentagon in Figure 9-11 as a degenerate hexagon.

---

[1] We call the line consisting of points that have congruent tangent segments to two circles the *radical axis* of the two circles. Furthermore, the radical axis of two intersecting circles is the common secant.

Geometric Surprises 337

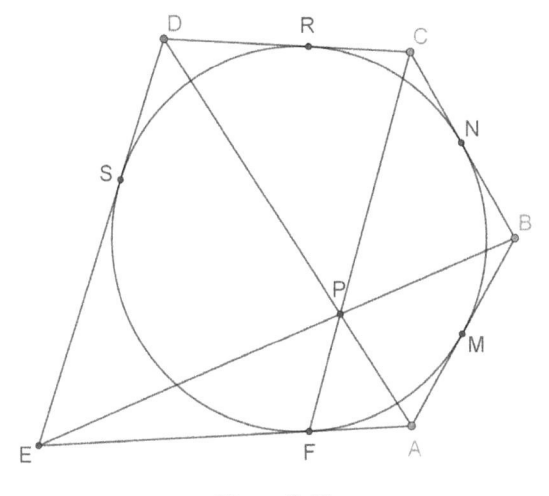

Figure 9-11

We then simply apply Brianchon's theorem to this degenerate hexagon to obtain our desired conclusion: *AD*, *BE*, and *CF* are concurrent at *P*, or *C*, *P*, and *F* are collinear.

Pappus' Theorem

Consider the vertices of a hexagon *AB'CA'BC'*, shown in Figure 9-12, being located alternately on two lines, as we can see in Figure 9-13. We draw the lines that were the opposite sides of the hexagon to locate their point of intersection. We find that the three points of

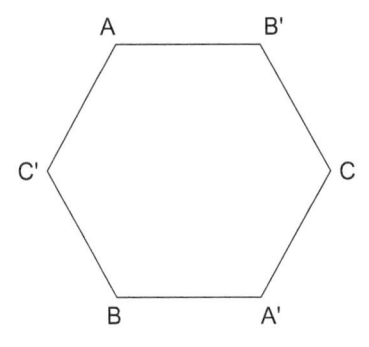

Figure 9-12

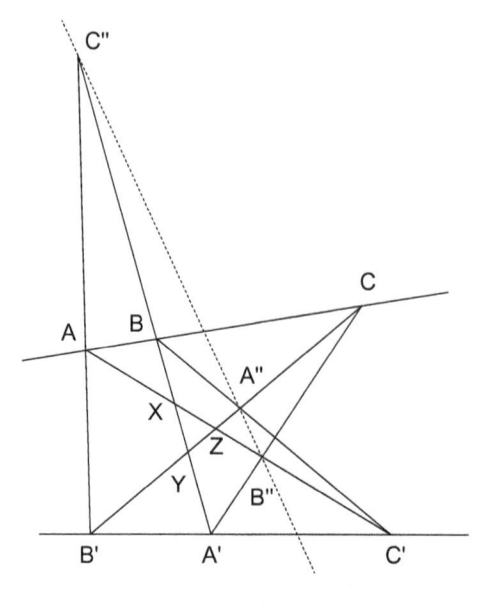

Figure 9-13

intersection of these pairs of "opposite sides" are collinear. This conclusion was first published by Pappus of Alexandria (290–350 CE) in his *Mathematical Collection* circa 300 CE.

Pappus' theorem states that points *A*, *B*, and *C* are on one line and points *A'*, *B'*, and *C'* are on another line (in any order). If *AB'* and *A'B* intersect at *C"*, while *AC'* and *A'C* intersect at *B"*, and *BC'* and *B'C* intersect at *A"*, then points *A"*, *B"*, and *C"* are collinear.

Proof

In Figure 9-13, *B'C* meets *A'B* at *Y*, *AC'* meets *A'B* at *X*, and *B'C* meets *AC'* at *Z*.

Consider *C"AB'* as a transversal of $\triangle XYZ$, by Menelaus' theorem:

$$\frac{ZB'}{YB'} \cdot \frac{YC''}{XC''} \cdot \frac{XA}{ZA} = 1 \qquad (\text{I})$$

Taking $A'B''C$ as a transversal of $\triangle XYZ$, by Menelaus' theorem:

$$\frac{YA'}{XA'} \cdot \frac{XB''}{ZB''} \cdot \frac{ZC}{YC} = 1. \tag{II}$$

Line $BA''C'$ is also a transversal of $\triangle XYZ$, by Menelaus' theorem:

$$\frac{YB}{XB} \cdot \frac{XC'}{ZC'} \cdot \frac{ZA''}{YA''} = 1. \tag{III}$$

Multiplying (I), (II), and (III) gives us equation (IV):

$$\frac{YC''}{XC''} \cdot \frac{XB''}{ZB''} \cdot \frac{ZA''}{YA''} \cdot \frac{ZB'}{YB'} \cdot \frac{YA'}{XA'} \cdot \frac{XC'}{ZC'} \cdot \frac{XA}{ZA} \cdot \frac{ZC}{YC} \cdot \frac{YB}{XB} = 1. \tag{IV}$$

Since points A, B, C are collinear, and A', B', C' are collinear, we obtain the following two relationships by Menelaus' theorem (when we consider each line a transversal of $\triangle XYZ$):

$$\frac{ZB'}{YB'} \cdot \frac{YA'}{XA'} \cdot \frac{XC'}{ZC'} = 1, \tag{V}$$

$$\frac{XA}{ZA} \cdot \frac{ZC}{YC} \cdot \frac{YB}{XB} = 1. \tag{VI}$$

Substituting (V) and (VI) into (IV), we get $\frac{YC''}{XC''} \cdot \frac{XB''}{ZB''} \cdot \frac{ZA''}{YA''} = 1$. Thus, points A'', B'', and C'' are collinear, by the converse of Menelaus' theorem.

Heron's Famous Formula for Finding the Area of a Triangle

There are various alternatives to finding the area of a triangle beyond the very well-known relationship that the area of a triangle is equal to one-half the product of the length of the base and the corresponding height $\left(A = \frac{1}{2}bh\right)$.

The famous Greek mathematician Heron of Alexandria (ca. 10 CE–70 CE) is credited with a technique for finding the area of a triangle ABC when we are not given the length of an altitude, but only the lengths of the sides of the triangle, a, b, and c. Heron's formula is area $\triangle ABC = \sqrt{s(s-a)(s-b)(s-c)}$, where s equals the semi-perimeter of the triangle, that is, $s = \frac{1}{2}(a+b+c)$, and is shown in Figure 9-14. We will provide a trigonometric derivation building on the two secondary-school-level relationships: the law of cosines ($c^2 = a^2 + b^2 - 2ab \cos C$), and the Pythagorean identity ($\sin^2 \theta + \cos^2 \theta = 1$). A well-known area formula for a triangle is area $\triangle ABC = \frac{1}{2} ab \sin C$.

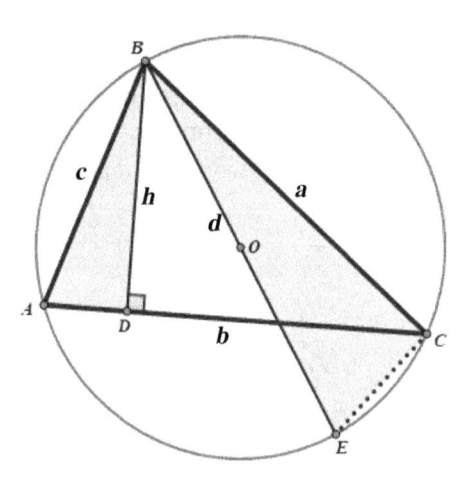

Figure 9-14

From the law of cosines, we can establish that $\cos^2 C = \frac{(a^2+b^2-c^2)^2}{4a^2b^2}$, and by substituting for $\sin C$ in this previous equation for the area, we get

$$\text{area}\triangle ABC = \frac{1}{2} ab \sin C = \frac{1}{2} ab\sqrt{1 - \cos^2 C} = \frac{1}{2} ab\sqrt{1 - \frac{(a^2+b^2-c^2)^2}{4a^2b^2}} =$$
$$\frac{1}{2} ab\sqrt{\frac{4a^2b^2 - (a^2+b^2-c^2)^2}{4a^2b^2}} = \frac{1}{4}\sqrt{4a^2b^2 - (a^2+b^2-c^2)^2}.$$

We now have to factor the term under the radical sign:

$4a^2b^2 - (a^2 + b^2 - c^2)^2 = -(a+b+c) \cdot (a+b-c) \cdot (a-b-c) \cdot (a-b+c) =$
$-(a+b+c)(a+b-c)[-(-a+b+c)](a-b+c) =$
$(a+b+c)(a+b-c)(-a+b+c)(a-b+c).$

With $a + b + c = 2s$, $a + b - c = 2(s - c)$, $-a + b + c = 2(s - a)$, and $a - b + c = 2(s - b)$, we get

$$\text{area}\triangle ABC = \frac{1}{4}\sqrt{2s \cdot 2(s-c) \cdot 2(s-a) \cdot 2(s-b)} =$$

$$\sqrt{s \cdot (s-a) \cdot (s-b) \cdot (s-c)}.$$

Another proof of Heron's formula uses Brahmagupta's formula for cyclic quadrilaterals, introduced in Chapter 6.

An Unexpected Triangle Area Formula

The area of a triangle can also be found by simply taking the product of the three sides and dividing that by twice the diameter of the circumscribed circle of the triangle. Symbolically, this is area $\triangle ABC = \frac{abc}{2d}$, where the lengths of the sides shown in Figure 9-14 are a, b, c, the altitude is h, and the diameter is d.

Proof

The common formula for the area of a triangle, which is shown in Figure 9-14, is area $\triangle ABC = \frac{1}{2}hb$. We begin by considering CE so that we have right angle BCE, which is inscribed in the semicircle. The two right triangles ADB and ECB can easily be shown to be similar, since both $\angle A$ and $\angle E$ are measured by the same intercepted arc $\overset{\frown}{BC}$. This enables us to set up the proportion $\frac{AB}{BE} = \frac{BD}{BC}$, which leads to our desired conclusion: $AB \cdot BC = BD \cdot BE$.

Thus, we have the following relationship:

$$ac = hd, \text{ or } h = \frac{ac}{d}.$$

Substituting for h in the area formula, we get

$$\text{area}\triangle ABC = \frac{abc}{2d}.$$

The Astonishing Morley's Theorem

Mathematics is full of astonishing relationships. In 1900, the mathematician Frank Morley (1860–1937) discovered an amazing relationship that can be applied to triangles regardless of their shape. Morley demonstrated (and proved) that if we trisect each of the angles of a triangle, the intersections of the adjacent trisectors will determine an equilateral triangle. Remember, what makes this so amazing is that this relationship holds true regardless of the shape of the original triangle. In Figure 9-15, we have each angle of triangle *ABC* trisected with the adjacent trisectors intersecting at points *D*, *E*, and *F*. When we join these three points, we will always arrive at an equilateral triangle *DEF*.

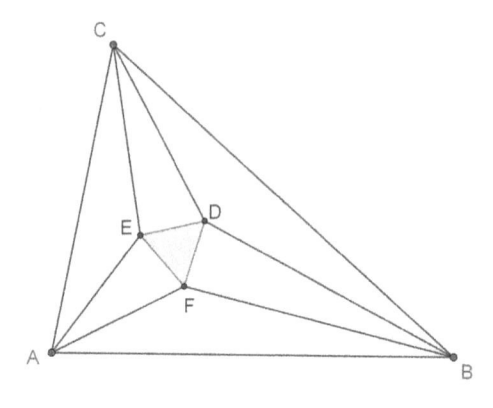

Figure 9-15

To demonstrate that Morley's theorem applies regardless of a triangle's shape, we offer a variety of triangles in Figures 9-16, 9-17, 9-18, and 9-19.

This theorem became an international challenge for mathematicians to prove. Consequently, there are now many proofs available, two of which we offer here.

Geometric Surprises **343**

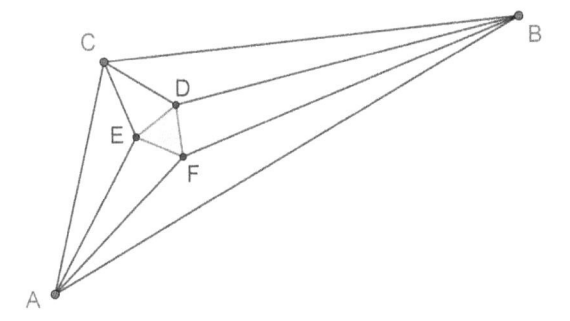

Figure 9-16

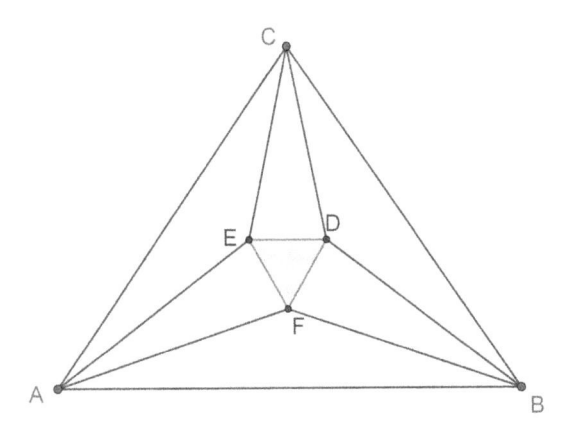

Figure 9-17

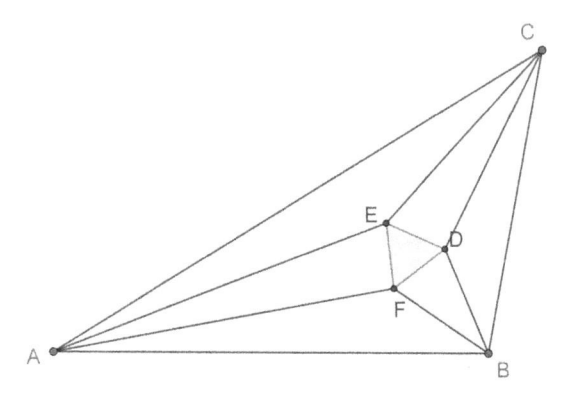

Figure 9-18

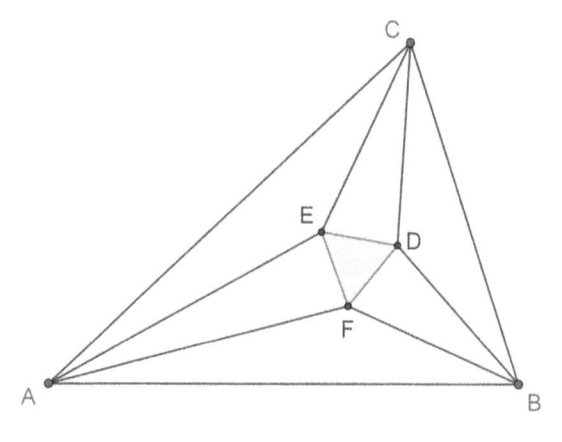

344 A Journey Through the Wonders of Plane Geometry

Figure 9-19

Proof 1

Let D be the intersection point of the two angle trisectors at B and C (adjacent to side BC) and the other two trisectors intersect at point P, as shown in Figure 9-20. Point D is clearly the incenter of $\triangle PBC$ and PD is the bisector of $\angle BPC$. From point D, two lines intersect CP and BP at points E and F, respectively, so that $\angle PDE = \angle PFD = 30°$, which is shown in Figure 9-20. We then have $\triangle PDE \cong \triangle PDF$ so that $DE = DF$.

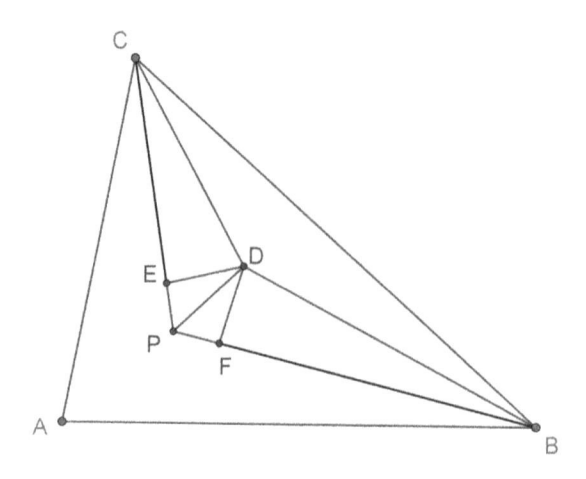

Figure 9-20

Because $\angle EDF = 60°$, we can conclude that $\triangle DEF$ is equilateral and $PD \perp EF$, as shown in Figure 9-21.

It remains to be shown that AE and AF are the trisectors of $\angle BAC$. Since CP and BP are bisectors of angle ACD and angle ABD, respectively, when we reflect D in CP and in BP, we get point R on side AC, and point S on side AB. The quadrilateral $REFS$ (which can be seen in Figure 9-21 and more clearly in Figure 9-22) consists of three equal line segments RE, EF, and FS, which are each equal to the

Figure 9-21

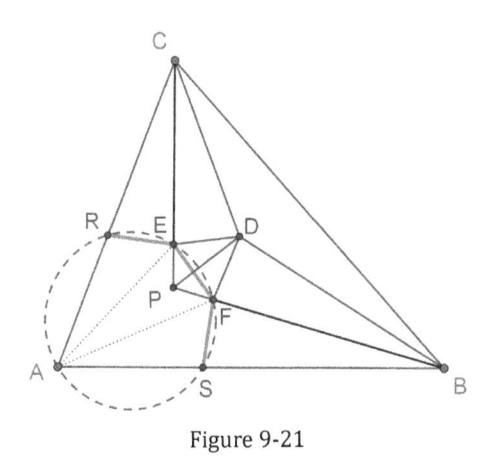

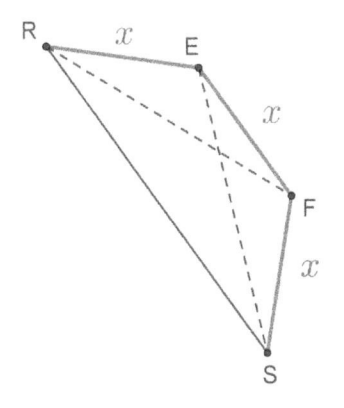

Figure 9-22

side length of the equilateral △DEF, shown in Figure 9-21. This creates equal angles at E and F, which are ∠PFS = ∠PFD = ∠PED = ∠PER and ∠PFE = ∠PEF. By addition, this results in ∠EFS = ∠REF. In addition to establishing that these two angles are equal, we can also calculate their measure as follows:

$$\angle BPC = \underbrace{180°}_{\angle A + \angle B + \angle C} - \frac{2}{3}\angle B - \frac{2}{3}\angle C = \underbrace{\frac{1}{3}(\angle A + \angle B + \angle C)}_{60°} + \frac{2}{3}\angle A = 60° + \frac{2}{3}\angle A,$$

so that

$$\angle PFE = 90° - \frac{1}{2}\angle BPC = 60° - \frac{1}{3}\angle A.$$

Then

$$\angle PFS = \angle PFD = 60° + \angle PFE = 120° - \frac{1}{3}\angle A,$$

and

$$\angle EFS = \angle PFE + \angle PFS = 180° - \frac{2}{3}\angle A.$$

We would get the same result for ∠REF, but it is not necessary to do the calculation again, because we know from the above ∠EFS = ∠REF.

Hence, REFS is an isosceles trapezoid, as shown in Figure 9-22, and is *cyclic*, with the supplementary angles ∠REF = 180° − ⅔∠A = ∠EFS, and ∠ERS = ⅔∠A = ∠FSA. In trapezoid REFS, the diagonals are angle bisectors of ∠R and ∠S since ∠RFE = ∠FRE, resulting from isosceles △RFE and ∠RFE = ∠FRS as RS∥EF. Hence, we have

$$\angle RFS = \angle EFS - \underbrace{\angle RFE}_{\frac{1}{2}\angle ERS} = 180° - \frac{2}{3}\angle A - \frac{1}{2} \cdot \frac{2}{3}\angle A = 180° - \angle A.$$

This establishes that ∠RFS is supplementary to ∠A. It therefore follows that point A lies on the circumcircle of REFS. Since the chords RE = EF = SF are equal, they determine equal inscribed angles

∠RAE = ∠EAF = ∠FAS, so that AE and AF are angle trisectors of ∠BAC. Thus, the trisectors of the three angles of triangle ABC form an equilateral triangle DEF.

Proof 2

We begin with a randomly drawn triangle ABC, shown in Figure 9-23, and define $\frac{1}{3}\angle A = \alpha$, $\frac{1}{3}\angle B = \beta$, and $\frac{1}{3}\angle C = \gamma$. We then have $\alpha + \beta + \gamma = 60°$. To simplify matters, we will denote with α' the angle $\alpha + 60°$ (analogously for β, γ).

We will employ a different proof strategy here; instead of beginning with △ABC and then showing that △DEF is equilateral, we will *work backwards*. That is, we will start with an equilateral △D'E'F' and construct △A'B'C', which will prove to be similar to △ABC. Moreover, we will see that in △A'B'C', the points △D',E',F' are nothing more than the intersection points of the angle trisectors of △A'B'C'. This will allow us, finally, to conclude the similarity △D'E'F' ~ △DEF, which means that △DEF is equilateral, as we seek to show. Working backwards is a clever principle.

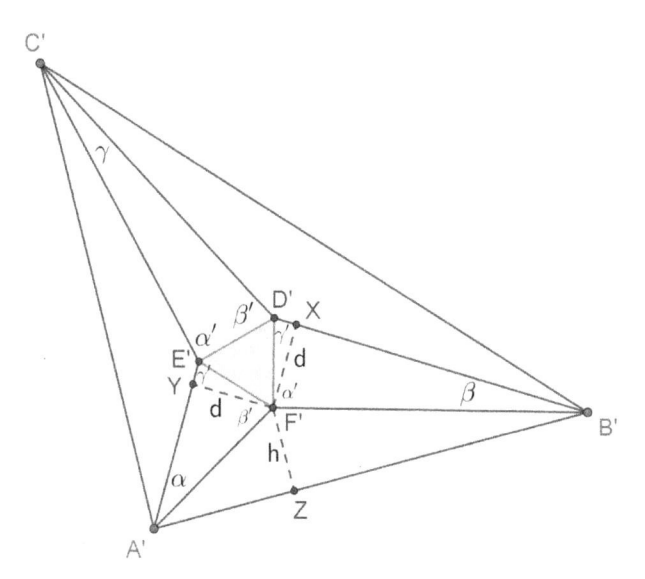

Figure 9-23

Let's describe this path of reasoning in more detail. As illustrated in Figure 9-23, we begin with the equilateral $\triangle D'E'F'$ and erect outwardly on its sides three triangles with the vertices A', B', C'. We will now show that $\angle F'A'B' = \alpha$ and $\angle F'B'A' = \beta$. To do this, we consider the perpendiculars $F'X$ and $F'Y$ to $B'D'$ and $A'E'$, respectively. Since $\angle A'E'F' = \angle F'D'B' = \gamma'$ and $E'F' = D'F'$, right triangles $\triangle F'D'X \cong \triangle F'E'Y$, and we then have $F'X = F'Y = d$.

First, we can calculate

$$\angle A'F'B' = 360° - \beta' - 60° - \alpha' = 180° - (\alpha + \beta). \qquad (*)$$

We then consider the perpendicular distance h from F' to $A'B'$, measuring this distance along line segment $F'Z$. If $h < d$, then the following holds (note that in a right triangle with fixed hypotenuse, increasing an angle yields longer opposite legs and vice versa): $\angle B'A'F' < \alpha$ and $\angle F'B'A' < \beta$, and thus, $\angle A'F'B' = 180° - \angle B'A'F' - \angle F'B'A' > 180° - (\alpha+\beta)$, which contradicts equation (*).

On the other hand, if $h > d$, it follows that $\angle B'A'F' > \alpha$ and $\angle F'B'A' > \beta$, and thus, $\angle A'F'B' < 180° - (\alpha + \beta)$, again a contradiction to equation (*). Hence, it follows that $h = d$, $\angle F'A'B' = \alpha$, and $\angle F'B'A' = \beta$.

In an analogous way, we can show that $\angle C'B'D' = \beta$, $\angle D'C'B' = \angle A'C'E' = \gamma$ and $\angle E'A'C' = \alpha$. We see that $A'E'$ and $A'F'$ trisect $\angle B'A'C'$, $B'F'$ and $B'D'$ trisect $\angle C'B'A'$, and $C'D'$ and $C'E'$ trisect $\angle A'C'B'$. Since $\angle B'A'C' = 3\alpha$, $\angle C'B'A' = 3\beta$, and $\angle A'C'B' = 3\gamma$, we know that $\triangle A'B'C'$ is similar to the originally given $\triangle ABC$. This similarity extends to all constructions, and the trisections of the interior angles of $\triangle ABC$ therefore yield an internal $\triangle DEF$ similar to $\triangle D'E'F'$, which must also be equilateral.

The Hidden Angle in an Isosceles Triangle

Now that we have marveled about the appearance of an equilateral triangle, we will solve a problem dependent on creating an equilateral triangle.

Geometric Surprises 349

We begin with the isosceles triangle *ABC*, shown in Figure 9-24, whose base angles are ∠*ABC* = ∠*BAC* = 80°. Point *P* is selected on *BC* so that *PC* = *AB*. The challenge here is to find the measure of ∠*PAC*. Of course, as with many geometric problems, there are a variety of procedures to reach the desired conclusion. However, in this case, the procedure is noteworthy because it employs an equilateral triangle.

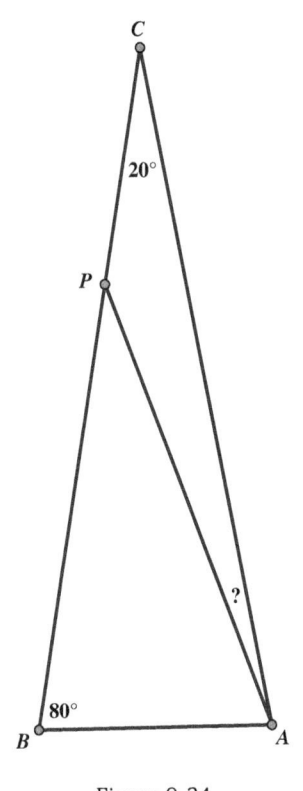

Figure 9-24

Solution

Most unusually, we begin by constructing an equilateral triangle ABQ, as shown in Figure 9-25, and we then draw PQ. Since $\angle QAB = 60°$, we have $\angle QAC = 20°$, and we already know that $\angle C = 20°$, so that $\angle C = \angle QAC$. Also, since $AB = PC$, we have $AQ = PC$. Therefore, we can show that quadrilateral $AQPC$ is an isosceles trapezoid, and PQ is parallel to AC. The corresponding angles of these parallel lines are $\angle BPQ = \angle C = 20°$. This determines that triangle PQB is isosceles, where $PQ = BQ$. Furthermore, since $\angle PQB = 140°$ and $\angle AQB = 60°$, we have $\angle PQA = 160°$. However, triangle PQA is also isosceles, and therefore, $\angle PAQ = 10°$. Thus, our sought-after angle measure is $\angle CAP = \angle QAC - \angle PAQ = 20° - 10° = 10°$.

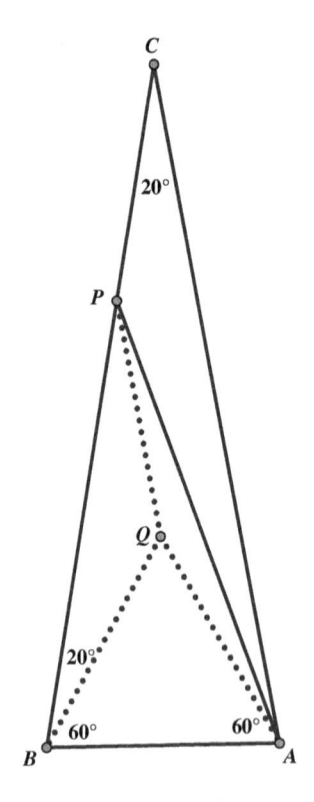

Figure 9-25

Japanese Geometry — Sangaku

For over 200 years, Japan was isolated from the rest of the world. During the Edo period (1603–1868), Euclidean geometry blossomed in Japan. Prior to this time, Japanese mathematics was influenced by the Chinese mathematicians. Beginning around 1600, European commercial interaction with the Netherlands and Portugal introduced European mathematics to Japan.

By 1631, when Japanese isolation intensified, Japanese mathematics was popularized as a form of recreation and challenge for the people. They focused on plane geometry, using triangles, polygons, and circles. In comparison to the European treatment of geometry, which was largely an axiomatic development of the subject, the Japanese were more concerned with the measurements of individual parts of the geometric configuration. A *Sangaku* (a mathematical tablet) was one such medium for spreading these geometric challenges to the populace. These tablets (Figure 9-26) were artistically presented and displayed in Shinto shrines and Buddhist temples. Unsolved problems were printed on these tablets, challenging viewers to seek solutions.

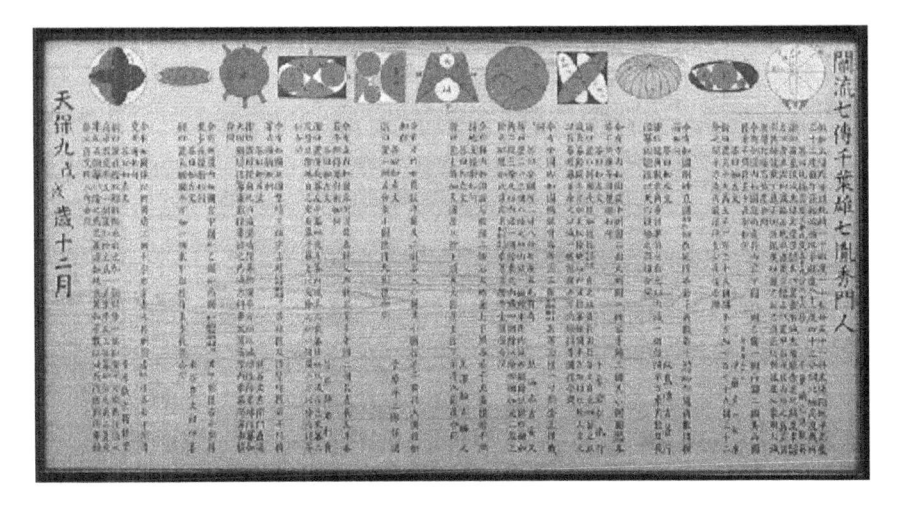

Figure 9-26

These problems dealt with applications of basic geometry. By solving some of these problems, new ideas evolved, although much of this was not documented until late in the 18th century. In 1790, the Japanese mathematician Fujita Kagen (1765–1821) published the first collection of these Sangaku problems in his book *Shimpeki Sampo*, followed by a sequel in 1806 entitled *Zoku Shimpeki Sampo*.

The complexity of the problems presented on the Sangaku tablets vary from simple to complicated. They require nothing more than high school level geometry, but demand a degree of cleverness. Over time, the problems presented on the Sangaku tablets became more complex. Through the destruction of some of the Japanese temples, many of these tablets were lost. Today, there exist about 900 Sangaku tablets.

The Sangaku tablet problems were accompanied by multicolored figures that detailed the problem and stated what specific answer was sought. This was followed by the actual answer. However, no method of solution was provided. They were written in *Kanbun*, which was a classical written language with Chinese characters. Most of the Sangaku drawings are sparse in the information given, as we show in Figure 9-27. However, we will provide auxiliary lines and label points as needed (Figures 9-28 and 9-29) so that our solutions can be easily followed.

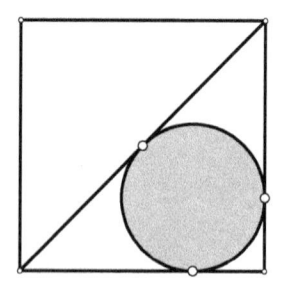

Figure 9-27

D a C

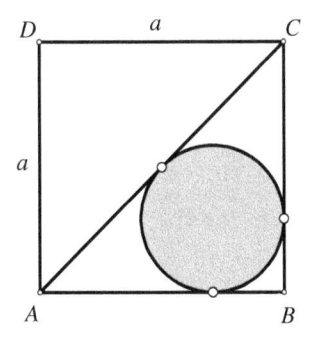

a

A B

Figure 9-28

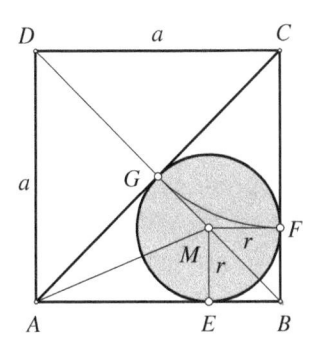

D a C

a G
         F
      M r
         r

A E B

Figure 9-29

Sangaku Problem 1

We are given a square with a circle inside the square, placed in such a way that it is tangent to two sides of the square and to the square's diagonal (Figure 9-27). We are asked to find the radius of the circle in terms of the side of the square. In other words, if we let the radius of the circle have length r, and the side of the square has length a, we seek an expression of r in terms of a.

Solution for Sangaku Problem 1

The center M of the circle must lie on the diagonal BD, where $r = ME = MF$, and $BG = BM + GM = BM + r$. Applying the Pythagorean theorem to

$\triangle BEM$, we get $BM = \sqrt{ME^2 + EB^2} = \sqrt{2r^2} = r\sqrt{2}$, so that $BG = BM + r = r\sqrt{2} + r = (\sqrt{2}+1)$. Applying the Pythagorean theorem to $\triangle BCD$, we get $BD = \sqrt{BC^2 + CD^2} = \sqrt{2a^2} = a\sqrt{2}$. Since the diagonals of a square bisect each other, we have $BG = DG = \frac{BD}{2}$. Combining equalities gives us $r(\sqrt{2}+1) = \frac{a\sqrt{2}}{2}$. Therefore, for the radii $ME = MF = MG = r$, and by rationalizing the denominator we can conclude that $r = \frac{\frac{a\sqrt{2}}{2}}{\sqrt{2}+1} = \frac{2-\sqrt{2}}{2}a \approx 0.29a$.

Naturally, there are other ways to solve this problem; we only offered one here.

Sangaku Problem 2

We are given a right triangle with an altitude drawn to the hypotenuse and a circle inscribed on each side of the altitude, so that each of these circles is tangent to the hypotenuse, to the circumscribed semicircle of the right triangle, and to the altitude of the right triangle (see Figures 9-30 and 9-31). We are asked to find the radius of each of the two circles in terms of the sides of the right triangle.

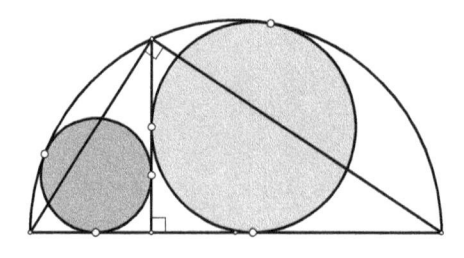

Figure 9-30

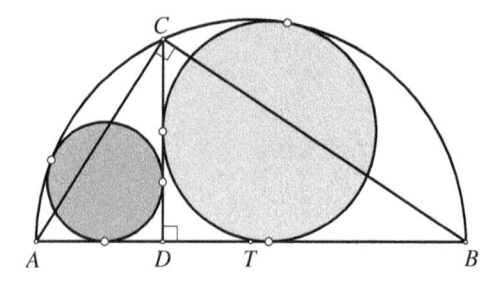

Figure 9-31

Solution for Sangaku Problem 2

As shown in Figure 9-32, point T is the center of the semicircle about $\triangle ABC$, whose sides have lengths a, b, and c.

We need to find the radius of the circle shown in Figure 9-33, since the circle on the other side of the altitude is analogous to this one. We will use segment lengths as shown in Figure 9-33, where in $\triangle ABC$, we have $BD = p$ and $AD = q$. Point G is the tangency point of the small circle with the large semicircular arc. A perpendicular line to the tangent at the point of tangency will contain the center of the circle, as well as the center of the semicircle. This implies that the points G, M, and T lie on the same line.

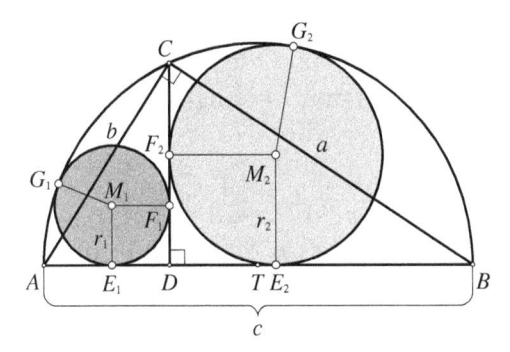

Figure 9-32

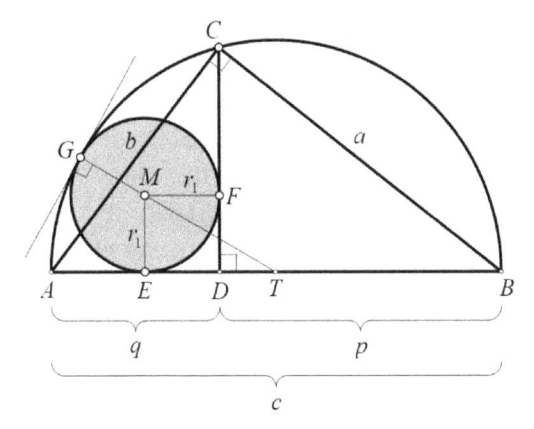

Figure 9-33

We are looking for the length of $EM = r_1$. We see in Figure 9-33 that $ET = ED + DT = ED + (BD - BT) = r_1 + p - \frac{c}{2}$, and $MT = GT - GM = \frac{c}{2} - r_1$. We apply the Pythagorean theorem to $\triangle EMT$ to get $EM^2 + ET^2 = MT^2$, or using the lengths of these sides, we get $r_1^2 + \left(r_1 + p - \frac{c}{2}\right)^2 = \left(\frac{c}{2} - r_1\right)^2$, which then can be written as $r_1^2 + r_1^2 + p^2 + \frac{c^2}{4} + 2pr_1 - cr_1 - cp = \frac{c^2}{4} - cr_1 + r_1^2$, and simplified as $r_1^2 + 2pr_1 + p^2 = cp$, or $(r_1 + p)^2 = cp$.

When the altitude is drawn to the hypotenuse of a right triangle, either leg of the right triangle is the mean proportional between the whole hypotenuse and the nearer segment. Therefore, $\frac{AB}{BC} = \frac{BC}{BD}$, or $BC^2 = AB \cdot BD$ and $a^2 = c \cdot p$. By substitution, $(r_1 + p)^2 = a^2$. Since we know the lengths are positive, we get $r_1 + p = a$, or $r_1 = a - p$. With $p = \frac{a^2}{c}$, it follows that $r_1 = a - p = a - \frac{a^2}{c} = a\left(1 - \frac{a}{c}\right) = \frac{a(c-a)}{c}$, which is what we sought, namely, to get the value of r_1 in terms of the sides of the triangle. Analogously, we can find the radius of the other circle. Thus, we now have the two radii that we sought:

$$r_1 = a - p = a - \frac{a^2}{c} = a\left(1 - \frac{a}{c}\right) = \frac{a(c-a)}{c},$$

$$\text{and } r_2 = b - q = b - \frac{b^2}{c} = b\left(1 - \frac{b}{c}\right) = \frac{b(c-b)}{c}.$$

Sangaku Problem 3

We are given a Reuleaux triangle[2] with three congruent circles inscribed within, as shown in Figure 9-34. The Reuleaux triangle is constructed by first drawing an equilateral triangle ($\triangle ABC$), as shown in Figure 9-35, using each vertex as a center of a circle with the side of the triangle as its radius, and drawing a circular arc joining the two remote vertices. The Reuleaux triangle has many fascinating properties, many of which result from its similarity to the properties of a circle.[3] For example, this odd-shaped figure can be placed tangentially

---

[2] Named for the German engineer Franz Reuleaux (1829–1905).
[3] Further characteristics of the Reuleaux triangle can be found in Alfred S. Posamentier and Ingmar Lehmann, *Pi: A Biography of the World's Most Mysterious Number*. Amherst (New York), Prometheus Books, 2004, pp. 158–170.

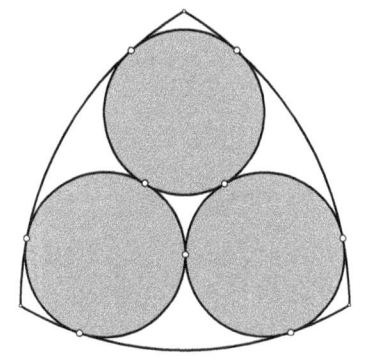

Figure 9-34

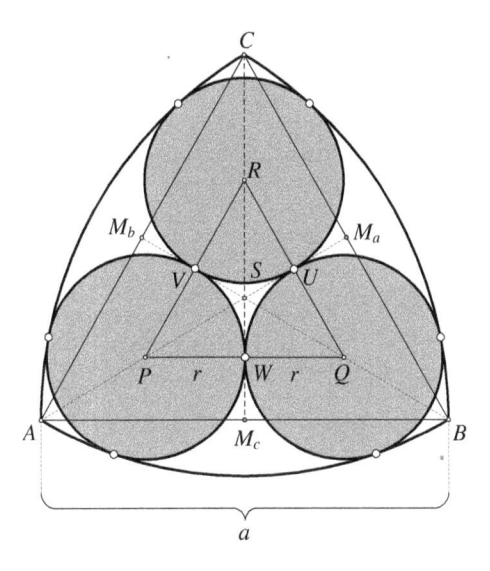

Figure 9-35

between two parallel lines and turned, all the while remaining tangent to the two parallel lines, just as a circle would under these circumstances. As an example, a circular button fits through a buttonhole regardless of which side of the button you press through buttonhole. The same is true for a button shaped as a Reuleaux triangle, which also fits through a buttonhole regardless of which side of the button is pushed through.

The problem, shown in Figure 9-35, is to determine the radius of these three congruent circles in terms of the side length (a) of the equilateral triangle ABC.

Solution for Sangaku Problem 3

We are given that the sides of equilateral triangle ABC are $AB = BC = AC = a$, their midpoints are M_c, M_a, and M_a, and their points of tangency are V, U, and W. The radii of the circles are $PW = QW = QU = RU = RV = PV = r$. The common center of both triangles is point S. By applying the Pythagorean theorem to $\triangle AM_cC$, we get $CM_c^2 = AC^2 - \left(\frac{a}{2}\right)^2 = a^2 - \left(\frac{a}{2}\right)^2 = \frac{3a^2}{4}$; therefore, $CM_c = \frac{a\sqrt{3}}{2}$.

Similarly, we can apply the Pythagorean theorem to $\triangle PWR$ so that $RW^2 = PR^2 - PW^2 = (2r)^2 - r^2 = 3r^2$, and then $RW = r\sqrt{3}$.

In Figure 9-36, we notice that $\triangle AM_cR$ is a right triangle with one leg $AM_c = \frac{a}{2}$ and the hypotenuse $AR = AA'' - RA'' = a - r$.

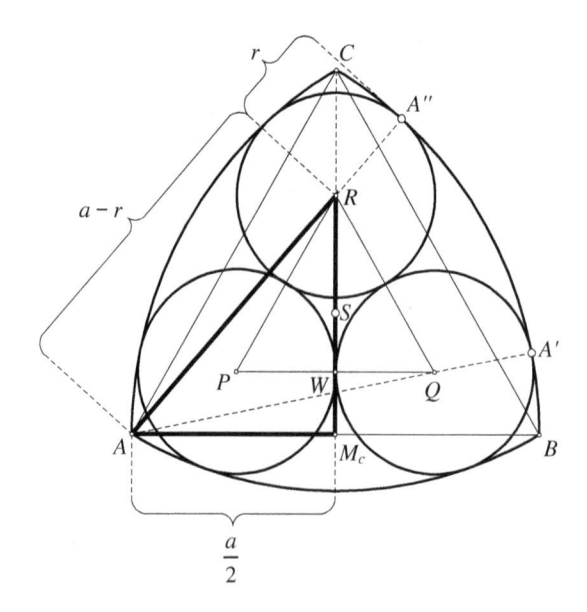

Figure 9-36

In Figure 9-36, since the medians of a triangle trisect each other, in $\triangle ABC$ we have $M_c S = \frac{1}{3} CM_c$, and in $\triangle PQR$ we have $SR = \frac{2}{3} RW$. Then applying the results we obtained from the Pythagorean theorem, we get $M_c R = M_c S + SR = \frac{1}{3} CM_c + \frac{2}{3} RW = \frac{a\sqrt{3}}{6} + \frac{2r\sqrt{3}}{3}$.

When we apply the Pythagorean theorem to $\triangle AM_c R$, we get: $AR^2 = AM_c^2 + M_c R^2$, and $(a-r)^2 = \left(\frac{a}{2}\right)^2 + \left(\frac{a\sqrt{3}}{6} + \frac{2r\sqrt{3}}{3}\right)^2$, which, in turn, gives us

$$a^2 - 2ar + r^2 = \frac{a^2}{4} + \frac{a^2}{12} + \frac{2ar}{3} + \frac{4r^2}{3}$$

$$a^2 - 2ar + r^2 = \frac{1}{3}a^2 + \frac{2}{3}ar + \frac{4}{3}r^2$$

$$\tfrac{1}{3}(2a^2 - 8ar - r^2) = 0$$

$$r^2 + 8ar - 2a^2 = 0.$$

Now solving this quadratic equation for r, we get $r_{1,2} = -4a \pm \sqrt{16a^2 + 2a^2} = -4a \pm 3a\sqrt{2}$, and ignore the negative solution $-4a - 3a\sqrt{2}$.

Therefore, the radius of each of the three circles in terms of the side of equilateral $\triangle ABC$ is $r = 3a\sqrt{2} - 4a = \left(3\sqrt{2} - 4\right)\cdot a \approx 0.24 \cdot a$.

Chapter 10

Geometric Fallacies

One may think that what we see in geometry is the truth. However, we justify "the truth" by using an established proof. There are times when an absurd statement can be "proved" by a procedure that has a subtle error. Sometimes these errors are well-hidden, and other times they are obvious. In this chapter, we will present absurd statements followed by a "proof," and challenge the reader to find the error in the proof. Not to worry, we will not leave the reader stranded, as we will provide an explanation of the fallacy in the proof.

How Can a Right Angle Equal an Obtuse Angle?

This geometric mistake points out a few properties that must hold and cannot be ignored. Furthermore, it shines a spotlight on a rarely recognized concept: the reflex angle. Follow along as we proceed to "prove" that a right angle can be equal to an obtuse angle (an angle that is greater than 90°).

In Figure 10-1, we begin with a rectangle $ABCD$, where $FA = BA$, R is the midpoint of BC, and N is the midpoint of CF. We will now "prove" that right $\angle CDA$ is equal to obtuse $\angle FAD$.

To set up the proof, we first draw RL perpendicular to CB, and draw MN perpendicular to CF. The rays RL and MN intersect at point O. If they did not intersect, then RL and MN would be parallel, and this would mean that CB would be parallel to, or coincide with, CF, which

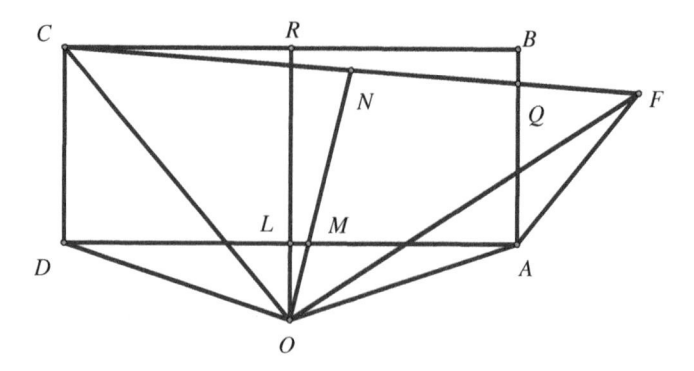

362 A Journey Through the Wonders of Plane Geometry

Figure 10-1

is impossible. To complete the diagram for our "proof," we draw the line segments *DO*, *CO*, *FO*, and *AO*.

We are now ready to embark on the "proof." Since *RO* is the perpendicular bisector of *CB* and *AD*, we know that *DO* = *AO*. Similarly, since *NO* is the perpendicular bisector of *CF*, we get *CO* = *FO*. Furthermore, since *FA* = *BA*, and *BA* = *CD*, we can conclude that *FA* = *CD*. This enables us to establish △*CDO* ≅ △*FAO* (SSS), so that ∠*ODC* = ∠*OAF*. We continue with *OD* = *OA*, which makes triangle *AOD* isosceles, and then the base angles ∠*ODA* and ∠*OAD* are equal. Now, ∠*ODC* − ∠*ODA* = ∠*ODF* − ∠*OAD* or ∠*CDA* = ∠*FAD*. This says that a right angle is equal to an obtuse angle. There must be some mistake!

Clearly, there is nothing wrong with this "proof;" however, if you use dynamic software or just a ruler and compasses to reconstruct the diagram, it will look like Figure 10-2.

As you will see, the mistake here rests with a reflex angle – one that is often not considered in plane geometry. For rectangle *ABCD*, the perpendicular bisector of *AD* will also be the perpendicular bisector of *BC*. Therefore, *OC* = *OB*, *OC* = *OF*, and then *OB* = *OF*. Since both points *A* and *O* are equidistant from the endpoints of *BF*, the line *AO* must be the perpendicular bisector of *BF*. This is where the fault lies; we must consider the reflex angle of ∠*BAO*. Although the triangles are congruent, our ability to subtract the specific angles no longer exists. Thus, the error with this "proof" lies in its dependence upon an incorrectly-drawn diagram.

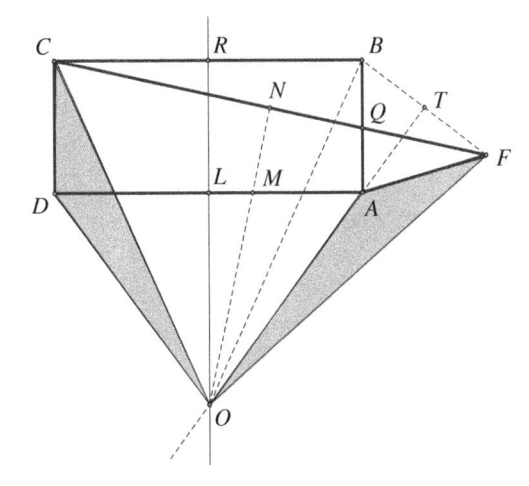

Figure 10-2

Every Angle is a Right Angle

We begin this demonstration with quadrilateral *ABCD*, where *AB* = *CD* and right angle ∠*BAD* = δ, as shown in Figure 10-3. We will consider ∠*ADC* = δ′ to be of random measure, but we will show that it is actually a right angle, which will then "prove" that any random angle is a right angle.

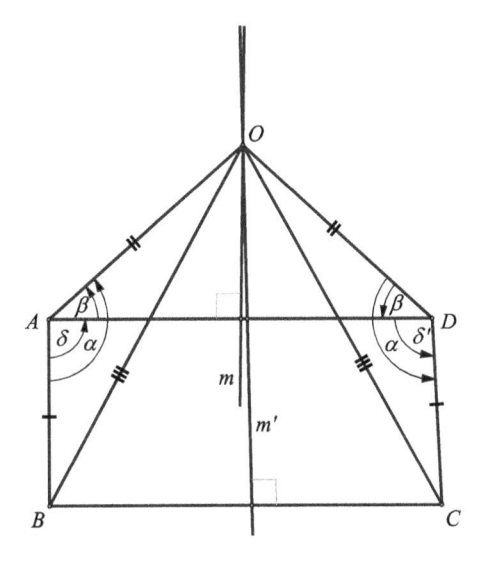

Figure 10-3

364 A Journey Through the Wonders of Plane Geometry

We construct m, the perpendicular bisector of AD, and m', the perpendicular bisector of BC. These perpendicular bisectors intersect at point O. The point O is then equidistant from points A and D, as well as from points B and C. Therefore, $OA = OD$ and $OB = OC$. We can then conclude that $\triangle OAB \cong \triangle ODC$, and it follows that $\angle BAO = \angle ODC = \alpha$. Since triangle OAD is isosceles, it follows that $\angle DAO = \angle ODA = \beta$. Therefore, $\delta = \angle BAD = \angle BAO - \angle DAO = \alpha - \beta$, and $\delta' = \angle ADC = \angle ODC - \angle ODA = \alpha - \beta$. Thus, $\delta = \delta'$. However, this result is silly, so there must be a mistake somewhere. To ferret out the mistake, let's revisit the original diagram.

In fact, the diagram presented in Figure 10-3 tricked us, and was intentionally false. The key error is the point where the two perpendicular bisectors meet, which must be further beyond the quadrilateral than what was indicated. The correct diagram would look like that shown in Figure 10-4. We then have $\delta = \alpha - \beta$; however, $\delta' = 360° - \alpha - \beta$. This then destroys the mistaken "proof."

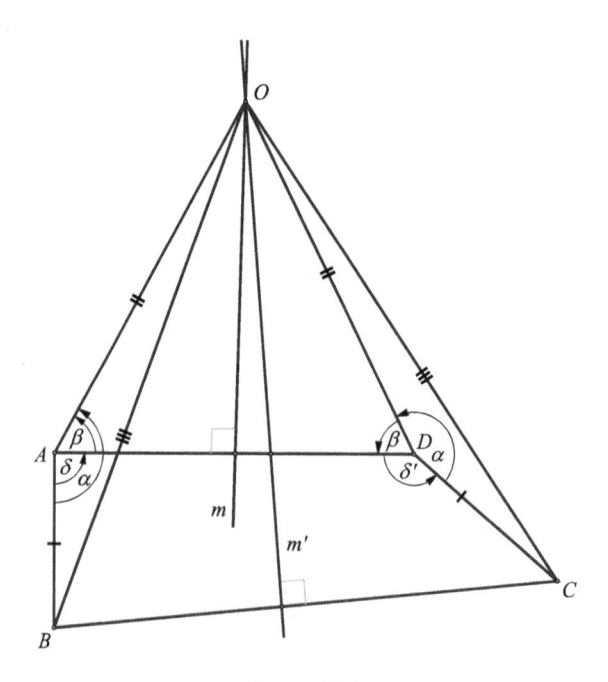

Figure 10-4

"Proving" That All Triangles are Isosceles: A Mistake?

Geometric fallacies tend to come from faulty diagrams resulting from a lack of definition. Yet, as we know, in ancient times some geometers discussed their geometric findings or relationships without diagrams. For example, in Euclid's work, the concept of "betweenness" was not considered. Ignoring this concept, we can prove that any triangle is isosceles – that is, that a triangle that has three sides of different lengths actually has two sides that are equal. This sounds a bit strange, but we present this "proof" and challenge the reader to discover where the mistake lies before we expose it.

We shall begin by drawing any scalene triangle (i.e., a triangle with no two sides of equal length) and then "prove" it is isosceles (i.e., a triangle with two sides of equal length). Consider the scalene triangle *ABC*, shown in Figure 10-5, where we draw the bisector of angle *C* and the perpendicular bisector of *AB*. From their point of intersection *G*, draw perpendiculars to *AC* and *CB*, meeting them at points *D* and *F*, respectively.

We now have four possibilities matching the above description for various scalene triangles:

In Figure 10-5, where *CG* and *GE* meet inside the triangle at point *G*.

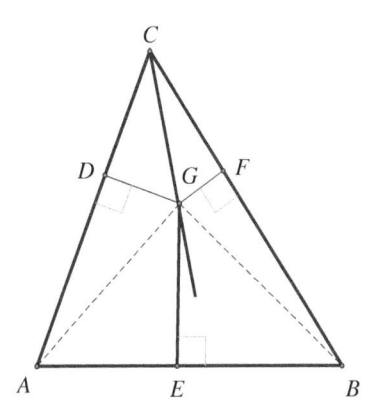

Figure 10-5

In Figure 10-6, where CG and GE meet on side AB. (Points E and G coincide.)

In Figure 10-7, where CG and GE meet outside the triangle at point G, but the perpendiculars GD and GF intersect the segments AC and CB at points D and F, respectively.

In Figure 10-8, where CG and GE meet outside the triangle, but the perpendiculars GD and GF intersect the extensions of the sides AC and CB outside the triangle at points D and F, respectively.

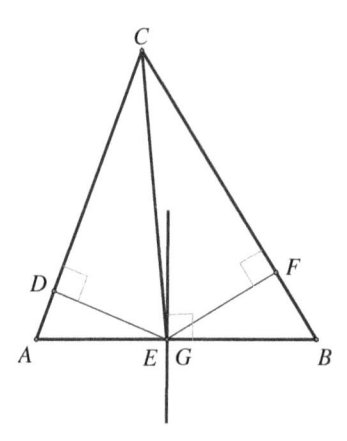

Figure 10-6

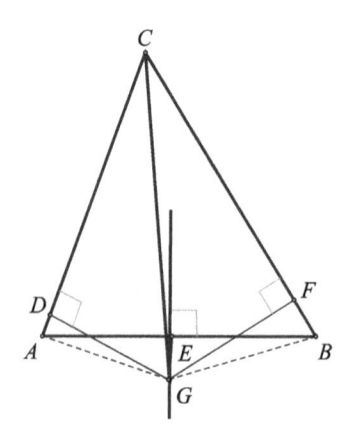

Figure 10-7

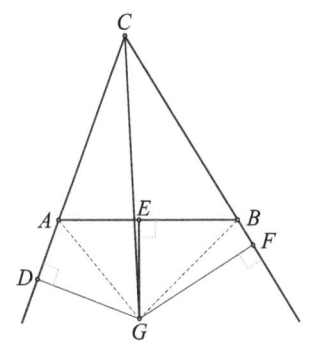

Figure 10-8

The "proof" of the fallacy can be completed with any of the above figures. Follow along and see if the mistake shows itself without reading further. We begin with a scalene triangle ABC. We will now "prove" that $AC = BC$, that is, that triangle ABC is isosceles.

From angle bisector CG, we have $\angle ACG \cong \angle BCG$. We also have two right angles, such that $\angle CDG \cong \angle CFG$. This enables us to conclude that $\angle CDG \cong \angle CFG$ (SAA). Therefore, $DG = FG$, and $CD = CF$. Since a point on the perpendicular bisector (EG) of a line segment is equidistant from the endpoints of the line segment, $AG = BG$. Also, $\angle ADG$ and $\angle BFG$ are right angles. We then have $\triangle DAG \cong \triangle FBG$ since they have respective hypotenuse and leg congruent. Therefore, $DA = FB$. It then follows that $AC = BC$.

At this point you may feel quite disturbed. You may wonder where the error was committed that permitted this mistake to occur. You might challenge the correctness of the figures. Well, by rigorous construction you will find a subtle error in the figures. We will now divulge the mistake and see how it leads us to a better and more precise way of referring to geometric concepts.

First, we can show that the point G *must* be outside the triangle. Then, when perpendiculars meet the sides of the triangle, one will meet a side *between* the vertices, while the other will not. We can "blame" this mistake on Euclid's neglect of the concept of betweenness. However, the beauty of this particular mistake lies in the proof of this betweenness issue, which establishes the mistake.

Begin by considering the circumcircle of triangle ABC, as shown in Figure 10-9. The bisector of ∠ACB must contain the midpoint, M, of arc $\overarc{AB}$, since ∠ACM and ∠BCM are congruent inscribed angles. The perpendicular bisector of AB must bisect arc $\overarc{AB}$ and therefore must pass through M. Thus, the angle bisector of ∠ACB and the perpendicular bisector of AB intersect on the circumscribed circle *outside* the triangle at M (or G). This eliminates the possibilities we used in Figures 10-5 and 10-6.

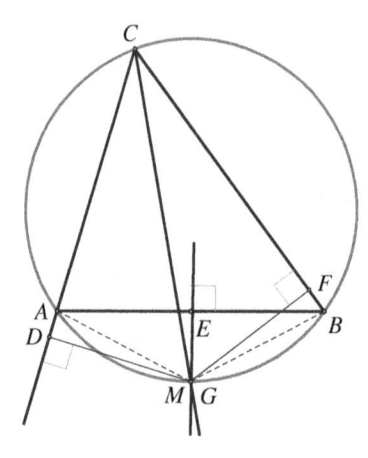

Figure 10-9

Now consider the inscribed quadrilateral ACBG. Since the opposite angles of an inscribed (or cyclic) quadrilateral are supplementary, ∠CAG + ∠CBG = 180°. If ∠CAG and ∠CBG were right angles, then CG would be a diameter and triangle ABC would be isosceles. Therefore, since triangle ABC is scalene, ∠CAG and ∠CBG are not right angles. In this case, one must be acute and the other obtuse. Suppose ∠CBG is acute and ∠CAG is obtuse. In triangle CBG, the altitude on CB must be *inside* the triangle, while in obtuse triangle CAG, the altitude on AC must be *outside* the triangle. The fact that one and *only one* of the perpendiculars intersects a side of the triangle *between* the vertices destroys the fallacious "proof." This demonstration hinges on the definition of betweenness, a concept unavailable to Euclid.

In Every Triangle the Base Angles are Equal

In Figure 10-10, where $BE = AD$, extend AC the length of BC to point D, and extend BC the length of AC to point E. We now have two isosceles triangles, $\triangle ACE$ and $\triangle BCD$, that share vertex angles at point C. Therefore, the triangles are similar that have equal base angles marked with ε.

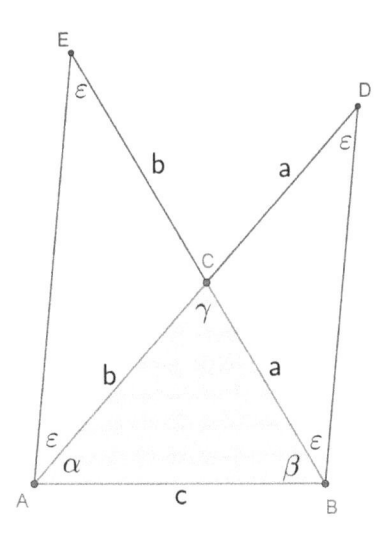

Figure 10-10

By applying the law of sines to the triangles ABD and ADE, we get

$$\triangle ABD : \frac{a+b}{\sin(\beta+\varepsilon)} = \frac{c}{\sin(\varepsilon)},$$

$$\triangle ABE : \frac{a+b}{\sin(\alpha+\varepsilon)} = \frac{c}{\sin(\varepsilon)}.$$

From that, we conclude first $\sin(\alpha+\varepsilon) = \sin(\beta+\varepsilon)$ and finally $\alpha = \beta$, thus indicating the triangle ACB is actually isosceles. But we knew from the start that ACB is not isosceles! Where is the mistake?

The mistake happens at the very end. From $\sin(\alpha+\varepsilon) = \sin(\beta+\varepsilon)$, it does not follow that $\alpha + \varepsilon = \beta + \varepsilon$ and $\alpha = \beta$! Because of the equal alternate interior angles marked ε at points A and D, we have

370 *A Journey Through the Wonders of Plane Geometry*

AE parallel to *BD*. Thus, the angles $\alpha + \varepsilon$ and $\beta + \varepsilon$ at points *A* and *B* are supplementary. Although such angles have the same sines $\sin(\varphi) = \sin(180° - \varphi)$, they are not equal, as we can see with the unit circle in Figure 10-11.

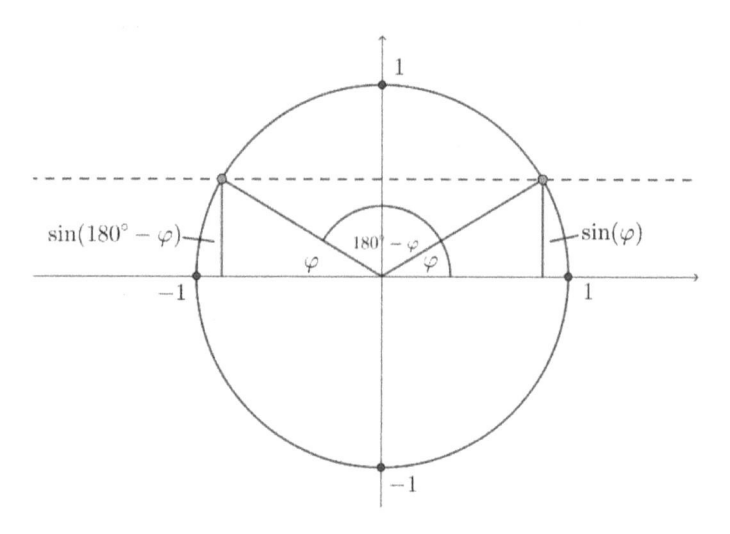

Figure 10-11

A Mistaken Proof That a Triangle Can Have Two Right Angles

The next geometric mistake is one that can truly upset an unsuspecting mathematician. With two intersecting circles of different or the same size, we will draw the diameters *AP* and *BP* from one of their points of intersection, *P*, and then connect *A* and *B*, as shown in Figure 10-12, where line *AB* intersects circle *O* at point *D* and circle *O'* at point *C*.

We find that ∠*ADP* is inscribed in semicircle *PNA*, and ∠*BCP* is inscribed in semicircle *PNB*, thus making them both right angles. We then have a dilemma: △*CPD* has two right angles! This is impossible. Therefore, there must be a mistake somewhere in our work.

The omission of the concept of betweenness in Euclid's work could lead us to this dilemma. When we draw this figure correctly, we

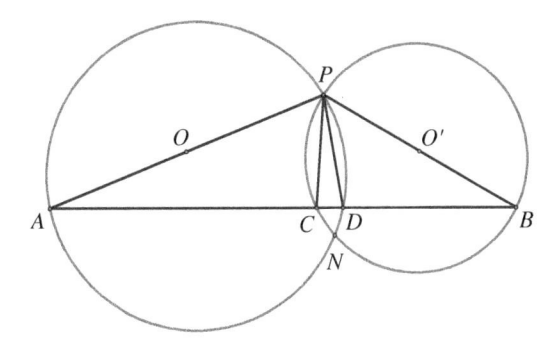

Figure 10-12

find that ∠CPD must equal 0, since a triangle cannot have more than 180°. That would make △CPD nonexistent. Figure 10-13 shows the correct drawing of this situation.

In Figure 10-13, we can easily show that △POO' ≅ △NOO', and then △POO' = △NOO'. Because of △AON, the exterior angle ∠PON = ∠A + ∠ANO. Since OO' bisects ∠PON, ∠ANO = ∠NOO', we have ∠POO' = ∠A, and then AN∥OO'. The same argument can be made for circle O' to get BN∥OO'. Since each of the two line segments AN and BN are parallel to OO', they must in fact be one line, ANB. This proves that the diagram in Figure 10-13 is correct and the diagram in Figure 10-12 is incorrect.

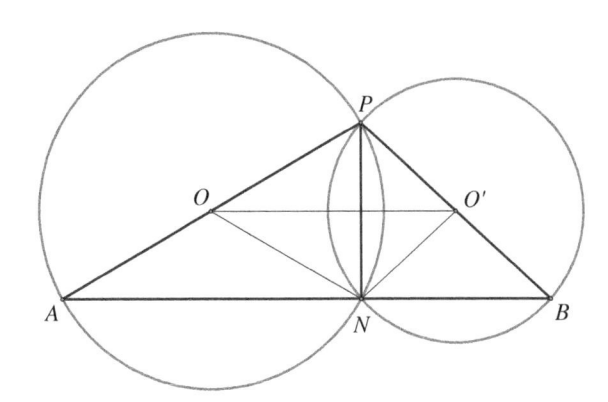

Figure 10-13

Two Unequal Lines Are Actually Equal

As we go through this "proof," see if you can spot the mistake. We will provide a clue: It has nothing to do with the diagram, as was the case in previous examples. We shall begin with $\triangle ABC$ and a line segment DE parallel to AB with endpoints on the other two sides, as shown in Figure 10-14.

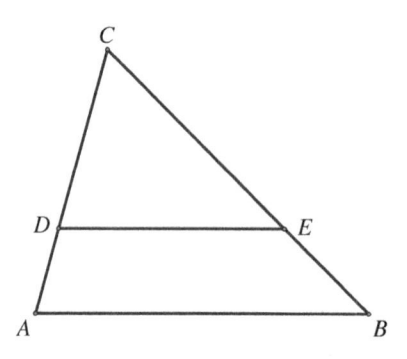

Figure 10-14

Then we have $\triangle ABC \sim \triangle DEC$. Therefore, $\frac{AB}{DE} = \frac{AC}{DC}$, or $AB \cdot DC = DE \cdot AC$.

Now we will multiply both sides of this equation by $AB - DE$ to get $AB^2 \cdot DC - AB \cdot DC \cdot DE = AB \cdot DE \cdot AC - DE^2 \cdot AC$.

Our next step will be to add $AB \cdot DC \cdot DE$, and subtract $AB \cdot DE \cdot AC$ from both sides of the above equation to get $AB^2 \cdot DC - AB \cdot DE \cdot AC = AB \cdot DC \cdot DE - DE^2 \cdot AC$.

Factoring the common term on each side of the equation yields $AB(AB \cdot DC - DE \cdot AC) = DE(AB \cdot DC - DE \cdot AC)$.

Now divide both sides by $AB \cdot DC - DE \cdot AC$ to get $AB = DE$. This is absurd, because we can see that $AB > DE$. There was no error in the diagram, so where does the error lie? Yes, we divided by zero, which is the forbidden division! This occurred when we divided both sides of the equation above by $AB \cdot DC - DE \cdot AC$, which is equal to zero, since $AB \cdot DC = DE \cdot AC$. One must be aware that there are times – such as this – when an algebraic mistake creates a geometric absurdity.

Every Exterior Angle of the Triangle Is Equal to One of Its Remote Interior Angles

We begin with the $\triangle ABC$ shown in Figure 10-15. We would like to demonstrate that the angles δ and α are equal.

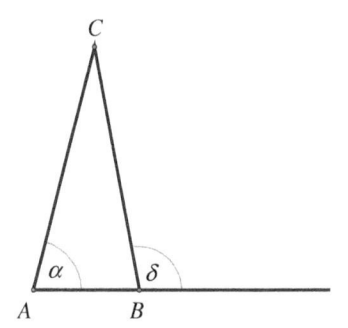

Figure 10-15

We now refer to Figure 10-16, where we have constructed quadrilateral $APQC$ so that $\angle CAP + \angle CQP = \alpha + \varepsilon = 180°$. We then construct a circle through the three points C, P, and Q. We will call the point where the line AP intersects the circle a second time point B. We draw BC to create a cyclic quadrilateral $BPQC$, where the following is true:

$$\angle CQP + \angle CBP = \varepsilon + \delta = \angle BCQ + \angle BPQ = 180°.$$

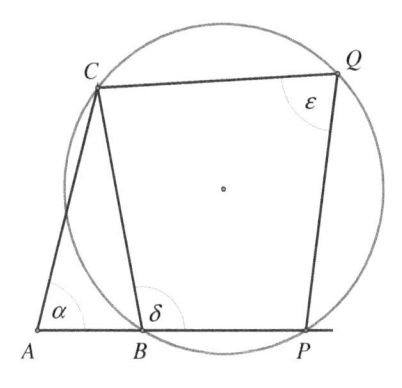

Figure 10-16

However, at the outset we drew $\angle CAP + \angle CQP = \alpha + \varepsilon = 180°$, so that we can now conclude that $\angle CAP = \angle CBP$, which is to say that $\alpha = \delta$. Something must be wrong! Where does the mistake lie?

If quadrilateral $APQC$ has the property that $\angle CAP + \angle CQP = \alpha + \varepsilon = 180°$, and the vertices C, P, and Q lie on the same circle, then the quadrilateral $APQC$ must also be cyclic, which implies that the point A must also lie on the circle. This implies that the two points A and B must be identical. In that case, the triangle ABC cannot exist. Thus, the mistake here has been revealed.

Two Non-Parallel Lines in a Plane That Do Not Intersect: A Paradox

We can also "prove" that if exactly one of two non-parallel lines is perpendicular to a third line, then the two non-parallel lines will not intersect. This paradox is attributed to Proclus Lycaeus (412–485 CE). There must be a mistake since only parallel lines never intersect, which is not the case here. Follow along and see if you can find the mistake. In Figure 10-17, we have $PB \perp AB$, while QA is not

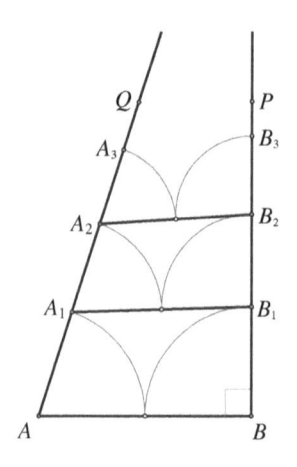

Figure 10-17

perpendicular to *AB*. Now we will "show" that the non-parallel lines *PB* and *QA* cannot intersect.

We begin by finding the midpoint of *AB*. Then, as we show in Figure 10-17, we mark off $AA_1 = \frac{1}{2}AB$ and $BB_1 = \frac{1}{2}AB$. The lines *AQ* and *PB* will not intersect anywhere along AA_1 or BB_1. If they did intersect at, say, a point *R*, then there would be a $\triangle ARB$, where the sum $AR + RB < AB$, which is impossible. We now consider the segment A_1B_1 and repeat the previous process so that we get $A_1A_2 = B_1B_2 = \frac{1}{2}A_1B_1$. As before, the lines A_1A_2 and B_1B_2 cannot intersect, and A_2 cannot coincide with B_2. We continue this process, bisecting A_2B_2 and marking off $A_2A_3 = B_2B_3 = \frac{1}{2}A_2B_2$. This process continues indefinitely, and we know that A_n will never coincide with B_n, since we would then have a right triangle where the hypotenuse AA_n would equal BB_n – clearly impossible! Therefore, at no step along this unending process will the oblique line intersect the perpendicular line. This is nonsensical! So where is the mistake?

Let's consider the two lines *AQ* and *BP*, intersecting as shown in Figure 10-18. Again, we will denote the similarly constructed

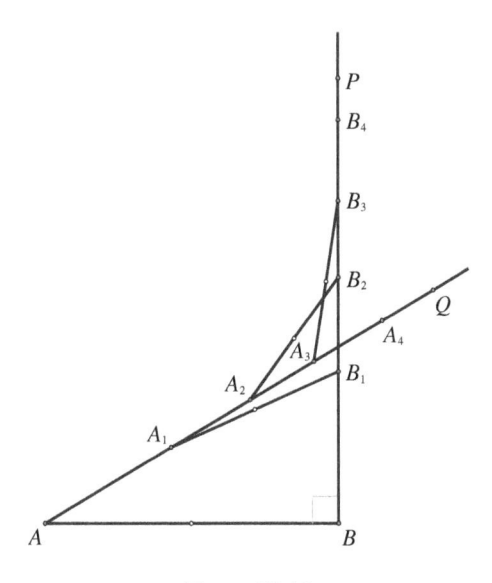

Figure 10-18

segments along AQ as we did before (AA_1, A_1A_2, A_2A_3, . . .), and do the same for the segments along BP (BB_1, B_1B_2, B_2B_3, . . .). We know that marking-off these segments along the two lines can continue indefinitely. Furthermore, segments with the same indices will not intersect. For example, A_1A_2 and B_1B_2 will not intersect. However, segments with different indices can intersect. For example, in Figure 10-18, A_3A_4 intersects B_1B_2. The mistake with our "proof" was to rest our argument on the idea that only certain segments – those with same indices – will not intersect, but that doesn't mean that other segments cannot intersect. This mistake is based on a limited form of reasoning.

Two Randomly Drawn Lines in a Plane are Always Parallel

We begin with the two randomly drawn lines l_1 and l_2, as shown in Figure 10-19. We then construct two parallel lines AD and BC that

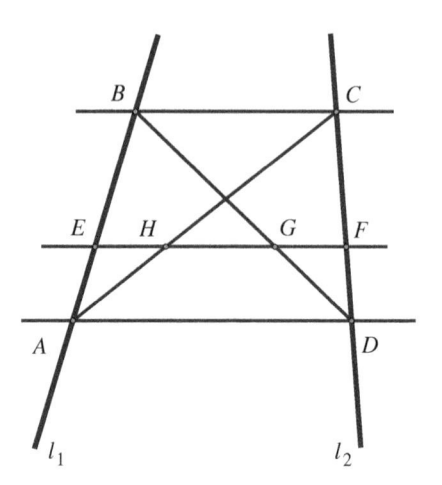

Figure 10-19

intersect our two given lines l_1 and l_2. By drawing $EF \parallel AD$, line EF intersects BD and AC in points G and H, respectively.

The triangles $\triangle AEH$ and $\triangle ABC$ are similar, as are the triangles $\triangle HCF$ and $\triangle ACD$. We therefore can establish the following proportions: $\frac{EH}{BC} = \frac{AH}{AC}$ and $\frac{HF}{AD} = \frac{HC}{AC}$. When we add the two proportions, we get $\frac{EH}{BC} + \frac{HF}{AD} = \frac{AH}{AC} + \frac{HC}{AC} = \frac{HF}{AD} = \frac{AH+HC}{AC} = \frac{AC}{AC} = 1$, which is to say that $\frac{EH}{BC} + \frac{HF}{AD} = 1$. Analogously, we can establish the similarity between the triangles $\triangle BGE$ and $\triangle BDA$, as well as between $\triangle BDC$ and $\triangle GDF$, and then get the following result: $\frac{EG}{AD} + \frac{GF}{BC} = 1$.

Since the last two equations are equal to 1, we get $\frac{EH}{BC} + \frac{HF}{AD} = \frac{EG}{AD} + \frac{GF}{BC}$, or $\frac{HF}{AD} - \frac{EG}{AD} = \frac{GF}{BC} - \frac{EH}{BC}$. Therefore, $\frac{HF-EG}{AD} = \frac{GF-EH}{BC}$.

From Figure 10-19, we find that $HF - EG = (EF - EH) - (EF - GF) = GF - EH$. This tells us that the numerators of the two equal fractions are equal. Consequently, the denominators must also be equal. Therefore, $AD = BC$. Since we began with $AD \parallel BC$, the quadrilateral $ABCD$ must be a parallelogram, and therefore, $AB \parallel CD$, or $l_1 \parallel l_2$.

Thus, we seem to have proved that two randomly drawn lines in the same plane are actually parallel. Clearly this is absurd, and so there must have been a mistake in this demonstration.

Let's take another look at what we have just done. From Figure 10-19, you can clearly see that $HF - EG = (HG + GF) - (EH + HG) = GF - EH$. From the parallel lines in the diagram, the following proportions follow immediately: $\frac{EH}{BC} = \frac{AE}{AB} = \frac{AH}{AC} = \frac{DF}{DC} = \frac{GF}{BC}$. Since $BC \neq 0$, we then have $EH = GF$. Therefore, $GF - EH = 0$, and $HF - EG$ must also equal 0. From the earlier equation $\frac{HF-EG}{AD} = \frac{GF-EH}{BC}$, by substitution, we have $\frac{0}{AD} = \frac{0}{BC}$.

This essentially tells us that we had no reason to state that $AD = BC$, since AD and BC can essentially take on any values to make this equation true. This explains where the mistake was made.

All Chords of a Circle are Equal

Let AB and CD be two arbitrary chords in a circle with different lengths, as shown in Figure 10-20. Since one can rotate chords around the circle's center without affecting their length, we can further assume that they are parallel without loss of generality. Let S be the intersection point of AC and BD.

The triangles ABS and CDS have equal angles and thus are similar, from which we can conclude $\frac{AB}{AS} = \frac{CD}{CS}$, or equivalently, $AB \cdot CS = CD \cdot AS$. Since we assume $AB \neq CD$, we can multiply $AB \cdot CS = CD \cdot AS$ by $AB - CD \neq 0$ and get

$$AB^2 \cdot CS - AB \cdot CS \cdot CD = CD \cdot AS \cdot AB - CD^2 \cdot AS,$$
or $AB^2 \cdot CS - CD \cdot AS \cdot AB = AB \cdot CS \cdot CD - CD^2 \cdot AS,$
or $AB \cdot (AB \cdot CS - CD \cdot AS) = CD \cdot (AB \cdot CS - CD \cdot AS).$

By dividing both sides of the equation by the common factor $AB \cdot CS - CD \cdot AS$, we finally conclude $AB = CD$.

So, we seem to have shown that if two chords are not equal, they are, in fact, equal! This is, of course, false, but where is the mistake in the proof? The mistake is at the very end. We neglected the common

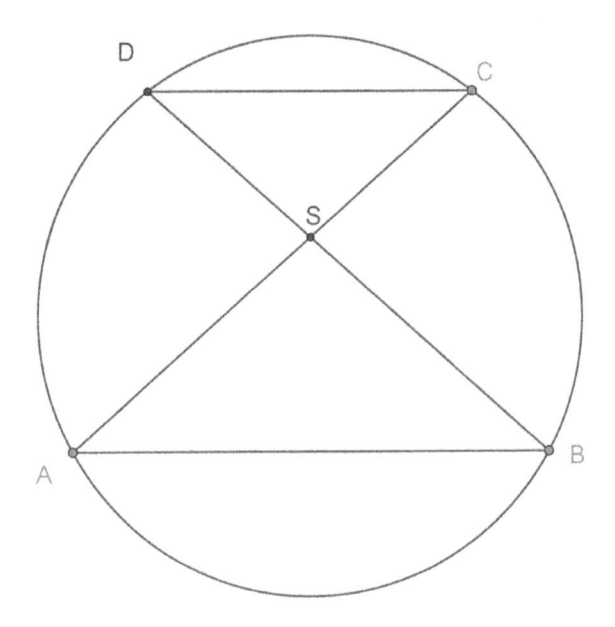

Figure 10-20

factor $AB \cdot CS - CD \cdot AS$, which equals 0 because $\frac{AB}{AS} = \frac{CD}{CS}$, from the earlier identified similar triangles ABS and CDS.

Geometrically it Can Be Shown That 64 = 65

Here is a mathematics mistake that was popularized by Charles Lutwidge Dodgson (1832–1898), who, under the pen name of Lewis Carroll, wrote *The Adventures of Alice in Wonderland*. In Figure 10-21, we notice that the square on the left side has an area of 8 × 8 = 64, and is partitioned into two congruent trapezoids and two congruent right triangles. Yet when these four parts are placed into a different configuration, as shown on the right side of Figure 10-21, we get a rectangle whose area is 5 × 13 = 65. How can 64 = 65? There must be a mistake somewhere!

When we correctly construct the rectangle formed by the four parts of the square, we find an extra parallelogram, as shown in Figure 10-22, where it is exaggerated in size.

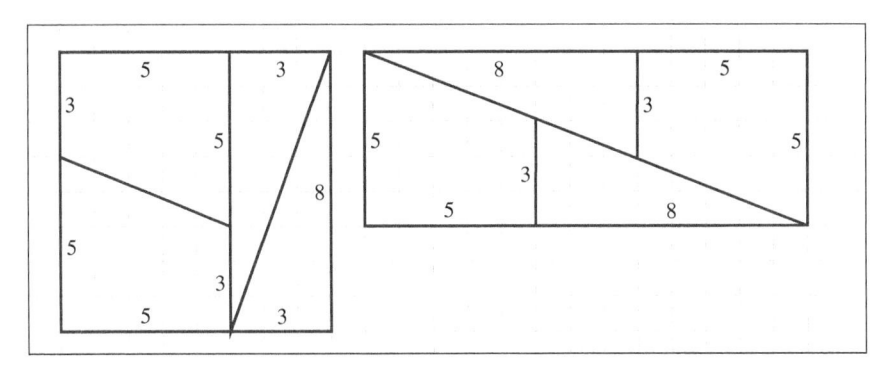

Figure 10-21

Figure 10-22

This parallelogram (shaded) results from the fact that the angles marked α and β are not equal. This is not obvious in the original diagram! Perhaps the easiest way to show this is to refer to the concept of "slope"; the lines CA and EC do not have equal slopes since $\frac{2}{5} \neq \frac{3}{8}$. In order for the line segment ACE to be a straight line – preventing a parallelogram being formed – the slopes, and equivalently the angles α and β, would have to be equal. With different slopes, this is not the case! Thus, the mistake – one easily overlooked – has been exposed.[1]

A Trapezoid Whose Bases Have a Sum of Zero!

The mistake in this proof is subtle and perhaps a bit difficult to find. Consider the proof first, and then we will reveal the mistake. We begin with a trapezoid $ABCD$, and extend the bases, as shown in Figure 10-23, to points E and F. To facilitate this demonstration, the segment lengths are marked in Figure 10-23.

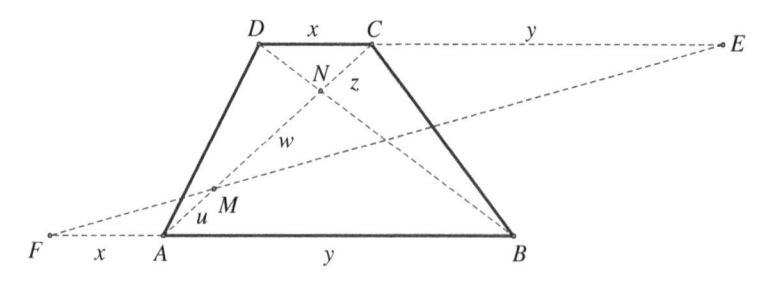

Figure 10-23

From the parallel lines (the bases of the trapezoid) we get some similar triangles:

$\triangle CEM \sim \triangle AFM$, which gives us the proportion

$$\frac{AF}{CE} = \frac{AM}{CM}, \text{ or } \frac{x}{y} = \frac{u}{w+z},$$

---

[1] More examples that relate to the Fibonacci numbers can be found in A. S. Posamentier and I. Lehmann. *The (Fabulous) Fibonacci Numbers*. Afterword by Herbert Hauptman, Nobel Laureate. Amherst (New York), Prometheus Books, 2007, 140–143.

and $\triangle ABN \sim \triangle CDN$, which gives us the proportion

$$\frac{CD}{AB} = \frac{CN}{AN}, \text{ or } \frac{x}{y} = \frac{z}{u+w}.$$

Therefore,

$$\frac{u}{w+z} = \frac{z}{u+w}.$$

There is a convenient (legitimate) operation on proportions that allows subtraction across the numerators and denominators, such as if $\frac{a}{b} = \frac{c}{d}$, then $\frac{a}{b} = \frac{a-c}{b-d}$.

We can now apply that process to the above proportions, so that $\frac{x}{y} = \frac{u-z}{(w+z)-(u+w)} = \frac{u-z}{z-u} = \frac{-(z-u)}{z-u} = -1$.

This would lead us to conclude that $x = -y$, or that $x + y = 0$. But how can the sum of the bases of the trapezoid be zero? Somewhere in this process there must have been a mistake. Let's review our development of this conclusion. Suppose we solve the earlier two equations $\frac{x}{y} = \frac{u}{w+z}$ and $\frac{x}{y} = \frac{z}{u+w}$ for u and z in terms of x, y, and w and get for the first equation

$$yu = x(w+z)$$
$$yu = xw + xz$$
$$xz - yu = -xw.$$

For the second equation, we get

$$yz = x(u+w)$$
$$yz = xu + xw$$
$$yz - xu = -xw.$$

Now, by adding these two newly obtained equations, we get $(xz - yu) + (yz - xu) = (xz + yz) - (yu + xu) = 0$. By rearranging terms, we can write this equation as $z(x+y) - u(x+y) = 0$. By factoring the $(x+y)$ term, we have $(x+y)(z-u) = 0$.

As usual, if either factor is zero, then the equation will be satisfied. In our "proof," we neglected the possibility that $z - u$ might be equal

382 *A Journey Through the Wonders of Plane Geometry*

to zero, and just used assumed that $x + y$ was equal to zero. However, logic tells us that $x + y$, which is the sum of the bases of the trapezoid, is clearly not equal to zero; therefore, $z - u$ must be equal to zero. With that, the fraction above, $\frac{-(z-u)}{z-u}$, becomes $\frac{0}{0}$, which is meaningless!

Any Point in the *Interior* of a Circle is Also *on* the Circle

Let's consider the conflicting statement that any point in the interior of a circle is also on the circle. It sounds ridiculous, but we can provide a "proof" of this statement. There must be a mistake, or else we are in a logical dilemma.

We shall begin our "proof" with a circle O, whose radius is r, as shown in Figure 10-24. We will then let A be any point in the *interior* of the circle distinct from O, and "prove" that the point A is actually *on* the circle.

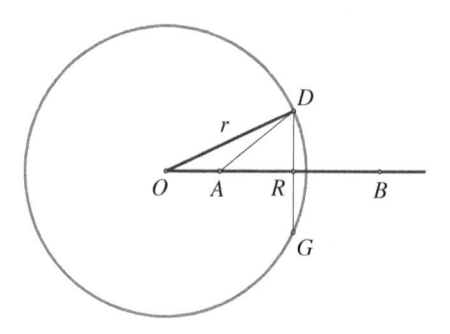

Figure 10-24

We will set up our diagram as follows: let B be on the extension of OA beyond A such that $OA \cdot OB = OD^2 = r^2$. (Clearly OB is greater than r, since OA is less than r.) The perpendicular bisector of AB meets the circle in points D and G, where R is the midpoint of AB. We now have $OA = OR - RA$ and $OB = OR + RB = OR + RA$. Therefore, $r^2 = OA \cdot OB = (OR - RA)(OR + RA)$, or $r^2 = OR^2 - RA^2$.

By applying the Pythagorean theorem to $\triangle ORD$, we get $OR^2 = r^2 - DR^2$, and applying it once again to $\triangle ADR$ gives us

$RA^2 = AD^2 - DR^2$. Therefore, since $r^2 = OR^2 - RA^2$, we get $r^2 = (r^2 - DR^2) - (AD^2 - DR^2)$, which reduces to $r^2 = r^2 - AD^2$. This would imply that $AD^2 = 0$; in other words, A coincides with D and thus lies on the circle. That is to say, point A inside the circle has been proved to be on the circle. There must be a mistake somewhere!

The fallacy in this proof lies in the fact that we drew an auxiliary line *DRG* with *two* conditions: it is the perpendicular bisector of *AB* and it intersects the circle. Actually, all points on the perpendicular bisector of *AB* lie in the exterior of the circle and therefore cannot intersect the circle. Follow along with the algebraic process:

$$r^2 = OA \cdot OB$$
$$r^2 = OA(OA + AB)$$
$$r^2 = OA^2 + OA \cdot AB. \qquad (I)$$

The "proof" assumes that $OA + \frac{AB}{2} < r$.

By multiplying both sides of the inequality by 2, we get $2OA + AB < 2r$.

By squaring both sides of the inequality, we have

$$4OA^2 + 4OA \cdot AB + AB^2 < 4r^2. \qquad (II)$$

By substituting 4 times equation (I), which is $4r^2 = 4OA^2 + 4OA \cdot AB$, into equation (II), we get $4r^2 + AB^2 < 4r^2$, or $AB^2 < 0$, which is impossible.

The mistake here alerts us to be cautious when allowing points to take on more properties than are possible. That is, when drawing auxiliary lines, we must take care that they use *one* condition only.

A Common Mistake Based on a Correct Principle

A basic concept in geometry is that the ratio of the areas of two similar figures is equal to the square of the ratio of two corresponding line segments. If we apply this principle to the following problem, we would run into a mistake.

384 A Journey Through the Wonders of Plane Geometry

We begin with two concentric circles whose radii are a and b, respectively, with $a > b$, as shown in Figure 10-25. Our task is to find the radius of a third concentric circle placed between the two given concentric circles such that the area of the outermost ring is twice the area of the next smaller ring. We will let the radius of the sought-after circle be x.

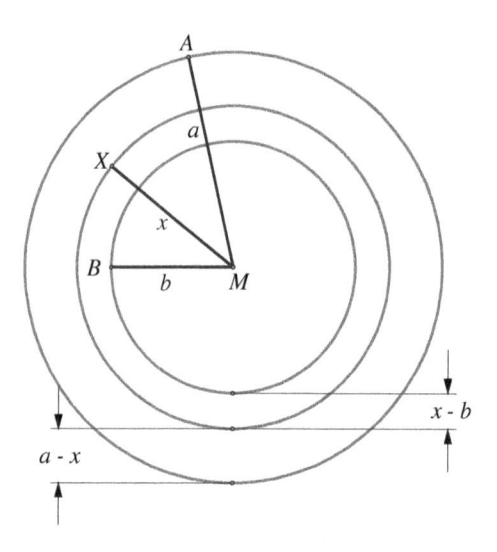

Figure 10-25

To apply the principle that we mentioned at the start, the two corresponding parts could be the distances between the circles, or the width of the rings. The outermost ring has a width of $a - x$, and the innermost ring has a width of $x - b$. We then get the following proportion:

$$\frac{(a-x)^2}{(x-b)^2} = \frac{2}{1}. \qquad (*)$$

Solving this equation for x, we get the following:

$$(a-x)^2 = 2 \cdot (x-b)^2$$

$$a^2 - 2ax + x^2 = 2 \cdot (x^2 - 2bx + b^2) = 2x^2 - 4bx + 2b^2$$

$$2x^2 - 4bx + 2b^2 - a^2 + 2ax - x^2 = 0$$

$$x^2 + 2x(a-2b) - a^2 + 2b^2 = 0$$

$$x = -a + 2b \pm \sqrt{(a-2b)^2 + a^2 - 2b^2} =$$
$$-a + 2b \pm \sqrt{a^2 - 4ab + 4b^2 + a^2 - 2b^2}$$
$$x = -a + 2b \pm \sqrt{2a^2 - 4ab + 2b^2} =$$
$$-a + 2b \pm \sqrt{2(a^2 - 2ab + b^2)}$$
$$x = -a + 2b \pm (a-b)\sqrt{2}.$$

Unfortunately, these two values of x are both wrong! We must have made a mistake somewhere. Our mistake occurred at the very first step, marked with (*). We mistakenly used the ring widths rather than the circles' radii to deal with the individual circles, whose areas would then be subtracted to get the ring areas.

If we let A_a, A_b, and A_x represent the areas of the circles whose respective radii are a, b, and x, with A_{a-x} and A_{x-b} representing the areas of the two required rings, we can set up the following:

$$A_{a-x} = A_a - A_x = \pi \cdot a^2 - \pi \cdot x^2 = \pi \cdot (a^2 - x^2),$$

and

$$A_{x-b} = A_x - A_b = \pi \cdot x^2 - \pi \cdot b^2 = \pi \cdot (x^2 - b^2).$$

With $A_{a-x} = 2 \cdot A_{x-b}$ we get $\pi \cdot (a^2 - x^2) = 2\pi \cdot (x^2 - b^2)$, which then yields $\frac{a^2 - x^2}{x^2 - b^2} = \frac{2}{1}$, which is considerably different from the initial "application" of the similarity principle.

Now proceeding *correctly*, we get $a^2 - x^2 = 2x^2 - 2b^2$. Therefore, $x^2 = \frac{a^2 + 2b^2}{3}$, which then gives us $x = \sqrt{\frac{a^2 + 2b^2}{3}}$ since the negative root is ignored as we are dealing with the length of a line.

A Rope Around the Equator: A Mistake of Our Intuition

A mistake in mathematics can also be one of judgment, where one makes a mistake because the correct answer is counterintuitive. Consider the planet earth, with a rope wrapped tightly around the equator. Let's assume that the earth is a perfect sphere, and that the equator is exactly 40,000 km long. Assume also that the earth has a smooth surface along the equator, just to make our work a bit easier.

We now lengthen the rope by exactly 1 meter. We position this (now loose) rope around the equator so that it is uniformly spaced off the globe, which we show in Figure 10-26. Will a mouse fit under the rope?[2] We might say "This is clearly not possible because 1 m is very small, practically negligible compared to 40,000 km, so the growth of the radius will also be very small." But we would be mistaken!

Figure 10-26

The traditional way to determine the distance between the circumferences of the earth and the rope is to find the difference between the radii. Let r be the radius of the circle formed by the earth (circumference = C), and R be the radius of the circle formed by the rope (circumference = $C + 1$).

Referring to Figure 10-27, we apply the familiar circumference formulas to yield $C = 2\pi r$, or $r = \frac{C}{2\pi}$, and then $C + 1 = 2\pi R$, or $R = \frac{C+1}{2\pi}$. We need to find the difference of the radii, which is $R - r = \frac{C+1}{2\pi} - \frac{C}{2\pi} = \frac{1}{2\pi}$. The "1" in the numerator means 1 meter. Therefore, we get $R - r = \frac{1m}{2\pi} = \frac{100cm}{2\pi} \approx 15.9$ cm = 0.159 m.

Wow! There is actually a space of approximately $6\frac{1}{4}$ inches for a mouse to crawl under. This result is all the more astonishing because the "intuitive" answer is wrong.

We might also have approached answering this question by using a very powerful problem-solving strategy called *considering extreme*

---

[2] This classic problem first appeared in the article "The Paradox Party. A Discussion of Some Queer Fallacies and Brain-Twisters" by Henry Ernest Dudeney. *The Strand Magazine. An Illustrated Monthly*. Edited by George Newnes, vol. 38 (No. 228) Dec. 1909, pp. 670–676.

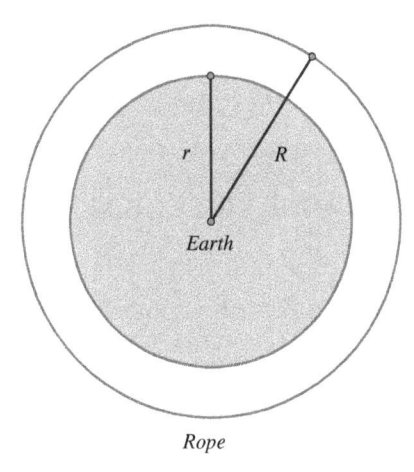

Figure 10-27

cases. The solution was independent of the circumference or radius r of the earth, since the end result did not include the circumference in the calculation. It only required calculating $\frac{1}{2\pi}$.

Here is a nifty solution using an extreme case. Suppose the inner circle in Figure 10-27 is very small, so small that it has radius zero, which means it is actually just a point. We were required to find the difference between the radii; in this case $R - r = R - 0 = R$. So, all we need to find is the length of the radius of the larger circle and our problem will be solved. We apply the circumference formula for the circumference of the larger circle:

$$C + 1 = 0 + 1 = 2p\,R, \text{ then } R = \frac{1}{2\pi}.$$

Our initial mistake lead us to two lovely little treasures. First, it revealed an astonishing result – one clearly unanticipated at the start – and second, it provided us with a useful problem-solving strategy.[3]

Why does the strategy of letting $r = 0$ yield the correct answer for the general case here? This can be explained easily by looking at the formula $C = 2\pi r$ with the *constant factor* 2π in another way: If the radius r changes by x, so $r_{new} = r + x$, the circumference changes by $2\pi x$

---

[3] For more cases and a discussion of similar problems, see A. S. Posamentier and I. Lehmann. *Pi: A Biography of the World's Most Mysterious Number*. Afterword by Herbert Hauptman, Nobel Laureate. Amherst (New York), Prometheus Books, 2004, pp. 222–243, 305–308

because $C_{new} = 2\pi(r+x) = \underbrace{2\pi r}_{C} + 2\pi x$, and this insight is completely independent of the value of r or C. Thus, we are free to choose $r = 0$.

Another Rope Around The Equator: A Further Counterintuition!

Up until now, the rope in the previous example was always tightened or loosened concentrically, that is, it was pulled away evenly on all sides. In this new situation, this is not the case. The rope will be pulled away at one single point, as if the earth were hung up on a hook. The rope length is again extended by 1 meter. But instead of being concentric, it is pulled away at one point so that a maximum distance from the earth's surface is achieved, as shown in Figure 10-28. How far away from the earth (x) can the rope be "pulled?"

Most people are astonished by the previous case, where the mouse has 15.9 cm to crawl under. The result of pulling the rope from one point will once again be quite surprising. The result is apparently a paradox that leads to mistaken conclusions. The 1-meter-longer rope

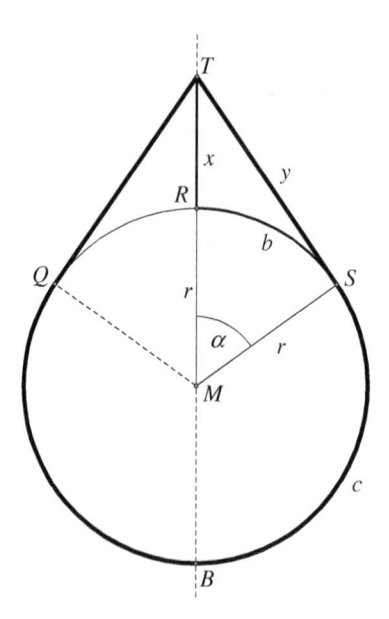

Figure 10-28

pulled taut from a point, where the rest of the rope "hugs" the earth's surface, reaches a point about 122 m above the earth's surface.

Let's see why this is so. This time, the answer is clearly dependent on the size of the earth, and not exclusively on π, although π will also play a role. From the exterior point T, the rope, which is 1 meter longer than the circumference of the equator, is pulled taut so that it hugs the earth's surface to the points of tangency, S and Q. We seek to find how high off the surface is point T. That means we will try to find the length of x, or RT. Remember, the length of the rope from B through S to T is 0.5 m longer than the circumference of the earth, so we have $\widehat{BS} + ST = \widehat{BSR} + 0.5$m. Our goal is to find the length of TR. Let's review where we are: The rope lies on the arc SBQ, and at points S and Q goes tangential to the point T. The lengths in Figure 10-28 are marked, and $\alpha = \angle RMS = \angle RMQ$. The length of the rope is $2\pi r + 1$, and we get the following relation: $y = b + 0.5$. This is equivalent to $b = y - 0.5$ because y is 0.5 m longer than b, since the rope is 1 m longer than the circumference.

In $\triangle MST$, the tangent function will be applied as follows: $\tan \alpha = \frac{y}{r}$, so $y = r \tan \alpha$.

We can form the ratio of arc length to central angle measure and get $\frac{b}{\alpha} = \frac{2\pi \cdot r}{360°}$, or equivalently, $b = \frac{2\pi \cdot r \cdot \alpha}{360°}$. Assuming that the equator is exactly 40,000,000 m, with $c = 2\pi r$, we can find the earth's radius to be $r = \frac{c}{2\pi} = \frac{40,000,000}{2\pi} \approx 6,366,198$ m.

Combining the equations we have above, we get

$$\underbrace{\frac{2\pi \cdot r \cdot \alpha}{360°}}_{b} = r \tan \alpha - 0.5.$$

We are now faced with a dilemma, namely, that this equation in α (obtained above) cannot be uniquely solved in the traditional manner. We will set up a table of possible trial values to see what will satisfy the equation $\frac{2\pi \cdot r \cdot \alpha}{360°} = r \tan \alpha - 0.5$. We will use the value of r we found above: $r = 6,366,198$ m.

Our various trials would indicate that our closest match of the two values occurs at $\alpha \approx 0.354°$.

For this value of α, we have $y = r \tan \alpha \approx 6,366,198 \cdot 0.006178544171 \approx 39,333.83554$ m, or about 39,334 m.

α	$b = \dfrac{2\pi \cdot r \cdot \alpha}{360°}$	$b = r\tan\alpha - 0.5$	Comparison of values (number of places in agreement – **bold**)
30°	**3**,333,333.478	**3**,675,525.629	1
10°	**1,1**11,111.159	**1,1**22,531.971	2
5°	**55**5,555.5796	**55**6,969.6547	2
1°	**111,1**11.1159	**111,1**21.8994	4
0.3°	**33,333**.33478	**33,333**.13940	5
0.4°	**44,444**.44637	**44,444**.66844	5
0.35°	**38,888.8**9057	**38,888.8**7430	6
0.355°	**39,444.4**4615	**39,444.4**5091	6
More exactly:			
0.353°	**39,222.22**392	**39,222.22**019	7
0.354°	**39,333,335**04	**39,333,335**54	8
0.3545°	**39,388.89**059	**39,388.89**322	7
0.355°	**39,444.4**4615	**39,444.4**5091	6

The rope is therefore almost 40 km long before it reaches its peak. Here, we intentionally ignore that, in reality, a 40 km long rope can never be modelled by a straight line because there would be a considerable slack span. But how high off the earth's surface is the rope? That is, what is the length of x? Applying the Pythagorean theorem to $\triangle MST$, we get $MT^2 = r^2 + y^2$. Then, $MT^2 = 6,366,198^2 + 39,334^2 = 40,528,476,975,204 + 1,547,163,556 = 40,530,024,138,760$. Therefore, $MT \approx 6,366,319.512$ m. We are looking for x, which is $MT - r \approx 121.512$ m, or about 122 meters.

This result is perhaps astonishing, because one intuitively assumes that by the circumference of the earth (40,000 km) an extra meter must almost disappear. But this is the mistake! The larger the sphere, the further the rope can be pulled away from it. Looking at the extreme case, where the radius of the equator decreases to zero, we have the minimum value for x, namely, $x = 0.5$ m.

Do All Circles Have Equal Circumferences?

Sometimes physical observations are difficult to explain and even paradoxical. For example, we know that when a circle rolls on a line and makes one complete revolution, it has traveled the distance equal to the length of its circumference. In Figure 10-29, when the larger circle travels from point A to point B, it will have traveled the distance AB, which is equal to the circumference of the larger circle. When we roll two concentric circles with unequal circumferences, we wonder how the smaller circle will have traveled one large-circle-circumference length at the same time as the larger circle traveled this distance, although both have made exactly one revolution. This may be seen in Figure 10-29, where AB is equal to CD. In other words, the small circle and larger circle have the same circumference. This paradox dates back to Aristotle (384–322 BCE). How is this possible? Where is the mistake?

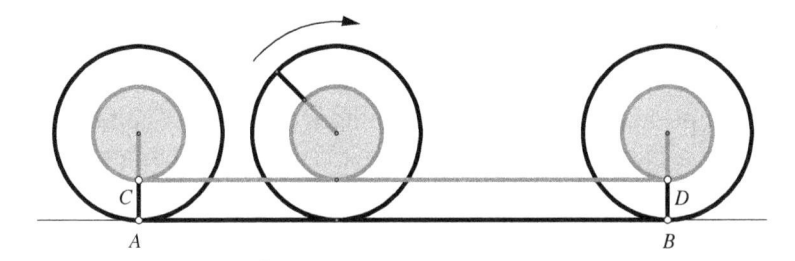

Figure 10-29

If we observe a fixed point on each of the two circles during this rolling exercise – in this case points A and C – we notice that the points travel on a cycloid path, as shown in Figure 10-30.

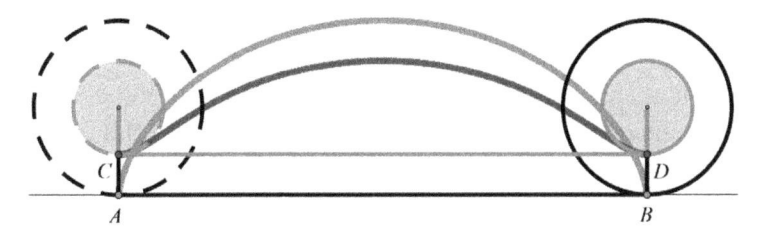

Figure 10-30

With additional revolutions this becomes even more clear, as you can see in Figure 10-31.

The curves describe the paths of points A and C for two complete revolutions of the circles. Yet, the paths along which they travel

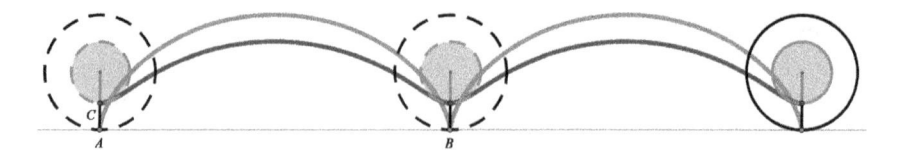

Figure 10-31

are not equal to the circumference of the circle on which they lie. The distance from A to B along the straight line is equal to $2\pi R$, where R is the radius of the larger circle. We can clearly see that the cycloid curve between points A and B is longer than the circumference of the larger circle. The length of the cycloid of points on each of the two circles is dependent on the circle's circumference. Interestingly enough, the length of the cycloid can be an integer if the radius is also an integer.[4]

The mistake that all circles have the same conference rests with the assumption that both circles simultaneously roll. The fallacy that we face here is actually not geometric; it is one of mechanics. Only one of the circles can roll at a time. It is crucial to see here that if the larger circle rolls, then the smaller circle additionally slides along. Were the smaller circle to roll, then the larger circle would supposedly slide somewhat backwards. The mistake here is to not have recognized that the wheels are not rolling together. While one wheel rolls, the other wheel slides along. Thus, the error is a mechanical one.

The Diagonals of The Hexagon: A Mistaken Count of Intersections

A common mistake when determining the number of intersections of the diagonals of a (convex) hexagon is to assume that the hexagon is a regular hexagon where all the angles and the sides are of equal measure. We are interested in the number of intersection points of the diagonals of *any* hexagon. In Figure 10-32, we can count the points of intersection of the regular hexagon shown. There are 13 such points.

---

[4] Circumference = $2\pi \cdot r$; using higher mathematics, we can show that a regular cycloid has length = $8r$, when r is the radius of the circle, which has made one complete revolution.

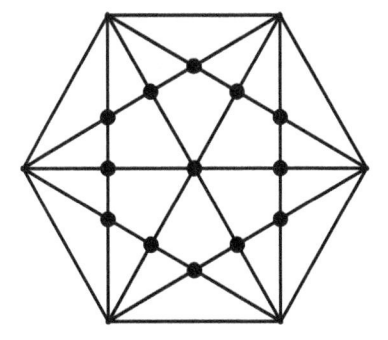

Figure 10-32

However, if we consider an irregular hexagon, as shown in Figure 10-33, we will find that the diagonals have two additional points of intersection. Thus, for a general hexagon, the number of points of intersection of the diagonals is 15. Unexpectedly, the mistake of using a regular hexagon when one isn't called for leads to a wrong answer.

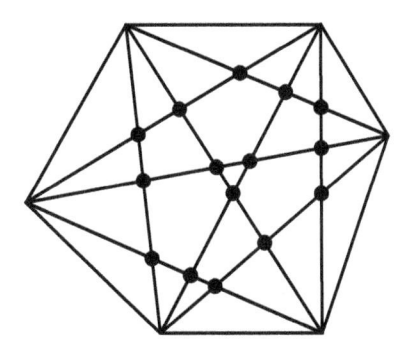

Figure 10-33

Confounding Constructions

We are going to take a slightly different tack now. Several different constructions for a regular octagon will be presented. They will all look correct. However, we will leave it to the reader to determine which of these produces a truly regular polygon, and which are erroneous constructions – despite the fact that they look correct.

394 *A Journey Through the Wonders of Plane Geometry*

Octagon Construction 1: In Figure 10-34, we begin with a square and then the midpoints of its four sides. At each of the four vertices of the square, an isosceles right triangle is formed. We bisect each of the acute angles of these isosceles right triangles to identify the remaining four vertices of the octagon.

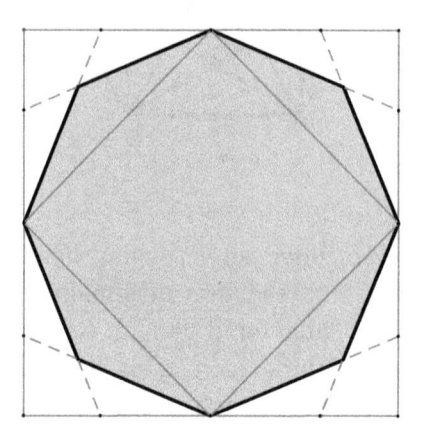

Figure 10-34

Octagon Construction 2: Again, we begin with a square, and then join the midpoints of each of the sides with the square's opposite vertices (See Figure 10-35).

Octagon Construction 3: We begin with four congruent tangent circles inscribed in a square, as shown in Figure 10-36. Next, we join

Figure 10-35

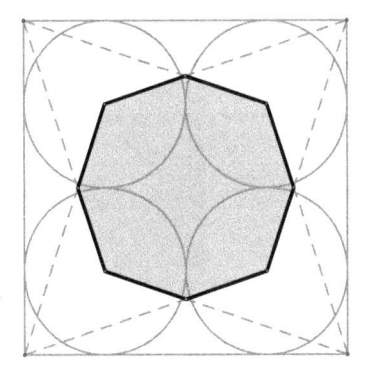

Figure 10-36

the centers of each of the circles with the vertices of the square. This determines the octagon.

Octagon Construction 4: In Figure 10-37, we once again begin with a square, and draw quarter-circles centered at each vertex of the square with radii half the length of the diagonal. The points at which these quarter-circles intersect the sides of the square determine the required octagon (see Figure 10-37).

Octagon Construction 5: Begin with a square, and construct quarter-circles centered at each vertex with radii the length of the side of the square. Mark the points at which the diagonals of the square intersect the quarter-circles. Through these four points, draw

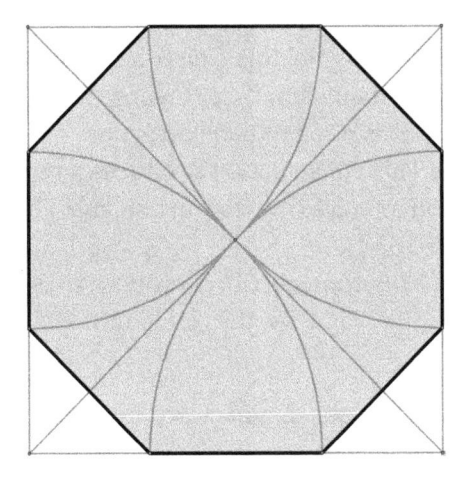

Figure 10-37

lines parallel to the sides of the square, which determine the vertices of the requisite octagon (see Figure 10-38).

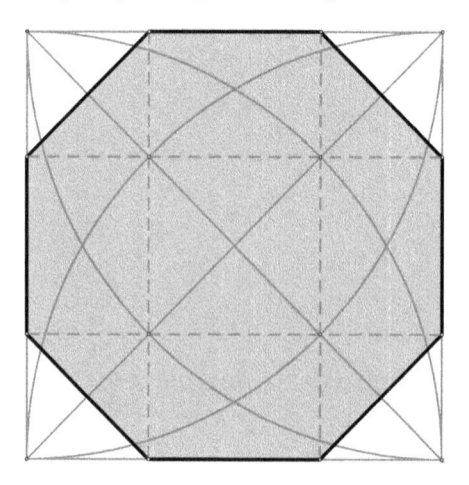

Figure 10-38

We now have five different constructions of an octagon. The question that remains is, which of these are regular octagons, and which of these are false constructions of regular octagons? Here are the results:

Construction 1: Is a correct construction of a regular octagon.

Construction 2: This construction was known to Archimedes (ca. 287–212 CE) and has equal sides, but not all angles are congruent. Therefore, it is *not* a regular octagon!

Construction 3: Once again, this construction produces an octagon whose sides are of the same length, but whose angles are not of equal measure. Therefore, it is *not* a regular octagon.

Construction 4: This construction produces a regular octagon and was first developed in 1543 by the artist and geometer Augustin Hirschvogel (1503–1553).

Construction 5: This construction produces a regular octagon and was first developed in 1564 by the goldsmith Heinrich Lautensack (1522–1568).

Therefore, constructions 2 and 3 are mistaken constructions of regular octagons.

For the ambitious reader, we provide the details of each of these constructions in the chart shown in Figure 10-39.

Geometric Fallacies **397**

(1)	(2) Archimedes	(3)	(4) Hirschvogel	(5) Lautensack
regular	equilateral, but not equiangular	equilateral, but not equiangular	*regular*	*regular*
$\varphi = \psi = 135°$	$\varphi \approx 126.9°$, $\psi \approx 143.1°$	$\varphi \approx 126.9°$, $\psi \approx 143.1°$	$\varphi = \psi = 135°$	$\varphi = \psi = 135°$
$b = \dfrac{a\sqrt{2-\sqrt{2}}}{2}$ $\approx 0.3827 \cdot a$	$b = \dfrac{a\sqrt{5}}{12}$ $\approx 0.1863 \cdot a$	$b = \dfrac{a\sqrt{10}}{12}$ $\approx 0.2635 \cdot a$	$b = a\left(\sqrt{2}-1\right)$ $\approx 0.4142 \cdot a$	$b = a\left(\sqrt{2}-1\right)$ $\approx 0.4142 \cdot a$
$A = \dfrac{a^2}{2}\sqrt{2}$ $\approx 0.7071 \cdot A_{\text{Sq}}$	$A = \dfrac{a^2}{6}$ $\approx 0.1667 \cdot A_{\text{Sq}}$	$A = \dfrac{a^2}{3}$ $\approx 0.3333 \cdot A_{\text{Sq}}$	$A = 2a^2\left(\sqrt{2}-1\right)$ $\approx 0.8284 \cdot A_{\text{Sq}}$	$A = 2a^2\left(\sqrt{2}-1\right)$ $\approx 0.8284 \cdot A_{\text{Sq}}$

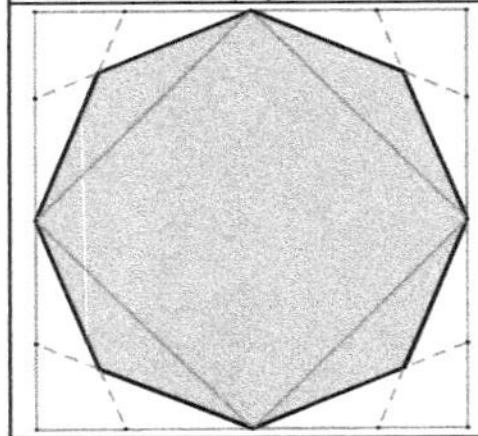

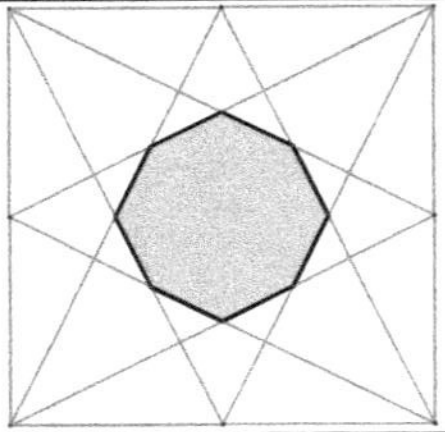

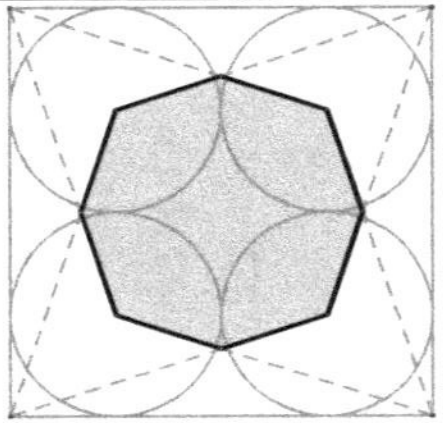

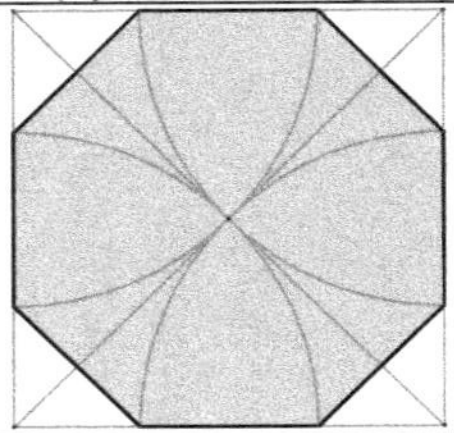

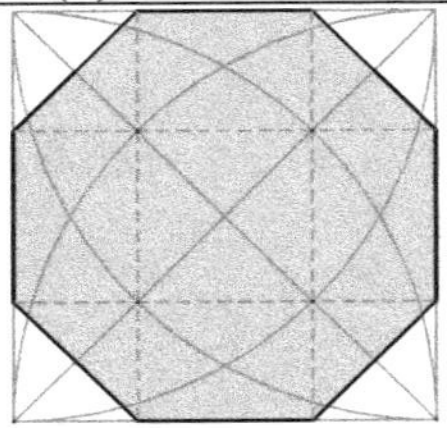

Figure 10-39   Comparisons of the five octagons

We denote the interior angles of each of the figures with the symbols (φ, ψ), the side length of the octagon with b, and the side length of the original square as a; A_{sq} represents the area of the square.

The Regular Pentagon That Isn't

One of the more difficult constructions to do using an unmarked straightedge and compasses is the regular pentagon. There are many ways to do this construction, none particularly easy. You might try to develop a construction on your own, and realize that the Golden Section is involved here (see Chapter 1).

For years, engineers have been using a method for drawing what appears to be a regular pentagon, but upon careful inspection is a tiny bit irregular.[5] This method, which we will provide below, was developed in 1525 by the famous German artist Albrecht Dürer (1471–1528).

In Figure 10-40, we begin with a segment AB. Five circles of radius AB are constructed as follows:

Circles with centers at A and B are drawn and intersect at Q and N. The circle with center Q is drawn to intersect circles A and B at points R and S, respectively. We have QN intersecting circle Q at P. Also, SP and RP intersect circles A and B at points E and C, respectively. Draw the circles with centers at E and C and radius AB to intersect at D. Joining the points in order, we get the (supposedly regular) pentagon $ABCDE$, as shown in Figure 10-41.

Although the pentagon "looks" regular, $\angle ABC$ is a bit more than $\frac{1}{3}$ of a degree too large. That is, for $ABCDE$ to be a regular pentagon, the measure of each angle must be 108°. Instead, we show that $\angle ABC \approx 108.3661202°$.

---

[5] For a discussion of where the error lies, see A. S. Posamentier and H. A. Hauptman. *101 Great Ideas for Introducing Key Concepts in Mathematics*. Corwin Press: Thousand Oaks, CA, 2001, pp. 141–146.

Geometric Fallacies 399

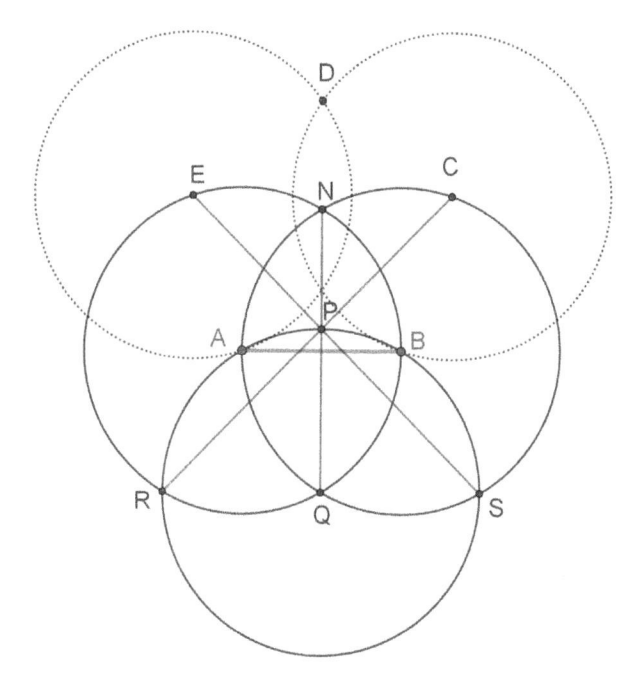

Figure 10-40

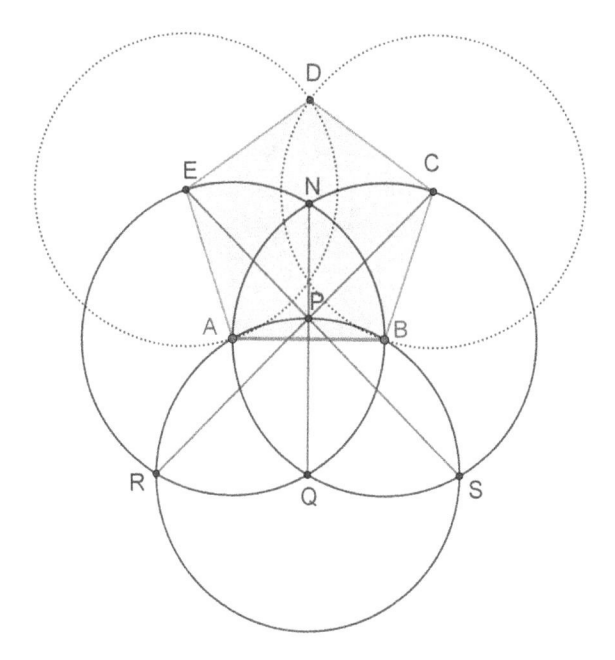

Figure 10-41

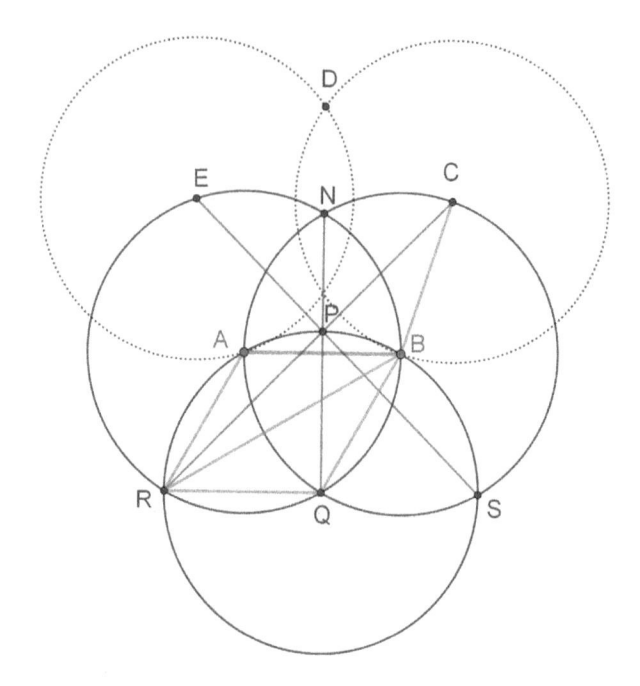

400 *A Journey Through the Wonders of Plane Geometry*

Figure 10-42

In rhombus *ABQR* of Figure 10-42, we have ∠*ARQ* = 60°, and *BR* = *AB*√3 because *BR* is actually twice the length of an altitude of equilateral △*ARQ*. Because △*PRQ* is an isosceles right triangle, ∠*PRQ* = 45°, and then ∠*BRC* = 15°. We apply the law of sines to △*BCR* and get $\frac{BR}{\sin \angle BCR} = \frac{BC}{\sin \angle BRC}$. That is, $\frac{AB\sqrt{3}}{\sin \angle BCR} = \frac{AB}{\sin 15°}$, or equivalently, sin ∠*BCR* = √3 sin 15°.

Therefore, ∠*BCR* ≈ 26.63387984. In △*BCR*, we then have ∠*RBC* = 180° − ∠*BRC* − ∠*BCR* ≈ 180° − 15° − 26.63387984° ≈ 138.3661202°.

Thus, because ∠*ABR* = 30°, ∠*ABC* = ∠*RBC* − ∠*ABR* ≈ 138.3661202° − 30° ≈ 108.3661202°. This is *not* 108°, as it should be in a regular pentagon.

Chapter 11

Homothety, Similarity, and Applications

Increasing and Decreasing the Size of Geometric Figures by Homothety

The process of *enlarging* or *reducing* the size of geometric figures has fascinated mathematicians for centuries. As this topic preceded the study of the concept of similarity, its allure rested in changing the size of figures and preserving their shape. It has been useful in many applications, such as:

- Using larger/smaller grid patterns to modify the sizes of various shapes. Modifying the size of figures is necessary at many levels of education.
- Using different scales for rectangular sports fields, for example, so that their internal lines are changed proportionally, that is, multiplied by a constant factor. The same holds true for building plans of houses.
- The concept of reduction and enlargement can be seen with models of toys, such as ships, cars, etc., as well as maps, cells, molecules, and the like.

Essentially, we want to keep objects the same shape and just modify their size. This type of enlargement or reduction is provided by *uniform scaling* or *homotheties*, also called *homogeneous dilations*, or *central dilations*, where scale factors work in all directions from a specified center, and where these scale factors can even be negative. In this chapter, we will consider a type of geometric mapping – primarily in the plane, but also with possible applications in the three-dimensional space.

A well-known example of this principle (in three-dimensional space) is the projection of photo slides onto vertical walls, as shown in Figure 11-1.

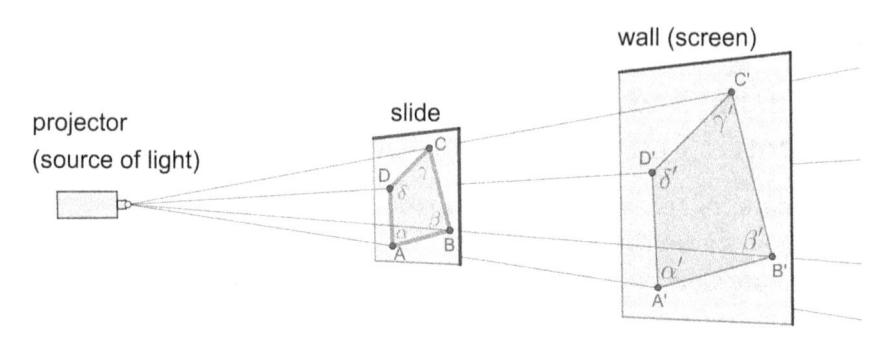

Figure 11-1   Projection of slides

For a homothety, we use the notation $S_{Z,k}$ to indicate that it has a center point Z and scale factor $k \neq 0$. Such a homothety does nothing more than multiply every (oriented) distance from the center point Z by a constant factor, k. Mathematically, the image points P' of P under a homothety $S_{Z,k}$ are constructed in the following way. In order to consider the orientation of the distances, we will introduce the concept of a *vector*. A vector is a quantity that has both magnitude and direction and is typically represented by an arrow whose direction is the same as that of the quantity and whose length is proportional to the quantity's magnitude.

1. For $P = Z$: $Z' = Z$ (fixed point)
2. For $P \neq Z$: P' lies on the straight line ZP, where $\overrightarrow{ZP'} = k \cdot \overrightarrow{ZP}$

When $k > 0$, point P' lies on the same side of point Z as point P, whereas when $k < 0$ it is on the other side of point Z, as shown in Figures 11-2 and 11-3.

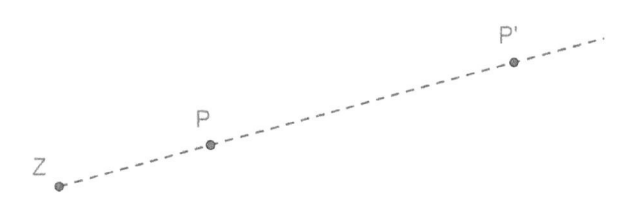

Figure 11-2 $k > 0$

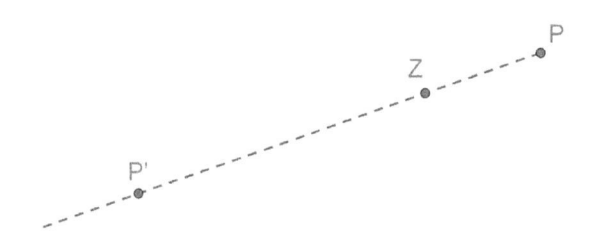

Figure 11-3 $k < 0$

The case of $k = 0$ is excluded because all the points of the plane would be mapped onto point Z itself, which would certainly not be a meaningful mapping. There are special cases: when $k = 1$, all points stay unchanged and the map is the *identity*; and when $k = -1$, we have a reflection in the point Z, or a half-turn around point Z.

In the following theorems and proofs, we mainly restrict ourselves to $k > 0$, while homotheties with $k < 0$ can be seen as homotheties with scale factor $|k| > 0$ joined by a reflection in the center Z (which is equivalent to a half-turn around point Z). Thus, all mentioned properties and theorems also apply for values $k < 0$.

Figure 11-4 shows a homothety of a triangle *ABC* with scale factor $k = 2$, which implies that all distances from point *Z* are doubled.

Such a mapping can be easily realized using dynamic geometry software, such as Geometer's Sketchpad or GeoGebra. By clicking an object (such as a triangle) and choosing a center point *Z* and a scale factor k, one immediately gets the enlarged/reduced figure. This is one of the advantages of modern technology.

In Figure 11-4, we can see how the distances of the points from *Z* and the distances between the points have been doubled: $A'B' = 2 \cdot AB$, $A'C' = 2 \cdot AC$, and $B'C' = 2 \cdot BC$. Furthermore, the images of the line segments seem to be parallel to the initial line segment, as $A'B' \| AB$. These observations and measurements are, of course, not a proof. We need to prove that this is always true, so that we will formulate these findings as a fundamental theorem about homotheties.

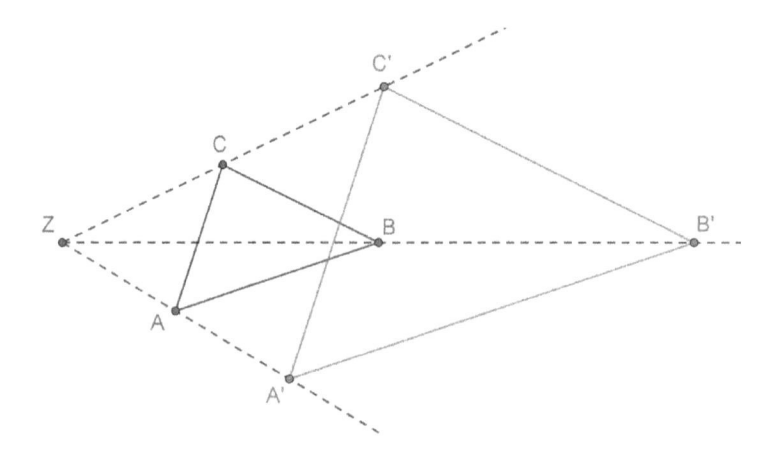

Figure 11-4 Homothety of a triangle *ABC* with center *Z* and $k = 2$

Theorem

Regarding a homothety with center *Z* and scale factor $k \neq 0$, the following holds:

1. $A'B' \| AB$: The image of a straight line is always parallel to the initial straight line.

2. $A'B' = |k| \cdot AB$: The length of a line segment is always multiplied by $|k|$.

From property 1 (invariance of the direction), homotheties preserve the angles of all figures.

Proof

Case 1: The straight line AB intersects point Z. Since all involved points and image points lie on one straight line through Z, the parallelism (or coincidence) $A'B' \| AB$ follows; furthermore, as shown in Figure 11-5, we have for $(k > 0)$: $A'B' = ZB' - ZA' = k \cdot ZB - k \cdot ZA = k \cdot (ZB - ZA) = k \cdot AB$.

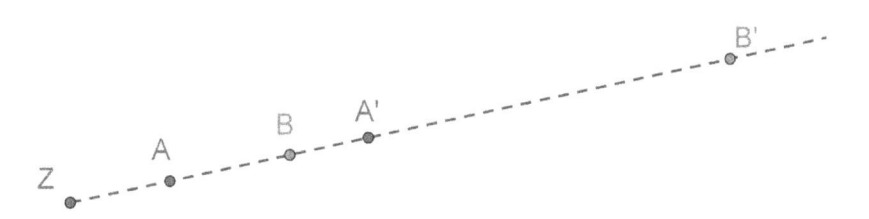

Figure 11-5 AB intersects Z

Case 2: The straight line AB does not intersect point Z. We consider the situation where $k = 3$, as shown in Figure 11-6 (analogous for other $k \in \mathbb{N}$).

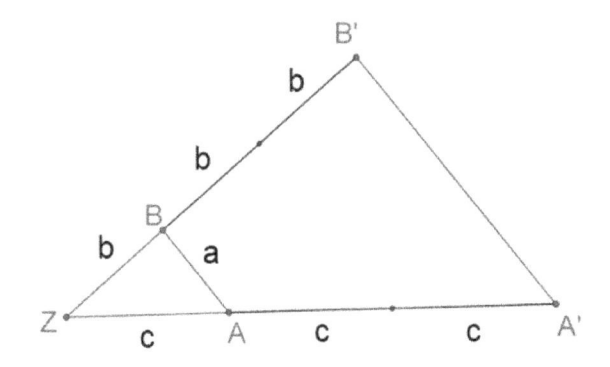

Figure 11-6 AB does not intersect Z, and $k = 3$

We now must show $A'B' \| AB$ and $A'B' = 3AB$. To simplify matters, we consider Figure 11-7, where ΔZA_2B_2 has $ZB = b$ three times on the ray h (originating at Z and passing through B) by starting at Z and getting to B_2, where $ZB_2 = 3b = ZB'$. If we drew everything in one diagram, we would have $B_2 = B'$. Let B_1 be the point after taking the segment $ZB = b$ twice. We translate (that is, move in a parallel manner) the straight line $c = ZA$ through B and B_1, and the straight line AB through B_1 and B_2, thus yielding the intersection points A_1 and A_2 with the straight line g.

From the congruence of angle-side-angle α, b, γ, three congruent triangles (shaded in Figure 11-7) and several congruent parallelograms arise, which enables the length c of the congruent triangles to transfer onto the straight-line g: $ZA_2 = 3 \cdot c = ZA'$. If everything were shown in one sketch, we would have $A_2 = A'$. Analogously, the length a is transferred three times onto AB so that $A_2B_2 = 3a$. From the side-angle-side congruence ($3b$, α, $3c$) we know that $\Delta ZA_2B_2 \cong \Delta ZA'B'$, and from $A_2B_2 \| AB$ and $A_2B_2 = 3AB$, respectively, we can conclude the original objective, which is an analogous statement with $A'B'$ instead of A_2B_2.

Conversely, if we take A_2B_2 as the initial line segment, we have a proof for $k = \frac{1}{3}$ (analogously, for $k = \frac{1}{n}, n \in \mathbb{N}$) with the result AB. If we

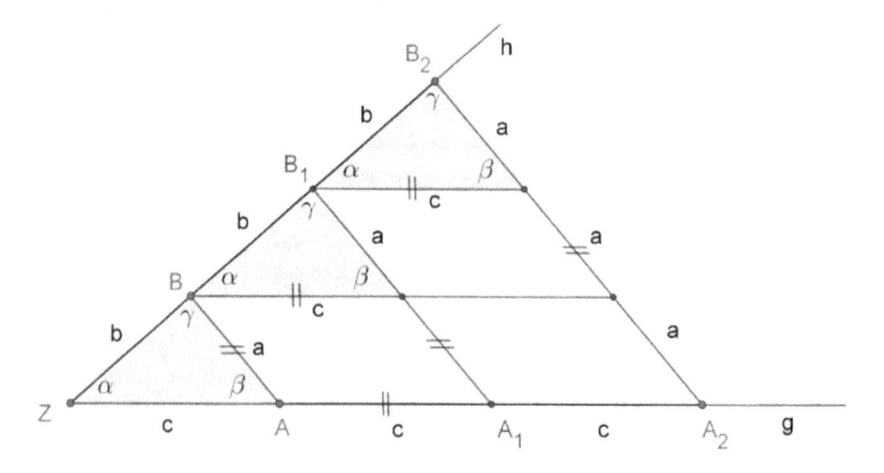

Figure 11-7   Taking b several times – translating straight lines

subsequently apply the same procedure to AB for $k = 2$ ($\to A_1 B_1$), we have – starting at $A_2 B_2$ – a proof for $k = \frac{2}{3}$ (analogously, for $k = \frac{m}{n}; m, n, \in \mathbb{N}$). This also applies for irrational values of k, since every real number can be approximated by rational numbers to arbitrary exactness. A homothety with scale factor $k < 0$ does the same as a homothety with scale factor $-k > 0$, just on the opposite side, and produces a half-turn around point Z, or equivalently, a reflection in point Z. Thus, we have established that the theorem holds for all real scale factors $k \neq 0$.

Actually, so far, we have only proved the following: The joining segment $A'B'$ of two image points A' and B' is parallel and $|k|$ times as long as the joining segment AB of the initial points A and B. We have not yet shown that the image of a line segment (or straight line) is again a line. Although this statement could be intuitively obvious, it still needs to be proved. When this is completed, the above theorem will be proved entirely. Fortunately, the proof is not complicated, as we merely have to show that the image point P' of any point P on AB lies on $A'B'$, that is, that $A'B'$ as a whole is the image of AB, being line segments or straight lines, as shown in Figure 11-8.

In Figure 11-8, we let P' be the image point of P. We have already shown that $A'B'$ is parallel to AB through A'. Analogously, $A'P'$ is parallel to AP (and also parallel to AB) through A'. Since there is only one parallel through point A' to AB and AP, respectively, point P' must lie

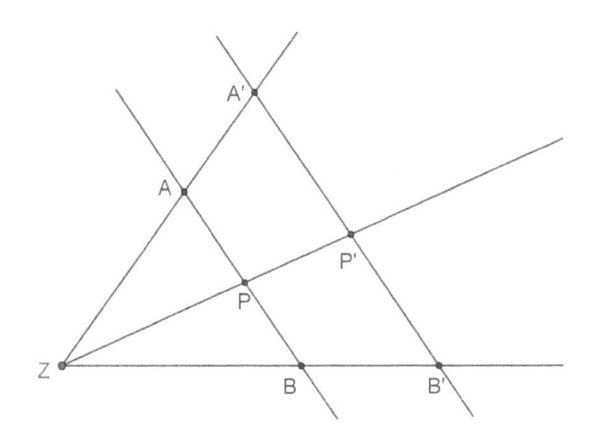

Figure 11-8 Point P' must lie on $A'B'$

on $A'B'$. This completes the proof that the image of a line segment (or straight line) is again a line segment (or straight line). We have finally completed the proof of this theorem, which states that: For *homotheties* with scale factor k, all angles are preserved, and all lengths are multiplied by $|k|$, which is what one would expect from enlargements/reductions. This property also means that homotheties preserve ratios in figures. That is, if a and b are two arbitrary lengths in the initial figure and a' and b' are the corresponding lengths in the image figure, then $\frac{a}{b} = \frac{a'}{b'}$ always holds, because $\frac{a'}{b'} = \frac{|k| \cdot a}{|k| \cdot b}$. This aspect finally leads to the concept of the slope m or, in other words, to the tangent function: If, and only if, two pairs of points have the same value of $\frac{\Delta y}{\Delta x} = \frac{y_1 - y_2}{x_1 - x_2}$ in the coordinate system, then the joining lines are parallel and have an equal elevation angle or slope.

In the practice of enlarging/reducing, only positive values $k > 0$ play a crucial role. Nevertheless, the geometrical map of *homothety* is defined more broadly, in that negative-scale factors $k < 0$ are also possible.

Homotheties Map Circles onto Circles and Map Tangents onto Tangents

Consider the following: Let M be the center of a circle with radius r and let X be an arbitrary point on this circle. A homothety with scale factor k yields $M'X' = |k| \cdot MX = |k| \cdot r$, that is, all images of circle points X have the constant distance $|k| \cdot r = r'$ from M'. Thus, these image points will create a circle with radius r' centered at M', as shown in Figure 11-9. Figure 11-9 also shows that the center Z of this homothety must lie on the line MM', which is a line of centers.

Any two circles are always *similar*. Furthermore, each of them can be brought to the size of the other by a homothety. In the plane, two circles are always *homothetic*. Referring to Figure 11-9, we consider the problem of how to find the center of a homothety when given two circles of different size. With concentric circles, the common center is the center of the homothety. With non-concentric circles, Z must lie on the straight line MM' containing the centers. One can use two

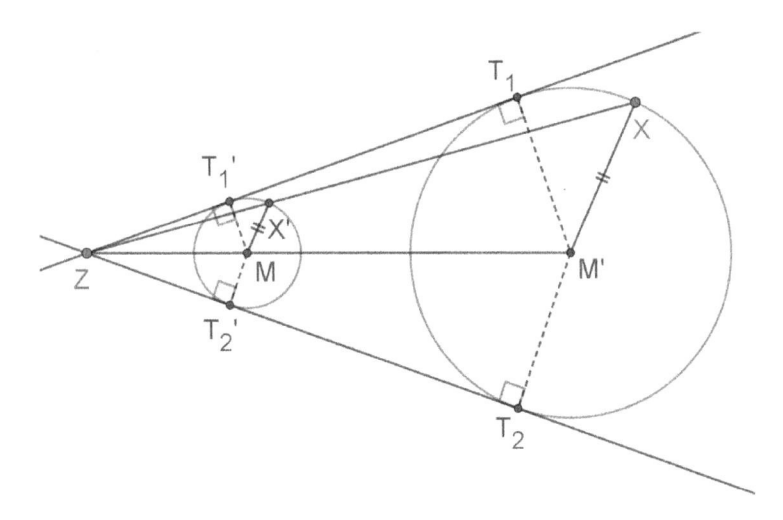

Figure 11-9 Homotheties map circles onto circles and tangents onto tangents

arbitrary parallel radii MX and $M'X'$ (but not on the straight line MM') to draw the straight line XX', on which the center Z must lie. The intersection of XX' with MM' must therefore be the sought-after center Z. Also, two of the common tangents of the circles must intersect each other at Z, as we can see in Figure 11-9. Notice that the two radii must be perpendicular to the tangents and thus parallel to each other. In general, tangents of circles map to tangents of circles under homotheties. Keep in mind that mapped segments and lines are parallel to the initial ones.

Two circles with equal radii are either identical, so that the center of their homothety can be chosen arbitrarily with scale factor $k = 1$, or not identical, so that we have $M_1 \neq M_2$ and the circle's homothety takes the midpoint of $M_1 M_2$ as its center and scale factor $k = -1$. This is equivalent to a reflection in this point, or a half-turn around this point.

General, Curved Figures with Homotheties

Since curves can be approximated "arbitrarily exact" by connected segments, and for segment lengths s we always have $s' = |k| \cdot s$, it is highly plausible for general (also curved) lengths ℓ that we have

$\ell' = |k| \cdot \ell$, especially for general perimeters p (also curved) $p' = |k| \cdot p$. Rather than provide a proper proof using limits, which is beyond the scope of this book, we will rely here on some plausible explanations.

Areas in Homothetic (Similar) Figures

Let us start with similar rectangles (a, b) and (a', b') with similarity factor k. Since $a' = ka$ and $b' = kb$, we have $a'b' = k^2 \cdot ab$. This means that the area of rectangle (a', b') is k^2 times the area of rectangle (a, b). The same is true for triangles with side a and altitude h: The area of triangle (a', h') is k^2 times the area of triangle (a, h). Since every polygon can be triangulated, for similar polygons P and P' with similarity factor k, we generally have area $P' = k^2 \cdot$ area P. This also holds for all similar *curved figures*, since curved lines can be approximated "arbitrarily exact" by connected segments. In case of circles or semicircles, this property can be seen via the corresponding area formula: For two circles with radii r and $r' = kr$, respectively, the area of the circle with radius $r' = kr$, which is $\pi \cdot (kr)^2$ or $k^2 \cdot \pi r^2$, is k^2 times the area of the circle with radius r.

Homothety is a special case of similarity. The concept of similarity is better known and wider spread in geometry textbooks than the concept of homothety, but homothety can be used to define similarity, which we will discuss in the next section. Moreover, the process of enlarging/reducing figures makes homothety a very useful and fundamental concept.

The concept of homothety can render geometric proofs shorter and more concise, as we will see in the following examples.

1. Consider the well-known theorem that the sides of the medial triangle $\triangle DEF$ (a triangle that is formed by joining the midpoints of the sides of a triangle) are parallel to the sides of the initial triangle $\triangle ABC$ and half as long. We will employ a dynamic method using homotheties, although there are other methods as well. In Figure 11-10, we have a homothety centered at C with scale factor $\frac{1}{2}$ that maps $A \mapsto E$ and $B \mapsto D$. Thus, ED is parallel to AB and half as long as AB. An analogous argument can be made for the other vertices as homothety centers.

Homothety, Similarity, and Applications **411**

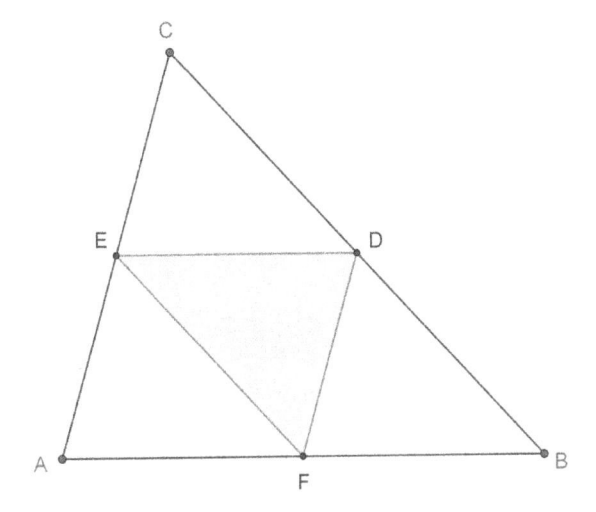

Figure 11-10 Medial triangle

2. Another example of the advantages of homotheties is presented in Figure 11-11, where we have $\triangle ABC$ and T is the point of tangency of the incircle to the side AB. We will consider T to be the "south pole" of the incircle, and denote its "north pole" as T', the point diametrically opposite T. We also have the points C, T', and T_1 collinear, where T_1 is the point of tangency of the excircle, which is a circle tangent to the three sides of the triangle externally. Our challenge here is to prove that this collinearity is true.

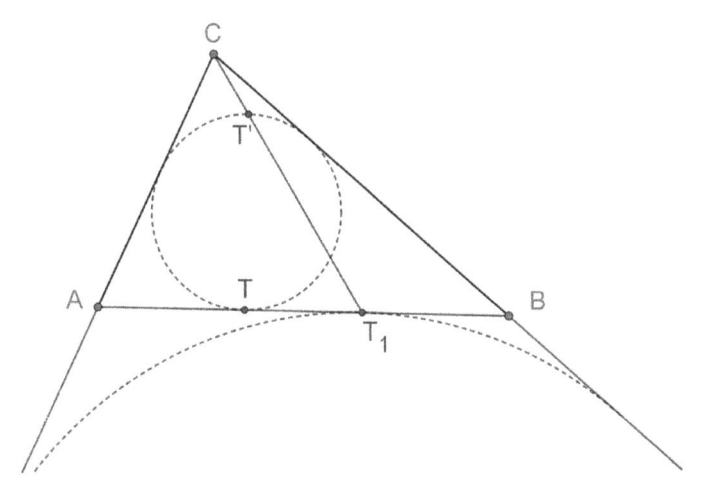

Figure 11-11 Points C, T', and T_1 are collinear

412 A Journey Through the Wonders of Plane Geometry

First, we will consider proof (**A**) using only similarity and then we will compare that to proof (**B**) using homothety.

A) We will use triangle similarity, and let T_1 be the intersection point of CT' with AB. We have to prove that T_1 is the point of tangency of AB with the excircle. From T_1 we draw a perpendicular to AB that intersects the angle bisector of angle C at M_1, as shown in Figure 11-12. We must show that M_1 is the center of the excircle. The triangles $\triangle CMT'$ and $\triangle CM_1T_1$ are similar, where M denotes the incenter of $\triangle ABC$. We define $k := \frac{CM_1}{CM}$. Then we have $T_1M_1 = k \cdot \underbrace{T'M}_{\rho}$ (where ρ represents the inradius), and the triangles CMS and CM_1S_1 are similar, as are triangles CMR and CM_1R_1. Hence, we have also $S_1M_1 = k \cdot \underbrace{SM}_{\rho}$, and $R_1M_1 = k \cdot \underbrace{RM}_{\rho}$. This means that $T_1M_1 = S_1M_1 = R_1M_1$ ($= k \cdot \rho$), and because we have right angles at R_1, T_1, S_1 we thus know that M_1 is the center of the excircle.

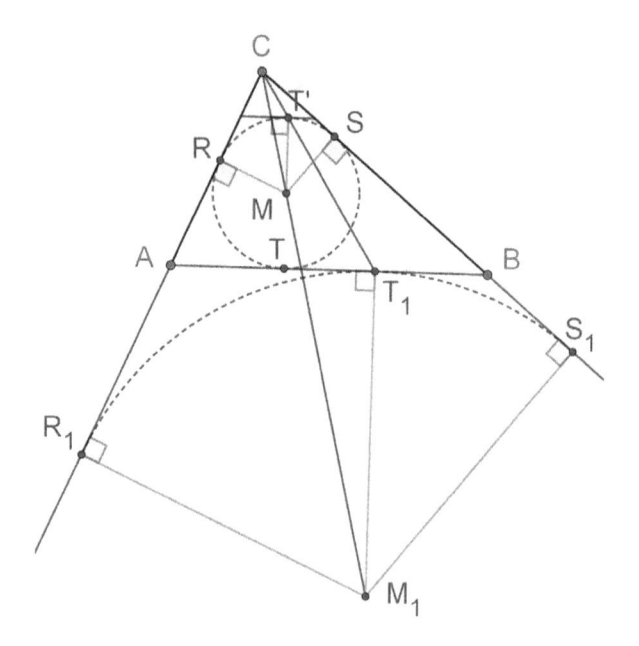

Figure 11-12

B) Here we present a much shorter proof using the concept of homothety. By extending *CA* and *CB* as shown in Figure 11-11, we see they are common tangents to the incircle and the excircle. Thus, there is a homothety centered at *C* that maps the incircle to the excircle. The scale factor would be the ratio of the two radii, but that is not relevant here. One could also consider expanding the incircle, emanating from the point *C*, until it becomes the excircle. During this process it gets increasingly larger and always stays tangent to *CA* and *CB*, or to their extensions. And since homotheties preserve angles, the north pole *T'* of the incircle maps to the north pole T_1 of the excircle (note: in both cases we have horizontal tangents, as shown in Figure 11-12). Hence, *C*, *T'*, and T_1 are collinear, which we sought to prove.

3. Consider, as next example, the famous Euler line of a triangle, which contains the circumcenter *O*, orthocenter *H*, and centroid *G*, where *G* is on the segment *HO* and $HG = 2GO$, as can be seen in Figure 11-13.

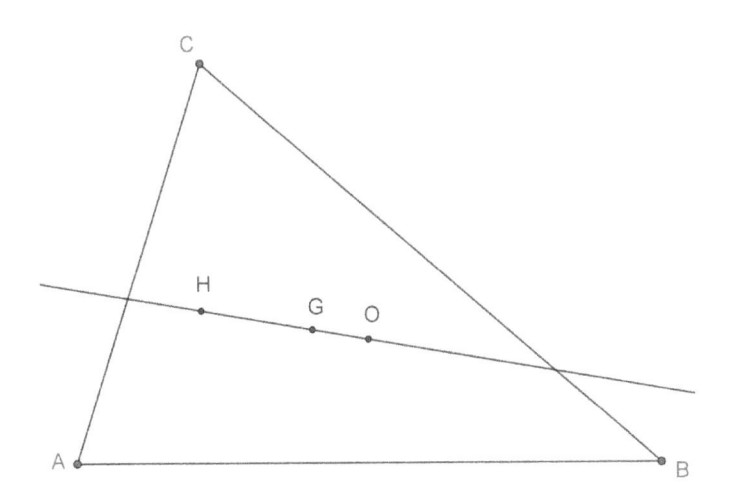

Figure 11-13 The Euler line

To facilitate the proof, we initially consider the medians, as their intersection point G, the centroid, divides them each in the ratio 2:1. The homothety with center G and scale factor $-\frac{1}{2}$ (note the negative scale factor) maps the vertices onto the midpoints of the opposite sides, as can be seen in Figure 11-14, where $A \mapsto D$, $B \mapsto E$, and $C \mapsto F$. The perpendicular from C to AB (altitude h_c) maps onto the perpendicular to AB at point F, which is the perpendicular bisector pb_c, because homotheties preserve angles. Analogously, with the other altitudes, we have $h_a \mapsto pb_a$ and $h_b \mapsto pb_b$. Therefore, this homothety maps the intersection point H of the three altitudes onto the intersection point O of the perpendicular bisectors of the three sides. Due to a crucial property of homotheties, initial point H, center G, and image point O lie on a common line, shown in Figure 11-13, where the center lies between initial point and image point because of the negative scale factor. Since the absolute value of the scale factor is $\frac{1}{2}$, the initial segment HG is twice the size of the image segment OG, which is what we sought to prove.

4. There is another famous line concerning triangles called the *Nagel line*, which is named for the German mathematician

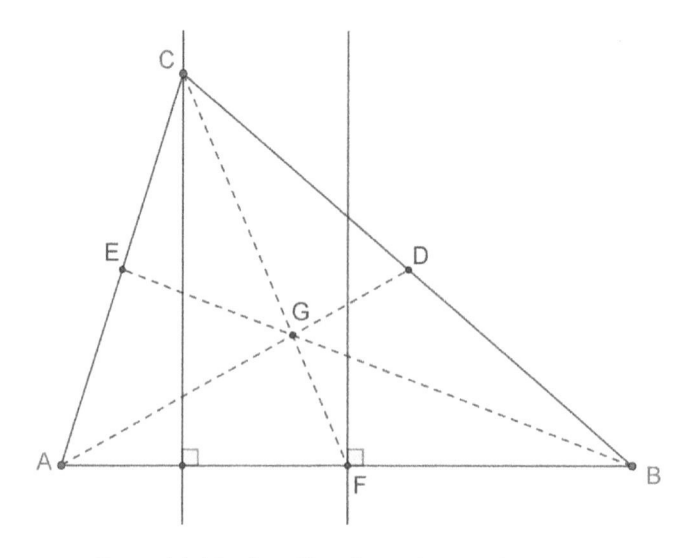

Figure 11-14   Proof by a homothety with center G

Christian Heinrich von Nagel (1803–1882). Although not as famous as the Euler line, it is important and has a similar property to the Euler line. The Nagel line contains: the Nagel point N, which is the intersection point of the cevians joining the vertices of a triangle to the tangent points of the opposite excircles; the centroid G, which is the intersection point of the medians of the triangle; and the incenter I, which is the point of intersection of the angle bisectors, as shown in Figure 11-14. Furthermore, point G divides NI in the ratio 2:1. This is somewhat analogous to the Euler line, and it has a similar proof. We employ the same homothety shown earlier in the case of the Euler line that is centered at G and has scale factor $-\frac{1}{2}$.

The plan for the proof is as follows: The above-mentioned homothety with center G and scale factor $-\frac{1}{2}$ maps the straight lines of the Nagel point cevians, such as CT_1, shown in Figure 11-15, onto lines through the incenter I. Thus, this homothety must map the Nagel point N onto the incenter I, which is our desired conclusion.

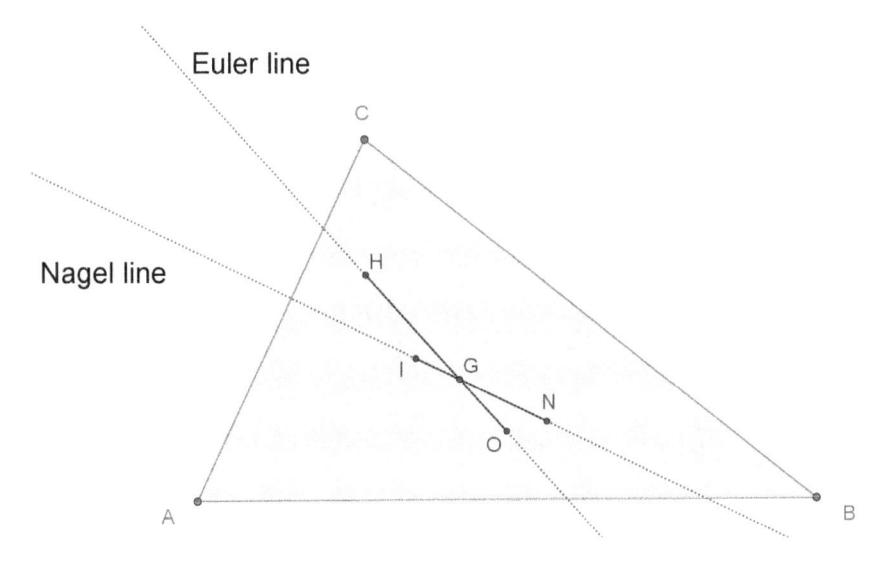

Figure 11-15

416 A Journey Through the Wonders of Plane Geometry

First, we will prove that the distances from vertices A and B of triangle ABC to the tangency points of the triangle's incircle and the excircle are equal, which we can see in Figure 11-16a, where $AT = BT_1$. An immediate consequence is that the midpoint M of AB is the midpoint of TT_1.

We have T and T_1 as the two tangency points on the side AB of the incircle and the excircle, respectively, as shown in Figure 11-16a. Since the tangent segments of a circle from an external point are equal, we have $CR = CS$ and $CP = CQ$, and furthermore by subtraction $PR = QS$.

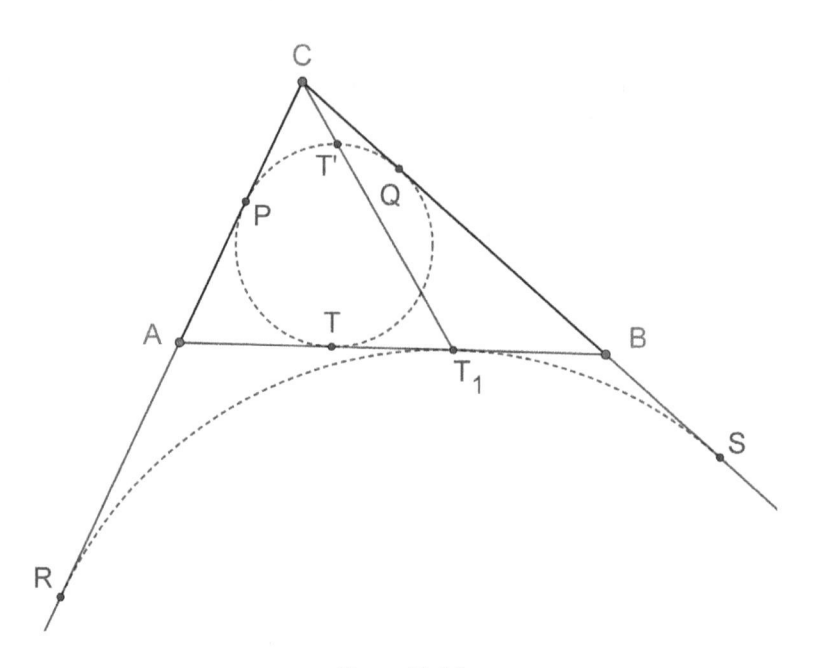

Figure 11-16a

We begin with $PA = AT$, and we also have $PR = PA + AR = PA + AT + TT_1 = 2AT + TT_1$. Similarly, we have $QS = QB + BS = TT_1 + T_1B + BS = 2BT_1 + TT_1$, and together with $PR = QS$ we can conclude that $AT = BT_1$.

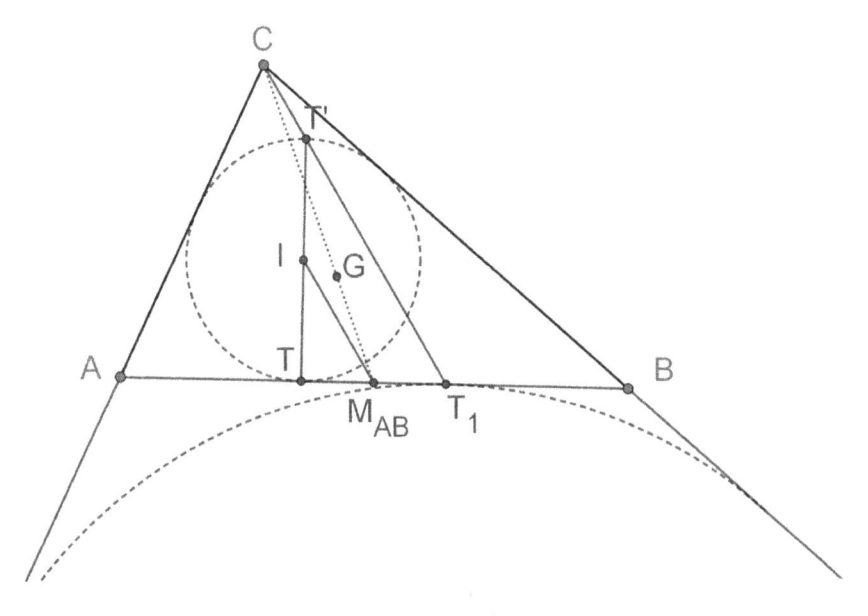

Figure 11-16b

Next, from example 2, above, we know that C and the two "north poles" T', T_1 are collinear (see Figure 11-11). Furthermore, since the incenter I is the midpoint of the diameter TT' and M_{AB} is also the midpoint of TT_1 (see Figure 11-16b), we know that IM_{AB} is parallel to $T'T_1$. The homothety centered at G with scale factor $-\frac{1}{2}$ maps C to M_{AB}. Since lines are mapped to *parallel* lines by homotheties, the image of the Nagel cevian CT_1 by the homothety must be the straight line IM_{AB}, which completes the proof.

5. Another problem that can benefit from the concept of homothety and also involves the Euler line is illustrated in Figure 11-17, where we have two distinct points $A \neq B$ and a straight-line g. We construct a point C such that g is the Euler line of $\triangle ABC$. The challenge is how to construct such a point and then to determine how many solutions there are.

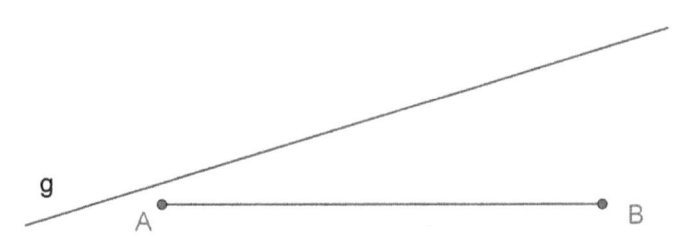

Figure 11-17 Line g is the Euler line of $\triangle ABC$

Solution: In Figure 11-18, we see the circumcenter O and the circumcircle c of $\triangle ABC$, where the point O is the intersection point of g and the perpendicular bisector of AB. We now admire a homothety playing a crucial role. The centroid G must lie on g, and G divides the line segment CM (where M denotes the midpoint of AB) in the ratio 2:1. Thus, we simply apply a homothety to g

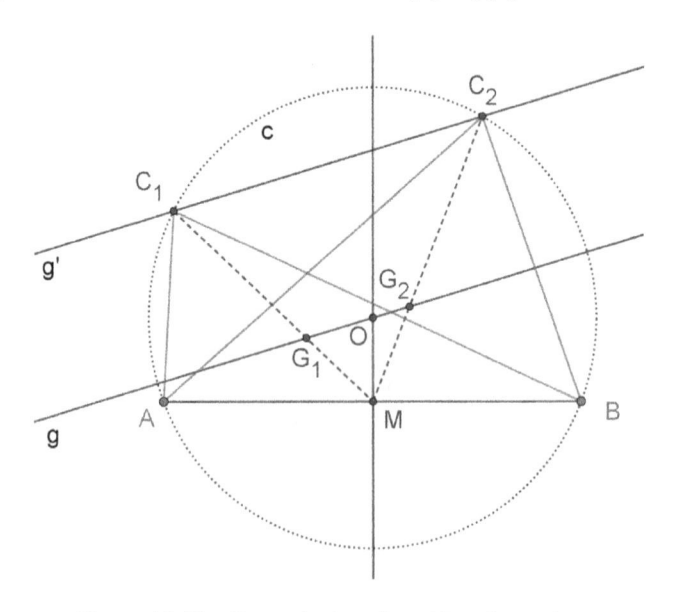

Figure 11-18 Two solutions found by a homothety

centered at M with scale factor 3, and the point C must lie on the image g'. Two possible positions of C are the intersection points of g' with the circumcircle, as can be seen in Figure 11-18. There may be only one or no solution, depending on how many intersection points g' and c have.

6. Vecten's theorem and the Vecten point,[1] named for the French mathematician who published this theorem in 1817, provide a further example of how a special homothety may shorten a proof. Although there are other proofs available, we present the following one because it cleverly uses homothety, which makes the proof vivid and clear. But first, let us establish Vecten's theorem, which is described in the footnote. On the sides of a triangle ABC, shown in Figure 11-19, squares are erected outwardly with centers J, K, and L. KJ and CL are equal and perpendicular. Analogous relationships can be shown at the other vertices A and B instead of C (see Chapter 4).

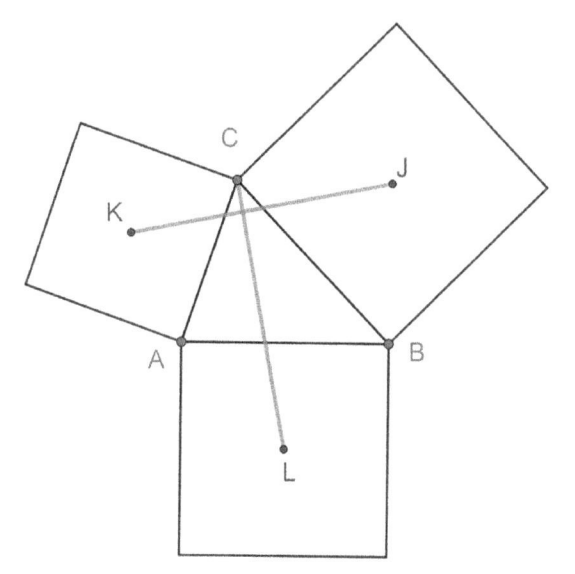

Figure 11-19   Vecten's theorem

---

[1] Vecten points are two triangle centers associated with any triangle. They are constructed by drawing three squares on the sides of the triangle, connecting each square's center by a line to the opposite triangle vertex, and finding the point where these three lines intersect. Drawing the squares outwardly produces the outer Vecten point, and drawing the squares inwardly produces the inner Vecten point.

We will present three more configurations in Figures 11-20, 11-21, and 11-22. In Figure 11-20, we see two shaded triangles, $\triangle ABM$ and $\triangle CBM$. In Figure 11-21, these triangles are reduced by homotheties, centered at A and C, respectively, with scale factor $\frac{1}{\sqrt{2}}$. Since before the homothety the triangles shared the side BM, both corresponding image sides are parallel and equal after the homotheties. Due to the Pythagorean theorem, we have the well-known relation $d = a \cdot \sqrt{2}$ between the sides a and diagonals d of a square. Thus, the scale factor $\frac{1}{\sqrt{2}}$ maps a diagonal of a square to its side, where in Figure 11-21 we have $AR = AC$. Furthermore, a side of a square is Mapped too one-half its diagonal, which in Figure 11-21 produces $CP = CK$, $CQ = CJ$, $AS = AL$. Therefore, we can rotate both triangles about ±45° around A and C, respectively, to get the end positions shown in Figure 11-22, which completes the proof, since the difference of +45° and −45° makes 90°, a right angle.

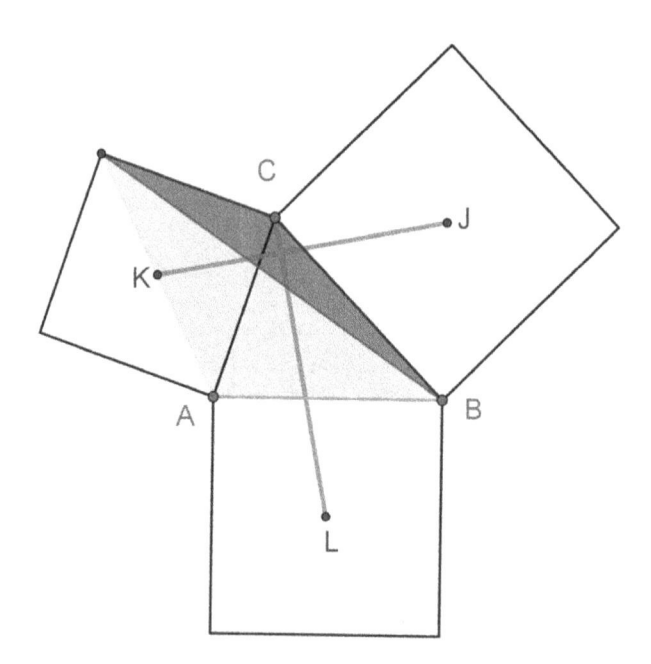

Figure 11-20

Homothety, Similarity, and Applications   421

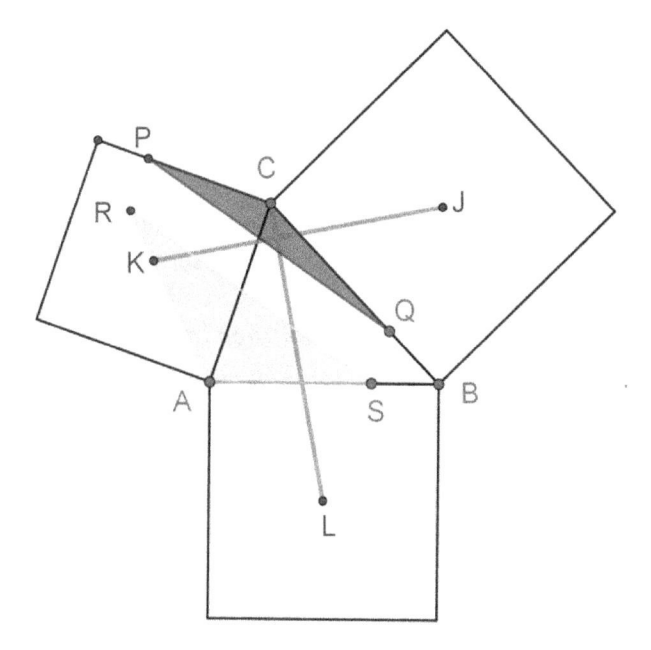

Figure 11-21

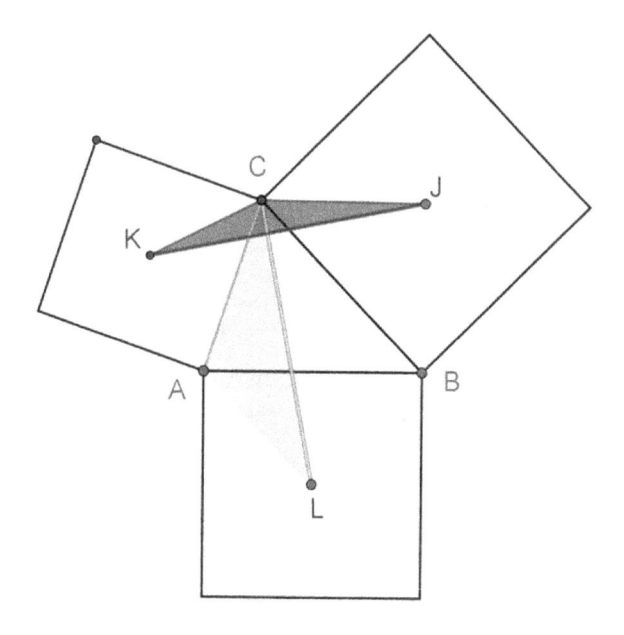

Figure 11-22

422 A Journey Through the Wonders of Plane Geometry

We now have a small step to find the outer Vecten point. If in the configuration of Figure 11-19 each vertex of the triangle is joined with the center of the opposite square, then these joining lines are concurrent at the outer Vecten point V, as shown in Figure 11-23.

This is immediately clear, since the mentioned lines are also altitudes in the triangle $\triangle JKL$, which is shown in Figure 11-23.

7. See the description of the *pantograph* on pp. 434 in this chapter.

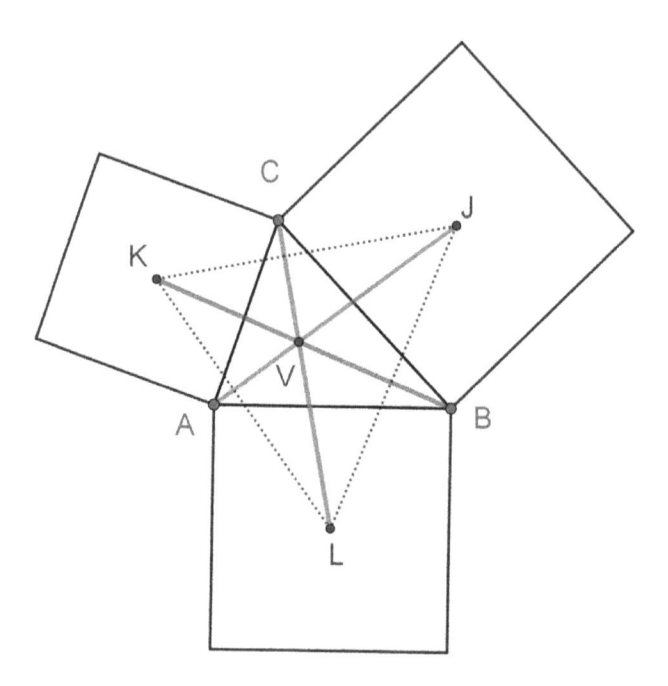

Figure 11-23   Vecten point

Similarity

As we know, figures that have the same shape and a (possibly) different size are called similar. There are two equivalent definitions for similarity of polygons.

Definition 1: Two polygons F and F', such as those shown in Figure 11-24, are called *similar* if

1. all corresponding angles are equal: $\alpha = \alpha'$, $\beta = \beta'$, ...
2. all ratios of lengths of corresponding sides are equal: $\frac{a'}{a} = \frac{b'}{b} = \frac{c'}{c} = \ldots = k$

Such **similarity factors** are always positive since they are ratios of lengths. Since homotheties have exactly these properties, where all lengths are multiplied by $|k|$ and all angles remain unchanged, they can be used to define similarity in an equivalent way.

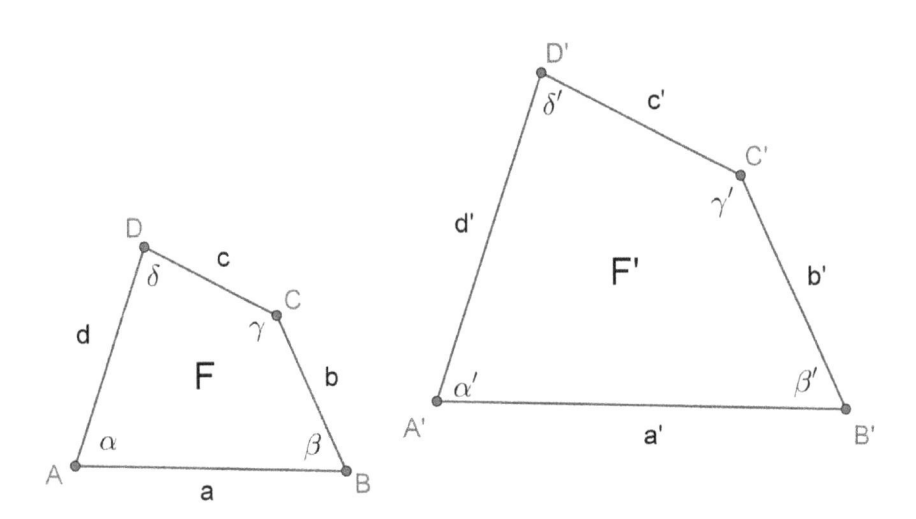

Figure 11-24   Similar polygons

Definition 2: Two arbitrarily placed polygons F and F' are called *similar* if there exists a homothety $S_{Z,k}$ that enlarges/reduces F in such a way that the image $S_{Z,k}(F)$ is congruent to F'. In other words, one figure is just an enlarged/reduced version of the other.

Some Remarks

- Definition 2 has the advantage of also being applicable to curved figures, and not only to polygons.
- Not all similar figures are homothetic. Additionally, they can be shifted, rotated, or reflected, as shown in Figure 11-25. Nevertheless, homothety is the prototype of all possible *similarities* (maps that produce similar figures).

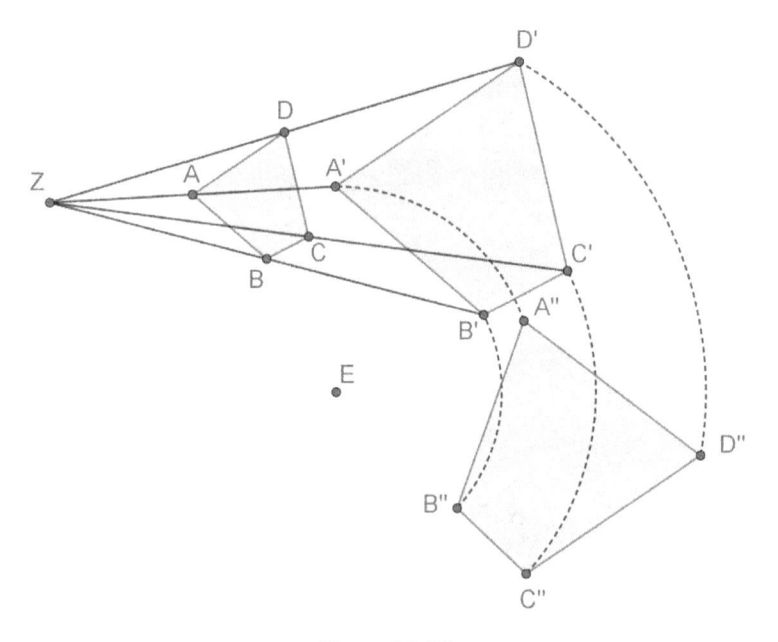

Figure 11-25

In Figure 11-25, we have $ABCD$ and $A'B'C'D'$, which are homothetic with center Z. $ABCD$ and $A''B''C''D''$ are not homothetic, just similar, as $A''B''C''D''$ is $A'B'C'D'$ rotated around point E.

Triangles have an important peculiarity regarding similarity. In the case of triangles, if one of the two conditions mentioned in definition 1 suffices for similarity (equal angles or equal ratios), the other condition is fulfilled automatically. In case of quadrilaterals, one condition is insufficient for similarity, as we can see in Figures 11-26 and 11-27. This is also true for other polygons (n-gons, where $n > 4$), where one condition is insufficient to prove similarity.

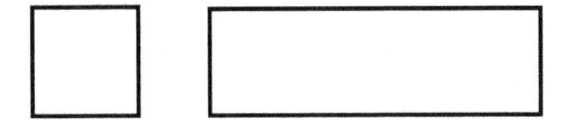

Figure 11-26 Both figures have four right angles but are not similar

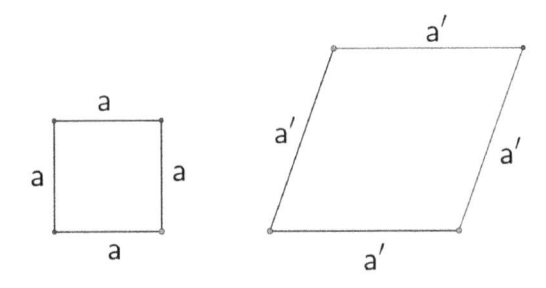

Figure 11-27 All four ratios are equal, $\frac{a'}{a} = \frac{a'}{a} = \frac{a'}{a} = \frac{a'}{a}$, but figures are not similar

In Figure 11-26, although the square and the rectangle have four right angles, they are clearly not similar in shape. Furthermore, a square is, in general, not similar to a rhombus, as we can see in Figure 11-27, where the square and the rhombus are clearly not the same shape despite having equal ratios.

One can argue that there cannot be an "AAAA" similarity theorem. In general, for quadrilaterals, one can take an arbitrary quadrilateral, where the sides are straight lines that can be translated in a parallel manner, and produce quadrilaterals with equal angles but very different shapes that are clearly not similar, as we can see in Figure 11-28.

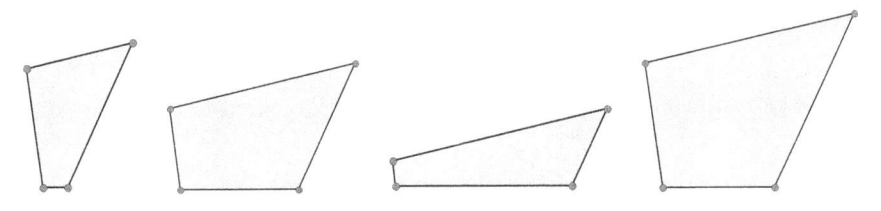

Figure 11-28 Not similar quadrilaterals with equal angles

It is also easy to see that there cannot be a "SSSS" (side-side-side-side) similarity theorem for quadrilaterals. For that purpose, imagine a quadrilateral with fixed bars as sides linked together at the vertices (a four-bar linkage). Without changing the lengths of the bars, one can vary the shape of the quadrilateral considerably.

Similarity of Rectangles

We now consider rectangles that are similar. All angles of rectangles are right angles, which indicates that the condition for similarity of rectangles considers only the side lengths. Two rectangles with side lengths a, b and a', b' are similar if and only if $\frac{a'}{a} = k = \frac{b'}{b}$, which is equivalent to $\frac{a}{b} = \frac{a'}{b'}$.

If the rectangles are placed with a common vertex A, long side upon long side, and short side upon short side, as shown in Figure 11-29, then in this position the diagonals of similar rectangles are on the same straight line. Without similarity, this would not be the case. Considering the situation of a homothety centered at A, it is clear that C' must lie on AC.

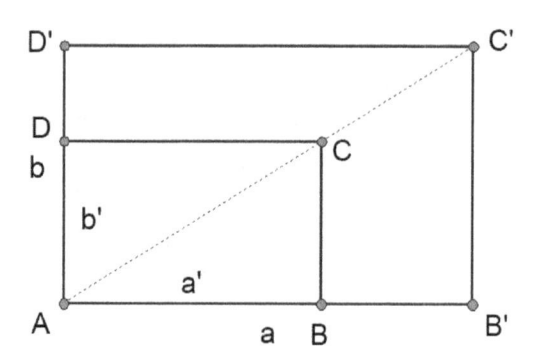

Figure 11-29 Similar rectangles

Similarity Theorems for Triangles

Similarity theorems are important for recognizing and proving similarities. One should not *define* similarity of triangles as "three (or two) equal angles" because this definition could erroneously be applied to polygons, where this does not hold. As mentioned above, it is a peculiarity of triangles that equal angles suffice for similarity.

Homothety and the aspect of similar triangles – including enlargements/reductions – focus more on dynamic situations. And with the help of dynamic geometry software such aspects become increasingly important.

Side-Angle-Side (SAS) Similarity Theorem

If two triangles ABC and $A'B'C'$ have one equal angle and equal ratios of the adjacent sides, that is, $\alpha' = \alpha$ and $\frac{c'}{c} = \frac{b'}{b} =: k$, then they are similar, as can be seen in Figure 11-30.

Our objective here is to show $\beta = \beta'$, $\gamma = \gamma'$, and $a' = k \cdot a$ so that all requirements of similarity are fulfilled.

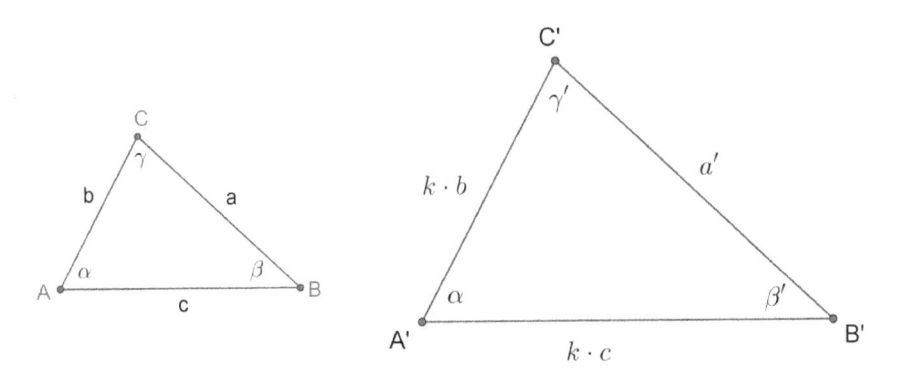

Figure 11-30 SAS similarity theorem

Proof

Consider Figure 11-31, where we apply a proper homothety with scale factor k to $\triangle ABC$. This yields the triangle $A_1B_1C_1$ with unchanged angles α, β, γ and with the sides $k \cdot a$, $k \cdot b$, and $k \cdot c$. Due to the SAS congruence theorem, we have $\triangle A'B'C' \cong \triangle A_1B_1C_1$, and this congruence certainly establishes similarity.

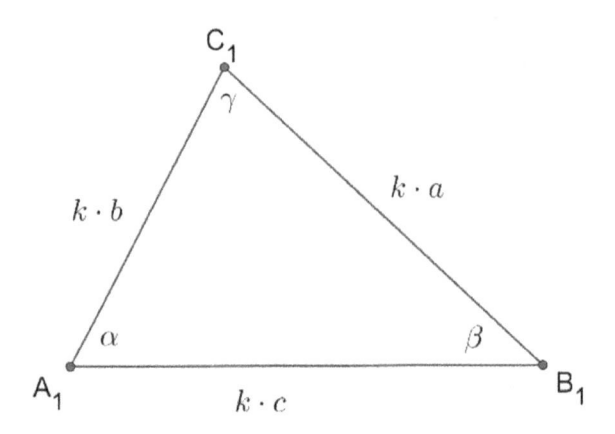

Figure 11-31 SAS similarity theorem, proof

Angle-Angle-Angle (AAA) Similarity Theorem

If all three corresponding angles of two triangles ABC and $A'B'C'$ are equal, that means $\alpha = \alpha'$, $\beta = \beta'$, and $\gamma = \gamma'$, then the triangles are similar, as can be seen in Figure 11-32. It should also be noted that since the angle sum of triangles is 180°, for similarity it suffices to show that two angles of one triangle are equal to two angles of the second triangle.

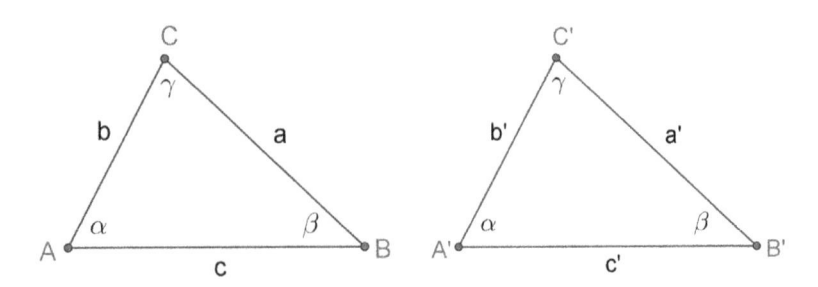

Figure 11-32 AAA similarity theorem

Homothety, Similarity, and Applications

To establish similarity through the AAA theorem, we have to show $\frac{a'}{a} = \frac{b'}{b} = \frac{c'}{c} = k$.

Proof

We determine the scale factor through $k := \frac{c'}{c} > 0$ and apply a proper homothety to $\triangle ABC$, so that the image triangle has the desired side length c'. We get the image triangle $A_1B_1C_1$ with $c_1 = k \cdot c = c'$ and unchanged angles α, β, γ. From the ASA congruence theorem, we have $\triangle A'B'C' \cong \triangle A_1B_1C_1$, as shown in Figure 11-33. This congruence, and $a_1 = k \cdot a$, and $b_1 = k \cdot b$, respectively, implies $a' = k \cdot a$ and $b' = k \cdot b$, which establishes our claim and, finally, the asserted similarity of the triangles ABC and $A'B'C'$.

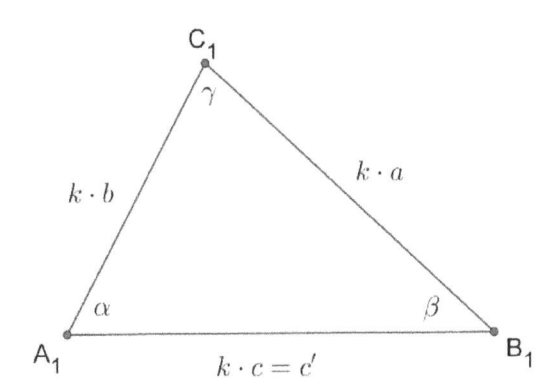

Figure 11-33 AAA similarity theorem, proof

Some Remarks

- The AAA similarity theorem is an essential aspect of geometry and is often realized through two equal corresponding angles. The SAS similarity theorem is important for further applications, in particular, to the concept of homothety.
- Of the four congruence theorems (SAS, ASA, SSS, SsA[2]) three have their immediate analog as a similarity theorem. The ASA

---

[2] The lowercase second letter s should indicate that the angle A is opposite of the uppercase first letter S, the longer side.

congruence theorem corresponds – see above – to the AAA similarity theorem.
- In each case, one can apply a proper homothety and use a corresponding congruence theorem for triangles. This is important to emphasize, and we provide explicit proofs for the three most important cases.

Side-Side-Side (SSS) Similarity Theorem

If the three ratios of corresponding sides of two triangles ABC and $A'B'C'$ are equal, that is, $\frac{a'}{a} = \frac{b'}{b} = \frac{c'}{c} (=:k)$, then they are similar. This is shown in Figure 11-34.

To establish the similarity between triangles ABC and $A'B'C'$, we need to show $\alpha = \alpha'$, $\beta = \beta'$, $\gamma = \gamma'$.

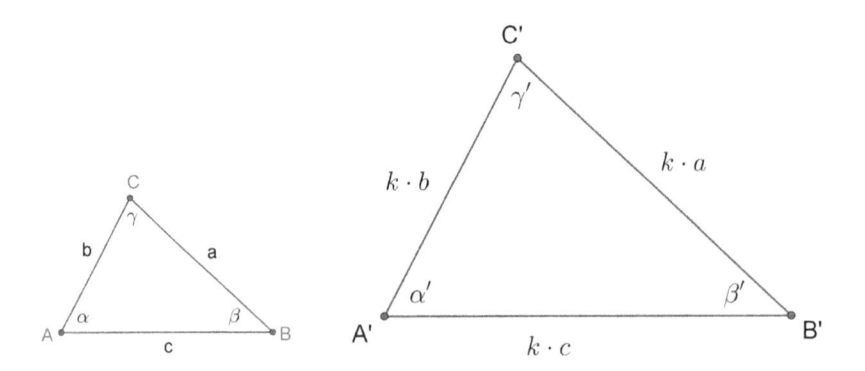

Figure 11-34   SSS similarity theorem

Proof

We can apply a proper homothety to $\triangle ABC$ with scale factor k so that the image triangle $A_1 B_1 C_1$ has the desired side length $c_1 = k \cdot c = c'$. We get $\triangle A_1 B_1 C_1$ with unchanged angles α, β, γ and with the same sides as $\triangle A'B'C'$. Because $(k \cdot a, k \cdot b, k \cdot c) \underset{\text{congruence}}{\overset{\text{SSS}}{\Rightarrow}} \triangle A'B'C' \cong \triangle A_1 B_1 C_1$, shown in Figure 11-35, $\triangle ABC$ and $\triangle A'B'C'$ have equal corresponding angles and are thus similar, which completes the proof.

Homothety, Similarity, and Applications **431**

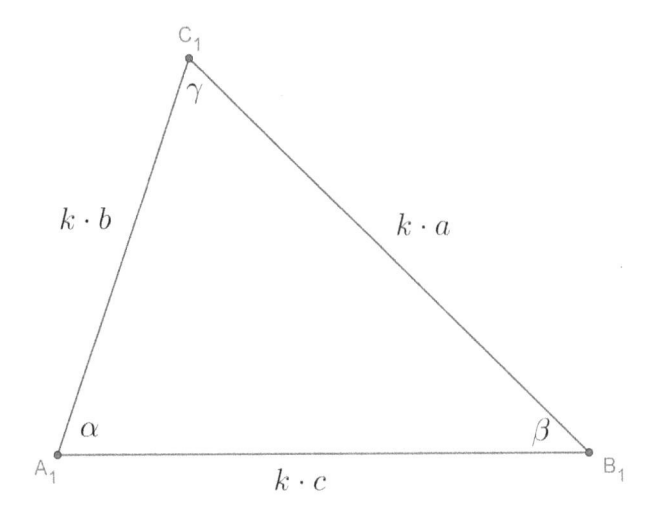

Figure 11-35 SSS similarity theorem, proof

Side-side-Angle (SsA) Similarity Theorem[3]

If two triangles $\triangle ABC$ and $\triangle A'B'C'$ have the same ratio of two sides and equal angles opposite of the longer side, so that $c > b$; $c' = k \cdot c$, $b' = k \cdot b$, $\gamma' = \gamma$, then they are similar, as can be seen in Figure 11-36. The reader is invited to find a proof with a proper homothety.

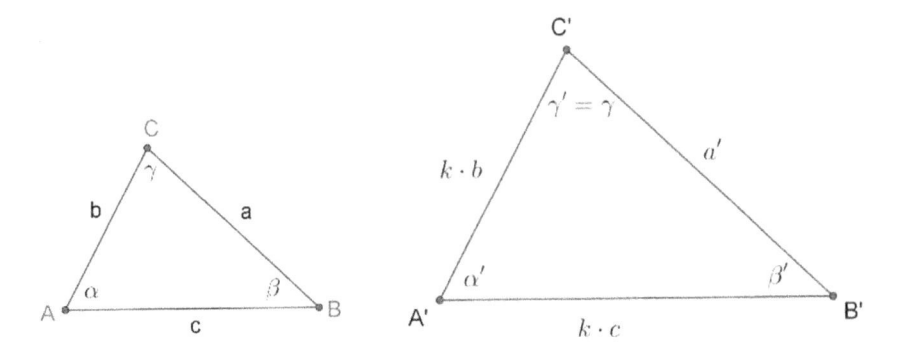

Figure 11-36 SsA similarity theorem

---

[3] The lowercase second letter s should indicate that the angle A is opposite of S, the longer side.

432 *A Journey Through the Wonders of Plane Geometry*

Closely related to homothety and similarity is the well-known *intercept theorem*, which states that if two non-parallel straight lines meeting at point S are intersected by two parallel straight lines not passing through S, then one of the configurations shown in Figure 11-37 can appear. One is an X-shaped figure, and the other is a V-shaped figure.

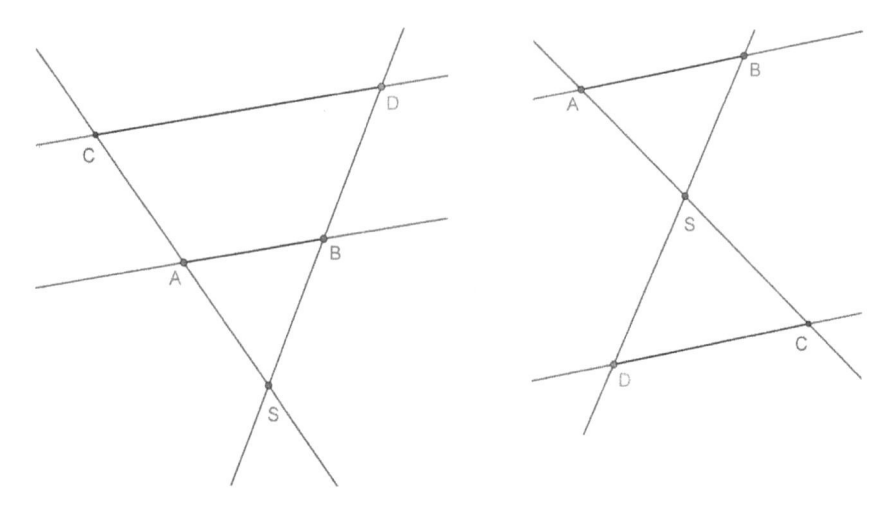

Figure 11-37   Intercept theorem (right: X-shaped figure, left: V-shaped figure)

In both cases, homothetic triangles SAB and SCD appear. As there are three equal angles, since the parallels make equal angles with a given straight line, the AAA similarity theorem can therefore be applied, which can be seen in Figure 11-37. This implies that the ratios of corresponding sides are equal:

1. $\dfrac{SA}{SC} = \dfrac{SB}{SD}$

2. a) $\dfrac{SA}{SC} = \dfrac{AB}{CD}$

   b) $\dfrac{SB}{SD} = \dfrac{AB}{CD}$

Equation 1 is often called the "first intercept theorem" (all the ratios are taken on the straight lines intersecting at S), and the equations of 2 are often referred to as the "second intercept theorem" (also involving the ratios on the parallel lines). As we see below, the converse of the first intercept theorem holds true, which is not the case for the second intercept theorem.

Converse of the First Intercept Theorem

If $\frac{SA}{SC} = \frac{SB}{SD}$ holds true, then AB must be parallel to CD, as can be seen in Figure 11-37.

Proof

If on the straight lines meeting at point S there are equal ratios, then the two triangles $\triangle SAB$ and $\triangle SCD$ must be similar because of the equal angles at point S (SAS similarity theorem). Thus, all the angles of the triangles are equal, and from that we can conclude $AB \| CD$.

A converse of the second intercept theorem would be: If $\frac{SB}{SD} = \frac{AB}{CD}$ holds true, then AB must be parallel to CD, as can be seen in Figure 11-37.

Figure 11-38 shows that the converse of the second intercept theorem does not hold. In spite of the proportion $\frac{SB}{SD} = \frac{AB}{CD}$, the parallelism $AB \| CD$ does not necessarily hold true, as one can see by starting

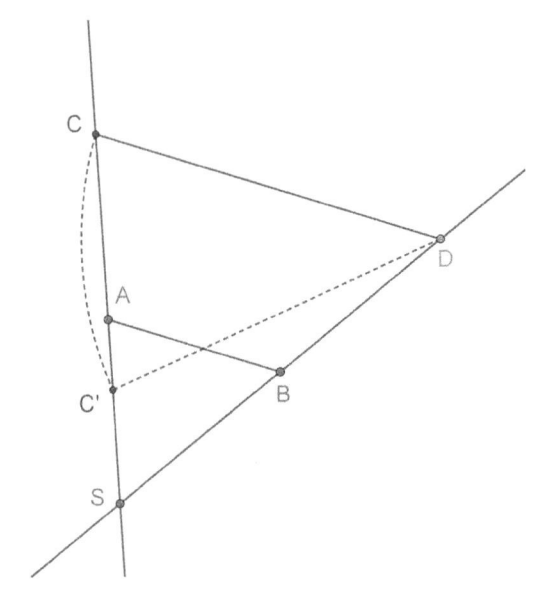

Figure 11-38 The converse of the second intercept theorem does not hold

with $AB \parallel CD$, where the circle with center at D and radius DC, intersecting with SC, yields point C'. This point C' satisfies $\frac{SB}{SD} = \frac{AB}{C'D}$ (note that the lengths did not change), but that obviously does not result in $AB \parallel C'D$.

All these contexts and relationships do not facilitate applying the intercept theorems, which focus primarily on static aspects; instead, we prefer the perception of similarity and the dynamic aspects of enlarging/reducing figures. In other words, we prefer the underlying *homothety*.

The Pantograph

Before computers became ubiquitous, drawings were scaled using a pantograph, which is a mechanical tool for enlarging/reducing the size of plane figures. At the website https://en.wikipedia.org/wiki/Homothety one can see an animation of scaling a heart figure. To make a pantograph, one links together two pairs of flat rods of equal length, such as AP and $A'Q$, and PQ and AA' (shown in Figure 11-39), to form a parallelogram with vertices A, P, Q, A'. The two rods meeting at A' are extended beyond A and Q by an arbitrary factor $k > 1$, yielding the points Z and P' at the other end, as shown in Figure 11-40. This determines $ZA' = k \cdot ZA$ and $P'A' = k \cdot \underbrace{QA'}_{=PA}$.

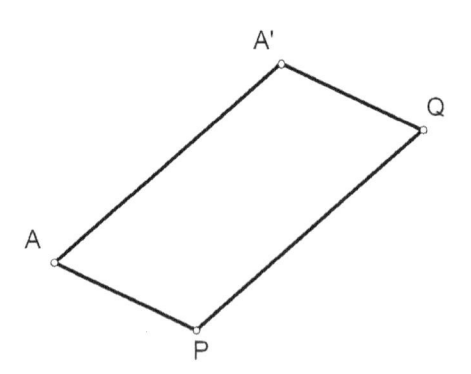

Figure 11-39 Parallelogram with two pairs of equal flat rods

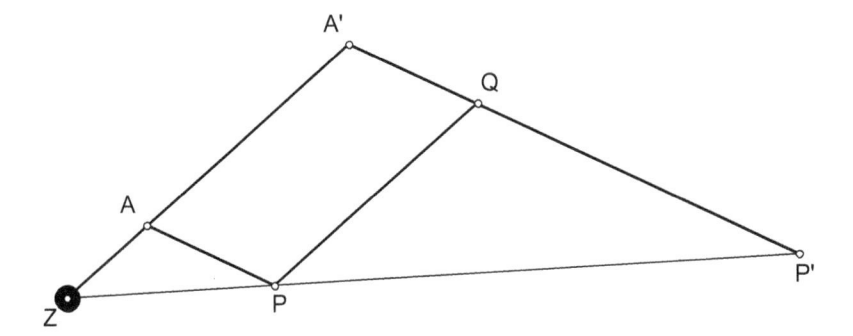

Figure 11-40 The whole pantograph

Since in a parallelogram opposite sides are parallel, we have $\angle ZAP = \angle ZA'P'$, and due to the ratios $\frac{ZA'}{ZA} = k = \frac{P'A'}{PA}$ we can use the SAS similarity theorem for the triangles and conclude the similarity $\triangle ZPA \sim \triangle ZP'A'$. Thus, we have $\angle PZA = \angle P'ZA'$, furthermore, Z, P, P' are collinear, and $\frac{ZP'}{ZP} = k$.

That means the map $P \mapsto P'$ is simply a homothety centered at point Z with scale factor k. One can use this in the following way:

1) Attach the mobile rods fixed and rotatable at point Z.

2) Place a pencil at point P'.

3) If, with point P, a figure (for example, a heart-shaped figure) is traced, then the pencil at P' draws this figure enlarged with scale factor k.

4) If you change the roles of P and P', then one can reduce the size of a figure with scale factor k.

In today's world, resizing geometric shapes is easily accomplished by using dynamic geometry. There, we choose a center of a homothety, an object (such as a triangle, quadrilateral, circle, ellipse, a 3-dimensional object, etc.), and a scale factor; the computer (the dynamic geometry program) does the corresponding work and immediately produces the result.

Homothety is not an advanced geometric concept, but a rather basic and fundamental one that is typically not part of the secondary school geometry curriculum. It can be used in many situations, for instance enabling proofs to be shorter and more concise. Furthermore, and even more importantly, homotheties can be used for describing the process of decreasing and increasing geometric objects dynamically. Dynamic aspects are increasingly more important for teaching and learning geometry since they make geometry vivid and promote the corresponding learning processes. Fortunately, there are powerful tools, such as dynamic geometry programs like Geometer's Sketchpad and GeoGebra that show and use these aspects.

Conclusion

Now we have completed our journey through the many aspects of plane geometry, focusing on the peculiarities that are often neglected or omitted during high school and at the university level. Curiosities and peculiarities involving concurrencies and collinearities often present aspects of geometry that leave the observer appreciating the wonders of plane geometry. We have also exposed quite a few geometric surprises, further enhancing the notion of the wonders of geometry. In order to fully understand what is appropriate and what may be inappropriate, we have guided the reader on a journey through fallacies that are often overlooked and sometimes ignored. We concluded our study of geometry by exposing the reader to an aspect of geometric analysis known as *homothety* or *central dilation*, which is not part of the American school curriculum and yet enhances geometric study. We hope that this geometric journey will be the beginning of further investigations by the motivated reader.

Index

A
affine-regular n-gon, 183
affine transformation, 183
altitudes, 214–215
    concurrency of, 58
    triangle, concurrency of, 66–67
angle-angle-angle (AAA) similarity theorem, 428–429
angle bisectors, 103–105
    concurrency of, 56
    triangle, concurrency of, 67–68
Anne's theorem, 121
arbelos, 286–289
    area of, 283
    rectangle in, 284
    semicircles forming, 282
architecture, Golden Ratio in, 24–37
art, Golden Ratio in, 46–51

B
betweenness, 365
Brahmagupta's theorem, 191–194
Brianchon's theorem, 333–337

C
central dilation, 436
centroid, 79–80
Ceva, Giovanni, 91
Ceva's theorem, 53, 60–63
    altitudes of triangle, concurrency of, 66–67
    angle bisectors of triangle, concurrency of, 67–68
    converse of, 64–65
    medians of triangle, concurrency of, 66
Cevian(s), 61–66, 70–71
    concurrent, aspects of, 71–74
    hidden length of, 70–71
circles
    centers, 224–225
    chords of, 310–311, 378–379
    generating equal lines, 257–258
    generating three equal lines with triangle, 253–254
    interior of, 382–383
    intersecting equal circles, 259–260

circumcenter, 79–80
circumscribed circles, 77–78, 95–96, 222–223, 276–277
collinear points, 93–95, 99–100, 103–105
collinearity, 91–125, 241–242
compound polygon, 297
concave n-gon, 294
concurrencies, 53–89
concurrency, by squares on sides of triangle, 149–152, 155–157, 160–161
concyclic points, 189–244, 216–221
congruence transformations, 183
congruent quadrilaterals, 261–263
congruent squares, octagons inscribed in, 307–309
Cosine law, 39, 331
cubit, 25
curved figures, 410
cyclic quadrilateral, 189–191

D
da Vinci, Leonardo, 49–50
Desargues, Gérard, 321
Desargues' theorem, 97–99, 321–323
directed angles, 294
Divina Proportione (Pacioli), 2, 41, 46
divine proportion, 2
dodecahedron, 40–41
Dodgson, Lutwidge, 379
Dürer, Albrecht, 50

E
Elements (Euclid), 28–29, 41, 111
equal circumferences, 390–392
equality, produced by circle and square, 248–249

equator, 385–390
equiangular point, 168
Euclid, 6, 28, 41, 111
Euler, Leonhard, 41
Euler six-point circle, 232
excenters, 57
excircles, 57
exterior angles, 294–296
eyeball theorem, 247–248

F
Fechner, Gustav, 17–19
Fermat–Torricelli point, 165–168
Feuerbach, Karl Wilhelm, 233
Fibonacci, 32
Fibonacci numbers, 19, 32
Finsler–Hadwiger theorem, 157–159
five concyclic points, 230–231
five-point circle, 228–229

G
Gergonne point, 74
Gergonne triangle, 74–75
  concurrency for, 75–76
Giza Pyramid, 27
Golden/Fibonacci Spirals, 21–24
Golden Parallelograms, 38–40
Golden Ratio, 1–51
Golden Rectangle, 1, 17–21
Golden Section, 2–17
Golden Triangles, 38
Grebe point, 151

H
Hemiunu, 24
Herodotus, 27
Heron of Alexandria, 5, 339
Heron's formula, 339–341

hexagon, 250-251
    diagonals of, 392-393
hexahedron, 40
hidden angle, 348-350
homotheties, 184
    areas in, 410-422
    general, curved figures with, 409-410
    increasing and decreasing size of geometric figures by, 401-408
homotheties map circles, 408-409
Honsberger, Ross, 85

I
icosahedron, 40-45
incenter, 56
incircle, 56
inscribed circle, 77-78, 95-96 251-253
    concurrency with, 81
    of triangle, 74-75
inscribed-triangle sides form, 102-103
intercept theorem, 432-434
intersecting circles, creating five concyclic points, 230-231
isogonal conjugation, 83-86
isosceles, 365-368
isosceles triangle, 348-350
    in regular octagon, 302-303
isotomic conjugation, 88-89

J
Japanese geometry, 351-359

K
Kagen, Fujita, 352
Kepler, Johannes, 1-2

L
lattice polygons, 316-320
Le Corbusier, 35, 37
Leonardo of Pisa, 32
Leybourn, Thomas, 111
Liber abaci (Fibonacci), 32
Lycaeus, Proclus, 374

M
medians, 100
    concurrency of, 59-60
    extensions of, 101
    triangle, concurrency of, 66
Menelaus' theorem, 91-92, 112-113, 322
    converse of, 93
    proof of, 92-93
minimum-distance point, 174, 178
Miquel point, 264
Miquel's theorem, 263, 266-272
Morley, Frank, 342
Morley's theorem, 342-348
mysterium hexagrammicum, 323

N
Nagel line, 110, 414-415
Nagel point of a triangle, 68-70
Napoleon's theorem, 168-185
Newton line, 120-125
nine-point circle, 233-240, 243-244
Niven, Ivan Morton, 301
Niven's theorem, 301
non-parallel lines, 374-376

O
octagon, 394-397
    with fixed area ratio within parallelogram, 305-306

inscribed in congruent squares, 307–309
octahedron, 40–42, 45
Ohm, Martin, 2
Orthic triangle, 99–100

P
Pacioli, Luca, 41
pantograph, 434–436
Pappus' theorem, 337–339
parallelogram, octagon with fixed area ratio within, 305–306
Pascal's theorem, 323–328
pentagram, 296
perpendicular bisectors, concurrency of, 55–56
perpendiculars, 103–105
Phidias, 3, 29
Pick's theorem, 316–320
Platonic solids, Golden Ratio in, 40–46
Pollio, Marcus Vitruvius, 46
polygon, sum of the interior angles of, 291–293
polygrams
  angles in, 296–300
  lengths of line segments in, 300–301
Prato, Giovanni di Gherardo da, 34
Ptolemaeus, Claudius, 194
Ptolemy's theorem, 194–206, 250
Pythagorean theorem, 1, 5–6, 11, 14, 16, 27, 127–129, 277–279, 312, 330–331, 420

Q
quadratic formula, 3
quadrilateral, with two intersecting circles, 226–227

R
rectangles, similarity of, 426
regular dodecagon
  determining area of, 301
  dissection of, 302
regular heptagons, astonishing property of, 303–304
regular n-gons, 311–316
regular octagon, isosceles triangle in, 302–303
regular pentagon, 398–400
regular polyhedra, 40–41
Reuleaux triangle, 356–357
right triangle
  altitude to hypotenuse of, 131
  placements of squares on, 136–137
  square on hypotenuse of, 138–143
  squares on all sides of, 145–146
  squares on hypotenuse of, 130–131
  squares on legs of, 129–130
  squares, placed on legs of, 131–135

S
Sangaku, 351–359
semicircles
  equal areas formed by, 281
  forming arbelos, 282
shear mappings, 184
shoemaker's knife, 282
side-angle-side (SAS) similarity theorem, 427–428
side-side-angle (SsA) similarity theorem, 431–433

side-side-side (SSS) similarity
  theorem, 430–431
similarity, 422–426
    of rectangles, 426
similarity theorems, for triangles,
    427–433
similar triangles, on triangle sides,
    165-188
Simson line, 114–115
    as bisector, 118–119
Simson, Robert, 111, 329
Simson's theorem, 111–114
    characteristic of, 116–120
sine law, 65, 301, 331, 369–370, 400
squares, 143–144
    on hypotenuse of right triangle,
      130–131
    on legs of right triangle, 129–130
    on quadrilateral sides, 162–164
    on sides of parallelogram,
      186–188
    on sides of random triangle,
      147–149
    on triangle sides, 127–164
Stewart, Matthew, 329
Stewart's theorem, 329–333
Sully, James, 2
sums of squared chords,
    310–311
symmedian point, 85, 152–154
symmedians, characterization of,
    86–88

T
tangent, 102–103, 285
tangent circles, 255–256
tetrahedron, 40
Thébault's theorem, 184, 186–188
Thorndike, Edward Lee, 19
triangle, 143–144
    exterior angle of, 373–374
    generating intersecting circles,
      264–265
    inscribed and escribed circle
      of, 274–275
    similarity theorems for,
      427–433
triangle area formula, 341–342
triangle centers, meeting of,
    109–110
triangle sides, similar triangles on,
    165–188

U
unit circle, 311–316

V
Vecten's theorem, 419
Vitruvian Man, 46–47

W
Wallace, William, 111

Z
Zeising, Adolf, 17

www.ingramcontent.com/pod-product-compliance
Lightning Source LLC
Jackson TN
JSHW011605180225
79191JS00002B/13